Samuel Edward Warren

Elements of Machine Construction and Drawing

Or, machine drawing, with some elements of descriptive and rational cinematics

Samuel Edward Warren

Elements of Machine Construction and Drawing
Or, machine drawing, with some elements of descriptive and rational cinematics

ISBN/EAN: 9783337164713

Printed in Europe, USA, Canada, Australia, Japan

Cover: Foto ©berggeist007 / pixelio.de

ELEMENTS OF MACHINE

CONSTRUCTION AND DRAWING:

OR,

MACHINE DRAWING,

WITH SOME ELEMENTS OF DESCRIPTIVE AND RATIONAL CINEMATICS.

A Text-book for Schools of Civil and Mechanical Engineering, and for the use of Mechanical Establishments, Artisans, and Inventors.

CONTAINING THE PRINCIPLES OF GEARING; SCREW PROPELLERS; VALVE MOTIONS, AND GOVERNORS; AND MANY STANDARD AND NOVEL EXAMPLES, MOSTLY FROM PRESENT AMERICAN PRACTICE.

BY

S. EDWARD WARREN, C.E.,

PROFESSOR IN THE RENSSELAER POLYTECHNIC INSTITUTE, AND AUTHOR OF A SERIES OF WORKS ON DESCRIPTIVE GEOMETRY AND STEREOTOMY.

NEW YORK:
JOHN WILEY & SONS,
15 ASTOR PLACE.
1883.

TROW'S
PRINTING AND BOOKBINDING CO.,
PRINTERS,
205-213 *East 12th St.*,
NEW YORK.

CONTENTS.

CLASS II.—RECEIVERS.

A—Point Receivers.

B—Line Receivers.

C—Surface Receivers.

a—Plane Receivers.

b—Developable Receivers.

c—Warped Receivers.

d—Double Curved Receivers.

CLASS III.—COMMUNICATORS.

A—Point Communicators.

B—Line Communicators.

Band, Cord, and Chain Wheels.

PAGE

THEOREM XIII.—In the rolling of three circles, with a common point of contact, any point of the inner circle will describe an epicycloid upon the circle on which it rolls, and a hypocycloid within the remaining circle. These curves will be the proper curves for teeth acting tangent to each other, to give a rolling motion to the circles to which they belong.......... 121

THEOREM XIV.—When any circle, less than either of two given pitch circles rolls on the exterior of both, and on the interior of both, it will, in the former case, generate the faces of the teeth of both wheels, and in the latter their flanks.......... 121

THEOREM XV.—Involutes are proper curves for the teeth of wheels.... 123

THEOREM XVI.—The teeth act by sliding contact, and their point of contact is on the generating circle.......... 123

THEOREM XVII.—Teeth formed by the preceding methods, give a constant angular velocity ratio to the wheels which carry them...... 123

THEOREM XVIII.—Within certain limits, the face of a driver acts best on the flank of a follower, and during the arc of recession; but, for either of a pair of wheels to be a driver, the teeth of each must have both faces and flanks.......... 124

I.—*General Solution*.......... 127

1.—Common spur wheels.......... 127

2.—Spur and annular wheels.......... 127

3.—Spur wheel and rack.......... 127

II.—*The describing circle equal to half the pitch circle*.......... 127

1.—Spur wheels.......... 127

THEOREM XIX.—Any two wheels of the same pitch, formed by the general solution, with a constant describing circle, will work together; but one made by the second solution will work perfectly only with those of *one* other number of teeth, and the same pitch.......... 128

2.—Spur wheel and rack.......... 128

III.—*The describing circle equal to one of the pitch circles*.......... 129

1.—Pin wheel and spur wheel.......... 129

2.—Pin wheel and rack.......... 130

3.—Annular pin wheels.......... 130

IV.—*Solution by involute teeth*.......... 131

1.—Spur wheels.......... 131

2.—Spur wheel and rack.......... 131

EXAMPLE XXXI.—To construct the projections of a spur wheel, seen first perpendicularly and then obliquely.......... 132

EXAMPLE XXXII.—To construct the projections of a bevel wheel, whose axis is perpendicular to the vertical plane.......... 138

EXAMPLE XXXIII.—To construct the projections of a bevel wheel, seen obliquely relative to the vertical plane.......... 140

Use of only three projections (101).......... 142

Practical forms of the teeth of wheels.......... 143

THEOREM XX.—Circular tooth curves, with centres on a line through the point of contact of the pitch circles, will give a sensibly constant velocity ratio to those circles.......... 145

CLASS IV.—REGULATORS.

A—Point Regulators.

B—Line Regulators.

C—Surface Regulators.

D—Volume Regulators.

CLASS V.—MODULATORS.

A—Point Modulators.

B—Line Modulators.

C—Surface Modulators.

a—Plane Modulators.

b—Developable Modulators.

c—Warped Modulators.

d—Double-curved Modulators.

CLASS VI.—OPERATORS.

A—Point Operators.

B—Line Operators.

C—Surface Operators.

a—Plane Operators.

b—Developable Operators.

c—Warped Operators.

THE SCREW PROPELLER.

D—Volume Operators.

BOOK II.

COMPOUND ELEMENTS, OR SUB-MACHINES.

SUPPORTERS.

COMMUNICATORS.

THE SLIDE VALVE AND ITS CONNECTIONS.

REGULATORS.

Governors.

MODULATORS.

PREFACE.

This book may be compared to an excursion train. Everything mechanical has called, or striven for a place in it, if only to cling to the platform of briefest mention. Yet not a tithe even of the *beauties* of mechanism have been admitted, for want of room. Indeed, one of the most arduous labors connected with the composition of this work has been to keep out the nearly irrepressible crowd of topics and examples that were pressing into it. Those that have been admitted have been selected with great care, after personal inspection in machine shops, and from valuable circulars, correspondence, and published authorities.

Other examples have been partially represented by woodcuts and brief notices, or have necessarily been excluded, and remanded back to the great world of mechanism.

In either case the governing idea has been, to develope the comprehensive scheme of the General Table proportionately, though briefly, with examples that should be American (mostly), new, and good.

There are at least six topics in this work, about which the troublesome problem has been to touch them at all, unless superficially, without devoting a volume to each. These are, *Turbines*, *Gearing*, *Propellers*, *Valve motions ; Governors*, with or without Variable Cut-offs; and *Trains* of gearing, as clock trains.

Of *turbines*, I have only taken one of each of the three essentially different kinds, with a simple description of its construction and action.

On *gearing*, I have been reasonably full, giving the substance of what need be known in behalf of proper practice, and with simple explanations.

As to *propellers*, their theory is so intricate, owing to the variety and indefiniteness of the data for calculations concerning them; and numberless experimental results are so full and accessible in Bourne and in similar works, that I have mostly interested myself in giving exact instruction, nowhere else accessible so far as I know, about making their projections; so as indirectly to correct grievous errors, and supply deficiencies, which have been found in print on this subject.

With the ample *geometrical* treatment of *valve motions* by Mr. Auchincloss,* and the masterly *analytical* work of Prof. Zeuner, to supplement the little that I have found room for on the same topic, I have had a narrowly limited and definite object in view in what I have had to say on that subject. The treatment of valve motions in the encyclopædias and the extended serial works, like Colburn's Locomotive Engineering, is generally unavailable. The works of Auchincloss and Zeuner, presuppose, expressly or impliedly, a good deal of familiarity with the subject. But many persons are wholly unfamiliar with it, and, unless apt to conceive readily of combined motions, find it a puzzling subject. My work has therefore been little more than to begin at the very beginning, and virtually to prepare for the beginner an introduction to those works. More, indeed, could not be attempted in a volume in which so many other topics have been introduced.

Governors, a plaything of American ingenuity, have been summarily, but with the most instructive variety attainable within small limits, passed over with the selection of the most marked varieties of governor and valve.

Trains of gearing, though very briefly noticed, have, it is hoped, been so treated as to afford some clear and accurate ideas on that subject, as a foundation for further study.

The classified table of machines has been prepared with great care, and compared with that in the Encyclopædia Britannica by Prof. Rankine, which is quite different, without material alteration in the result. I have endeavored, in the paragraphs immediately preceding the table, so to distinguish machines from instruments, as to rationally exclude engineering, astronomical and musical instruments from the province of machines, in which he includes them; contrary to those common usages of speech, which I believe will be found upon analysis to be grounded on real differences.

* Graduate of the R. P. I., 1862.

Still, the number and uses of machines are so endless, that I cannot profess to have found a strictly scientific, and therefore exhaustive classification of them.

A word now as to the intended use of this book in the class-room may be considered seasonable. Previous to its appearance, the subject of it was taught orally, and with no small labor, to classes, which, in their turn, could progress neither so rapidly nor pleasantly as if provided with a text-book. The present volume is naturally much fuller than an oral course could well be, and is intended as a text-book upon which daily interrogations and black-board exercises are to be held, as well as a manual, to be constantly open before the student for a guide in the preparation of his drawings.

With the plates of uniform size for each student, so that they can be agreeably bound together, but with a choice as to that size, from quarter super-royal to semi-super-royal, until the best size can be experimentally determined, it may be well, wherever practicable, to require that one of them should be constructed from actual measurements, made by the student, and accompanied by a plate containing an inked copy of the sketches and measurements.

Some of the plates should have the measurements recorded substantially in the style of office practice, and they should generally be titled, in addition to the general title-page of the collection. Or there should be a separate plate containing the several titles.

The "heavy lines" are omitted, or displaced in some plates, as an exercise for the student in supplying or correcting them.

Finally, the following lists will show to what helping friends I am indebted, and what sources of information I have diligently consulted. Also the signatures of student draftsmen, of the classes of '70 and '71, R. P. I., on many of the plates will always happily remind me how kindly my labors in that direction were lightened.

TROY: November, 1870.

ESTABLISHMENTS VISITED OR DRAWN UPON FOR MATERIALS USED IN THIS WORK.

American Saw Co.	Trenton, N. J.
Andrews Bros.	New York.
Atlantic Works	Boston.
Babcock and Wilcox Eng. Works	New York.
Bement and Dougherty	Philadelphia.
Bessemer Steel Works	Troy, N. Y.
Boston and Albany R. R. Shops	Boston.
Boston, Hartford and Erie R. R.	"
Bond's Chronometer Rooms	"
Brooklyn Water Works	Brooklyn.
Brown's Machine Works	Troy, N. Y.
Bullard and Parsons	Hartford, Conn.
Cambridge Machine Works	Cambridge, N. Y.
Chubbuck & Sons	Boston.
Collins' Turbine Works	Norwich, Conn.
Delamater Iron Works	New York.
Gurley W. & L. E.	Troy, N. Y.
Harmony Mill	Cohoes, N. Y.
Hopedale Mach. and Furnace Co	Hopedale, Mass.
Hinckley & Williams' Loc. Works	Boston.
Horton E. Machine Works	Hartford, Conn.
Hotchkiss, Power Hammers	New York.
Huntoon Governor Co	Boston.
Jones and Laughlin	Pittsburgh, Pa.
Judson Governor Works	Rochester, N. Y.
Lowell Machine Shop	Lowell, Mass.
Ludlow Valve Co.	Troy, N. Y.
Lyall Positive Motion Loom Co	New York.
Mason V. W. Friction Clutches, etc.	Providence, R. I.
McMurtrie & Co., Machine Agency	Boston.
Milwaukee & St. Paul R. R., E. M. Hall, Supt. Power	Milwaukee, Wis.
Morgan Iron Works	New York.

New York Central R. R. Shops................	Albany, N. Y.
Novelty Iron Works..........................	New York
Pennsylvania R. R. Car Shops................	Altoona, Pa.
Putnam Machine Works........................	Fitchburg, Mass.
Rensselaer Iron Works.......................	Troy, N. Y.
Ruggles' Machine Works......................	Poultney, Vt.
Sault M. & T. Co............................	New Haven, Conn.
Sellers, Wm. & Co...........................	Philadelphia.
Shaw & Justice, Hammers.....................	"
Starbuck Bros., Engineers...................	Troy, N. Y.
Steere, E. N., Cotton Machinery.............	Providence, R. I.
Swain Turbine Co............................	Chelmsford, Mass.
Tremain, Balance Valves.....................	Chicago, Ill.
Troy & Boston R. R. Machine Shop............	Troy, N. Y.
U. S. Navy Dep't............................	Washington, D. C.
Washington Iron Works.......................	Newburgh, N. Y.
Wheeler, N. W., Eng. Office.................	New York.

PERIODICALS AND WORKS OF REFERENCE CONSULTED.

American Artisan.
Auchincloss, Link and Valve Motions.
Belanger, Cinematique.
Borgnis, Composition of Machines, 1818.
Bourne on the Screw Propeller.
" Catechism of the Steam Engine.
Brown, H. T., 507 Mech'l Movements.
Burn, R. S., Mech. and Mechanism.
Colburn, Locomotive Engineering.
Engineer, The.
Engineering.
Engineer and Mach. Drawing Book.
Fairbairn, Mach. of Transmission.
Francis, Hydraulic Experiments.
Hughes on Water Works, Weale's Series.

Imperial Cyclopædia of Machinery.
Jour. of Franklin Inst., 1860, 1864, 1867–70.
Joynson, Gearing.
King, W. H., Notes on Steam.
Leroy, Géométrie Descriptive, Applications.
Long & Buel, Cadet Engineer.
Olivier, Géométrie Descriptive, Applications.
R. P. I. Collections of Mechanical Lithographs.
Roebling, Wire Rope Transmission.
Scientific American.
Sellers on a System of Screw Threads.
Weisbach, Mechanics.
Weissenborn, Amer. Eng'g illustrated.
Willis' Principles of Mechanism.
Zeuner, on Valve Gears.

FUNCTIONAL CLASSI

Fixed Elements. Supporters. Local.	Fixed Elements. Supporters. General.	Receivers.	Communicat
Pillow Blocks. * Axle Box. * Hangers. * Footsteps. *		Chain hooks.	Crank Pins. Block Cross He Fixed Coupling
Guides. Straight. * Curved. Arms.	Drill posts. Standards. * Solid beam-frames or beds. *	Winches. Treadles. Engine hand levers. Levers of Horse powers.	Flexible { Band, Cord, Chai Rigid. { Crank, Rocke, Conne rods, Links, Excen rods, Worki boar
Planer tables. Face plates. Flat brackets.	Plane bed plates. Flat Frames, { Web. *, Open beam. *	Flat pistons. * Endless platforms in Horse powers.	Excentrics. *
Cylindrical brackets. Local prismatic beds. * Stuffing boxes. *	Prismatic beds. * Corliss Vert. Eng. Frames.	Cylindrical water wheel floats. Fourneyron Turbine	Band wheels. * Spur wheels. * Bevil wheels. *

TABLE.

ION.			
MOVING ELEMENTS.			COMPOUND ELEMENTS OR SUB-MACHINES.
REGULATORS.	MODULATORS.	OPERATORS.	SUPPORTERS.
Governor balls.*	Idler pulley. Escape wheel.	Saw teeth. Shuttles.	Compound Chucks. " Slide Rests. " Tool Holders.
Fly wheels. *	Band shifter arms. Pin clutches. Simple slide rests.	Drills. Cutters, Helical as in Hay cutters and Ruggles' Slate trimmers.	COMMUNICATORS. Beam Engine Connections.* Tumbling Beam Movements.* Clock Trains.*
Throttle valves. Puppet valves. Flat slide valves. Flat oscillating valves. (Corliss.)	Variable crank.	Air pump buckets.* Printing press platens.	REGULATORS. Valve Motions.* Governors. Ball*, Fan*, Oil* } for { Throttle Valve or Steam Valve.
Cage valves. Cylindrical throttle valves.	Cone pulleys. Speed " Dead " Sectoral motions. Elliptic gears.	Bending, polishing, and shaping rolls.	Gauges.
	Helicoidal clutch.	Screw propellers.*	MODULATORS. Feed Motions. Band Shifters. Quick Returns.* Friction Clutches. Compound reversing gear.* Escapements.*
Ball valves.	Paraboloidal pulleys.	Clock bells.	

ELEMENTS

OF

MACHINE CONSTRUCTION AND DRAWING.

BOOK FIRST.

SIMPLE OR SINGLE ELEMENTS OF MACHINES.

PART I.

INTRODUCTION.

GENERAL PRINCIPLES.

1. Bodies, in addressing the eye, exhibit not only the attributes of color, transparency, or opacity; polish, or roughness; but the two *fundamental geometrical attributes* of *Form* and *Size.*

2. FORM is a determinate arrangement of an assemblage of points, according to some law. It depends upon the *relative* lengths and *directions* of the bounding lines of a body.

3. SIZE is the amount of space occupied by a body, and is due to the *extent* of its bounding lines, as compared with a unit of measure.

4. Drawings may represent objects, in respect to their *size*, as larger, or smaller than they really are; or, in their real size.

5. Drawings which represent the *apparent forms* of bodies as presented to the eye, are called *perspective drawings*, or *pictures;* and are intended chiefly for ornament, or for popular illustration.

6. Drawings which represent the *real forms* of objects, as determined by the sense of *touch*, in taking measurements, are called *projections.* Since such drawings show the *real* propor-

tions of objects, they constitute a graphic language, by which the thoughts of a designer can be most clearly conveyed to a workman, who can thence construct the objects represented. Hence projections are often called *working drawings.*

7. While working drawings represent the *real forms* of objects; they represent them, in a majority of cases, in less than their *real size.* But to preserve the true proportions in the drawing, all the parts of the object must be similarly reduced in the drawings. This is what is called *drawing by scale.* That is, each distance, as a foot, on the object, is represented by some less distance, as an inch, or a quarter inch, etc., on the drawing.

Scales.

8. The only scales necessary to be understood by students of the present work, are the *linear scale,* and the diagonal scale *of equal parts,* which we will now explain.

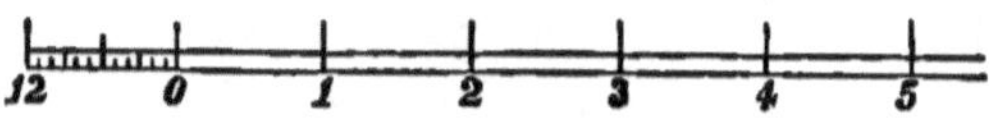

FIG. 1.

Fig. 1 represents a plain linear scale of *three feet to the inch,* each of its units as from 0 to 1 being *one-third of an inch.* The equal space from 0 to 12, is divided into twelve equal parts, representing inches. Thus, from 4 to the fifth mark to the left of 0, *represents* four feet and five inches, and is therefore *called* four feet and five inches. For brevity this is written 4′:5″.

9. Other linear scales of equal parts, being similarly constructed, can readily be understood from this example. The ivory scales used by draftsmen, contain a variety of such single scales, with the left hand unit divided both into tenths and twelfths. It also contains others, expressed in inches to the foot, as one inch to one foot, three-fourths of an inch to a foot, etc., and numbered accordingly IN = inch; $\frac{3}{4}$; $\frac{5}{8}$; $\frac{1}{2}$; etc.

FIG. 2.

10. Fig. 2 represents a simple diagonal scale of units, 5ths;

and 4ths of the 5ths. The space 0—5′, which is equal to 0—1, etc., is then divided into five equal parts; and so is *ac*. The five equidistant horizontal lines afford four equal spaces. We then reason thus: If in coming down four spaces on 0*b*, to *b*, we depart from the vertical 0*a* by the space *ab*, which is one-fifth of the unit 0′1, in coming down one space, we should depart one-fourth of *ab*, which equals one-twentieth of 01. We thus have the rule for reading the scale: proceed to the left of 0 as many spaces as there are 5ths required, and then down on the diagonal thus reached as many spaces as there are 4ths of 5ths required. Thus the distance between the stars is 3 units, 3-fifths and 2-fourths of a fifth, or

$$3 + \tfrac{3}{5} + \tfrac{2}{4} \text{ of } \tfrac{1}{5} = 3\tfrac{14}{20} = 3\tfrac{7}{10}.$$

All other diagonal scales, including the more familiar decimal diagonal scale, are made and used in a similar way; so that if any one of them be rationally and fully comprehended, all others may easily be understood.

If, as is sometimes done, the diagonals were drawn in the direction *ad*, the numbers 0, 1, 2, etc., would be on the lower line *ac*.

11. In regard to the manual operations of machine drawing, the proper standard of precision should be carefully observed. That is, the student should always imagine himself in a drafting office, working as if his compensation, or position, depended upon the accuracy of his work. And the latter should be the same as if his plates were to form working drawings for the actual construction of finished machinery. To this end, all *points* should be *accurately located*, and *finely marked*; and all lines should be *finely drawn* with none but the *hardest pencils*, and *exactly through* the *proper points*.

In much of machine drawing, the distances to be laid off are numerous, and quite small, hence the fine spacing dividers, pens, and pencils, are of especial use, as well as the most accurate scales.

12. *The instruments called scales*, are simply pieces of metal, ivory, wood, or paper containing a variety of linear and other scales.

The leading forms of scales are *edge scales* and *surface scales*. An *edge*-scale is a scale whose graduations are on the edge of the substance containing it. This form of scale is most convenient, because a distance can be transferred from it to the paper, *directly*, by laying the scale on the paper and pricking off, with a needle-point, the extremities of the given distance. The *best form of edge scale* is the *triangular scale*, which contains *six* linear edge scales. The other form, or *flat-edge scale*, having its edges chamfered on one side to ensure greater accuracy in its use, can conveniently carry but *two* edge scales, except as two or more, each of which is just double the other, may lie against the same edge.

13. *Surface scales* are flat pieces of some hard material, usually ivory, containing a set of various linear scales, side by side, and, all together, covering the surface of the instrument. These give more scales on a single instrument than edge scales; but to transfer a distance from them to paper, we must proceed *indirectly* by taking up this distance in a pair of dividers, and then laying it down on the paper.

Elements of Projections.

14. The following brief rehearsal of the elements of projections may assist many, or all who make use of this volume.

A solid has three dimensions, at right angles to each other. Therefore if a horizontal plane, as RQ, Fig. 3, be placed parallel to two of the dimensions, as AB and BC, of a solid, and if the latter be then viewed in a direction, Aa, perpendicular to the plane RQ, those dimensions can be seen, correctly represented, in *length* and *direction*, upon that plane, as at *ab* and *bc*.

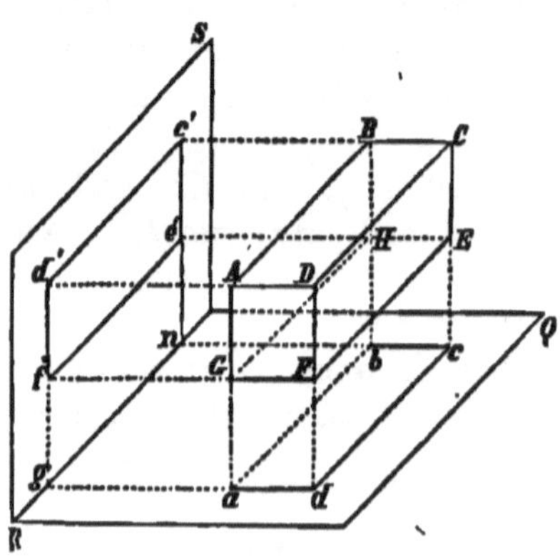

FIG. 3.

The figure *abcd* is thus equal to the visible top of the given body, and is called its *horizontal projection;* or, in the language of practice, its *plan*.

15. In like manner, if a vertical plane, RS, be taken parallel to the dimensions, AB and AG, and if it be viewed

perpendicularly, as in the direction D*d'*, these dimensions can be correctly shown on that plane. The figure *d'c'e'f'*, equal to DCEF, is called the *vertical projection*, or the *elevation* of the given body.

Thus we see that the two projections of a body, taken together, show its three dimensions, when the latter are parallel to the planes RQ and RS, which are called the *planes of projection.*

16. Observe now that *d'g*, the height of the vertical projection of D above the ground line, is equal to D*d*, the height of D, itself, above the horizontal plane. In like manner, *cn*, the distance of the horizontal projection of C from the ground line, is equal to the perpendicular distance of C, itself, from the vertical plane.

That is: *The perpendicular distance of a point in space from either plane of projection, is equal to the distance from the ground line to its projection on the other plane.*

17. Analyzing the figure a little, we see that when a line, as AB, is parallel to a plane of projection, its *projection ab*, or *c'd'*, upon such plane, is *equal* and *parallel* to itself. Also if a line, as DF, perpendicular to the horizontal plane, or DA, perpendicular to the vertical plane, is perpendicular to a plane, its projection on *that* plane, as *d* or *d'* respectively, is a point; and on the other plane, as at *d'f'* and *da*, respectively, is perpendicular to the ground line, and parallel to the line in space. With this suggestion, the reader can make out the projections of lines in other positions, as the diagonals, AC, DE, AF and AE, not shown.

18. Summary of definitions.

A *plane of projection* is one on which an object is represented.

A *projecting line* is a line from any point of an object, perpendicular to a plane of projection; and it represents the direction in which the object is looked at.

The *projection of any point*, is the intersection of the projecting line of that point with a plane of projection.

The *projection of any object* is the figure formed by joining the projections of the bounding points of that object.

The intersection, R*n*, of the two planes, is called the *ground line*. So simple an apparatus as a folding slate or sheet of stiff paper, with the leaves placed horizontally and vertically, and a few straws, will serve to illustrate the principles here stated.

19. The planes of projection, which are at right angles to each other in space, coincide upon paper. This is accomplished by supposing the vertical plane, RS, Fig. 3, to revolve backward about the ground line, until it coincides with the horizontal plane produced backwards.

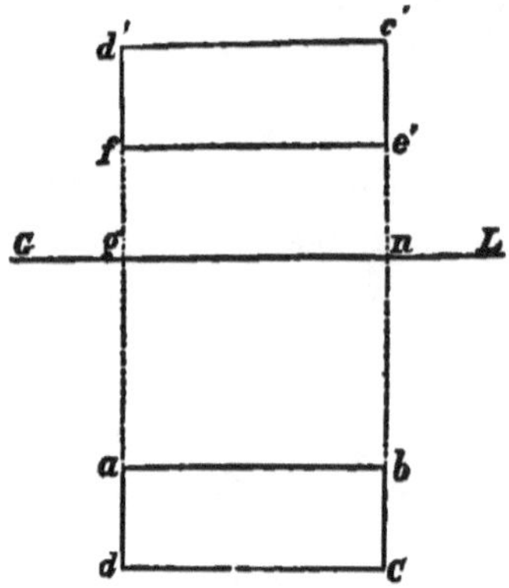

Fig. 4.

Supposing now the planes to be of indefinite extent, Fig. 3 would be thus transformed into Fig. 4, which shows the projections *abcd* and *d'c'e'f'*, as they really are, instead of pictorially, as in Fig. 3. GL, the ground line, represents R*n* in Fig. 3.

The two projections, as *d* and *d'*, *of the same point, are* thus *in the same perpendicular to the ground line.* This should be carefully remembered.

20. *A point is named by naming its projections.* Thus the point *dd'*, Fig. 4, means the point itself, D, Fig. 3, whose projections are *d* and *d'*. The like is true of lines. Thus the line *dc—d'c'*, Fig. 4, means the line DC, Fig. 3, whose projections are *dc* and *d'c'*.

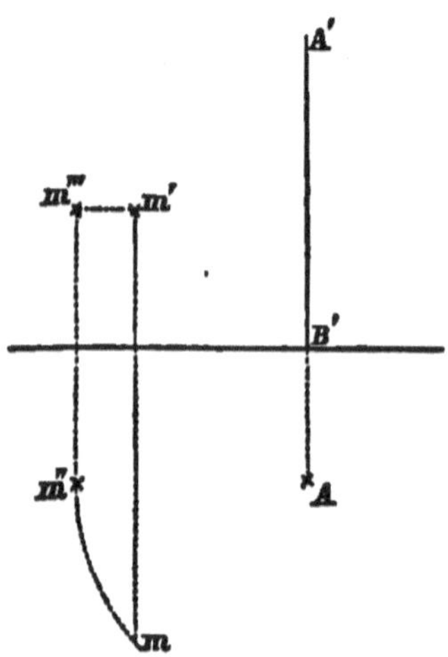

Fig. 5.

21. Resuming, now, the conclusion of (15) if the dimensions of a body are not parallel to the planes of projection, they may be made so either by *turning the body*, or by taking a *new plane of projection.* In turning a body, it is sufficient to study the motion of one of its points.

This understood, the following principles pertain to the revolutions of points.

a—If a point as *mm'*, Fig. 5, revolve about a *vertical* axis, as A—A′B′ (see DF, Fig. 3), it will describe a *horizontal arc*, as *mm″—m′m‴*, whose horizontal projection, *mm″*, will be *an equal arc*, with centre at A, and whose vertical projection, *m′m‴*, will be *a straight line parallel to the ground line.*

b—Similarly, if a point *mm′*, Fig. 6, revolve about an axis, AB—A′, which is *perpendicular to the vertical plane* (See DA, Fig. 3), it will describe an *arc parallel to the vertical plane;*

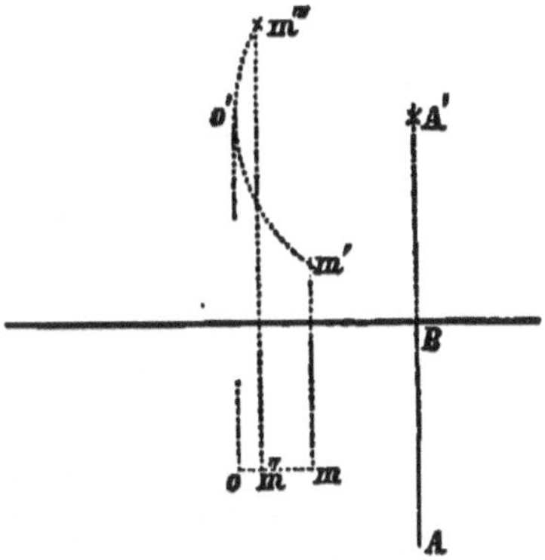

FIG. 6.

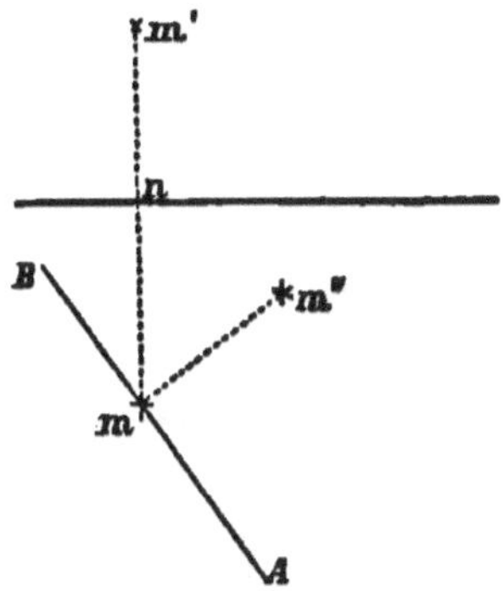

FIG. 7.

whose vertical projection, *m′o′m‴*, will be *an equal arc*, with centre at A′, and its horizontal projection, *mom″*, *a straight line, parallel to the ground line.*

c—If a point *mm′*, Fig. 7, which is vertically over a horizontal axis, A B, revolves 90°, it will appear, as at *m″*, on a perpendicular, *mm″*, to AB, and equal to its height *m′n above the axis.* (16.) For the arc of its revolution is in a vertical plane, perpendicular to AB, and its horizontal projection is therefore straight, and perpendicular to AB.

Here the axis is *in* the horizontal plane. If it had been merely *parallel to* that plane, *m′n* would have been estimated from its vertical projection, which would have been parallel to the ground line.

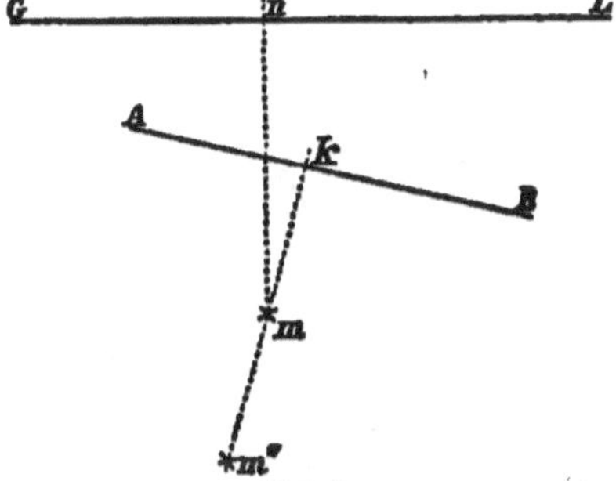

FIG. 8.

Like results would be true for a revolution about an axis *in*, or *parallel to* the *vertical* plane. The student should construct figures to represent these cases.

d—If a point *mm′* not vertically over an axis, AB, in the horizontal plane, be revolved about that axis, into that plane, Fig. 8, it will appear at a perpendicular distance, *km″*, from that axis, equal to its true perpendicular distance, in space, from AB. This distance, as may be made evident by the simplest model, will be the hypothenuse of a right-angled triangle, whose base equals *mk*, and whose altitude equals *m′n*, which last is the true height of the point itself above its horizontal projection, *m* (16.)

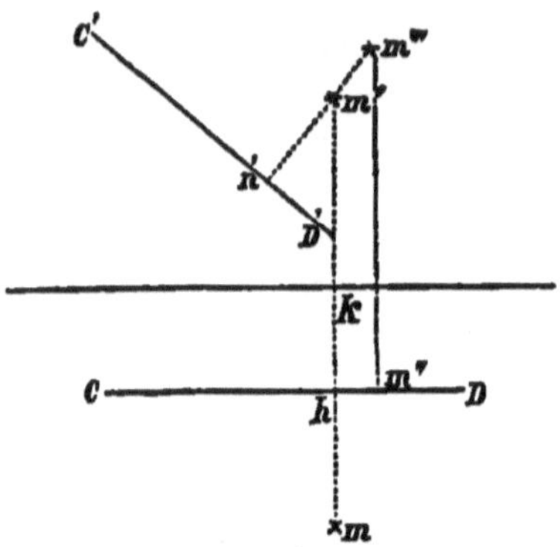

Fig. 9.

e—Similarly, in Fig. 9, the axis, CD—C′D′, is *parallel* to the *vertical* plane, at a perpendicular distance from it, equal to *hk*, and the point *mm′* is revolved about it, into a vertical plane containing CD—C′D′. After such revolution, *mm′* will appear at *m″m‴*, where *m‴n′*, perpendicular to C′D′, is equal to the hypothenuse of a right-angled triangle, whose base is *m′n′*, and altitude, *mh*, the perpendicular distance of the horizontal projection of the point from that of the axis, or from the vertical plane through the axis.

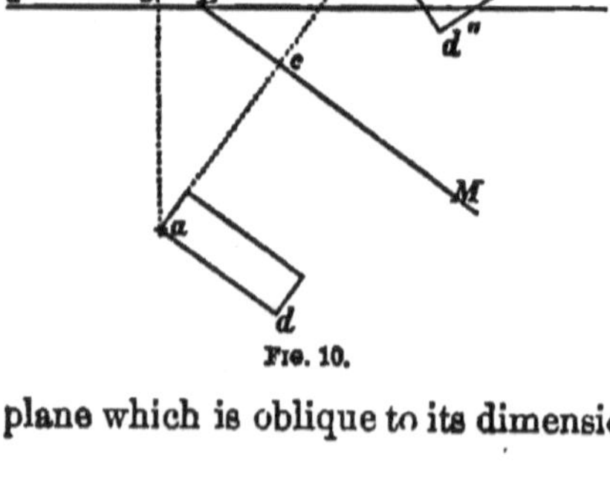

Fig. 10.

22. These, being examples of the main principles and operations relating to the *revolution of points*, the following relates to the *selection of new planes of projection.*

a—When it is desired to represent an object on a plane which is oblique to its dimensions, it is obviously necessary to begin with a projection made on a plane which is parallel to *two* of its dimensions. Thus, in Fig. 10, the plane ML is vertical, and parallel to two dimensions of the rectangular block, *ad*—*a''d''*.

b—The principle is also to be observed, in these operations, that any number of different elevations of the same point are at equal heights above the common horizontal plane. Thus *a'b* is made equal to *a''c*, and on a perpendicular to the ground line through *a* (19), in order to find *a'*, the projection of *a*, on that vertical plane whose ground line is GL.

c—To avoid the use of vertical planes which are oblique to each other, as they are in Fig. 10, conceive the body, as *ad*—*a''d''*, to be turned horizontally about any vertical axis, till it is brought parallel to the one vertical plane used, and begin with its projections in that position. Thus, in Fig. 11, after making

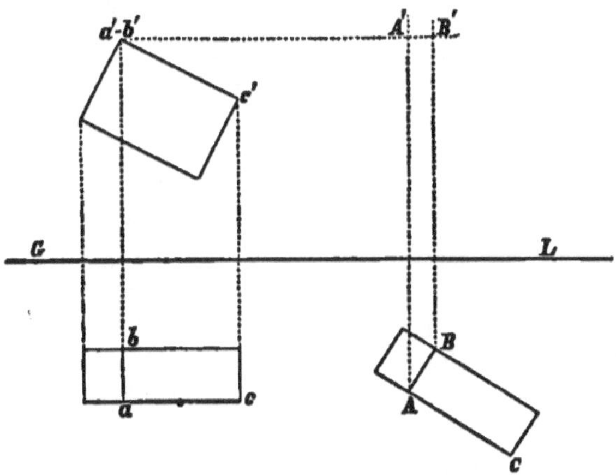

FIG. 11.

first the elevation, *a'b'c'*, and second the plan, *abc*; third, make the new plan, ABC, of the same *form* as *abc*, but turned to represent the desired position of the body relative to the vertical plane of projection. As the position of the body, relative to the horizontal plane, has not changed, all its points will be at the same height as before. Therefore any point, as A', of the desired elevation, is at the intersection of the projecting line AA', perpendicular to the ground line, GL, with the line *a'*A', which is parallel to the ground line.

Constructions of the Ellipse.

23. It not unfrequently happens, that some of the wheels or circular parts of a machine are situated in planes which are oblique to each other.

All such parts, when oblique to the plane of projection, will be projected in ellipses. For the further preliminary information of self-instructors, especially, some convenient constructions of the ellipse are therefore added.

24. An *ellipse*, Fig. 12, is a plane curve, such that the sum of the distances, as PF+PF′, from *any* point of the circumference, to the fixed points F, F′, within the curve, is always equal to AB; the longest line within the curve, and which is called the *transverse axis.* F and F′ are called *foci.*

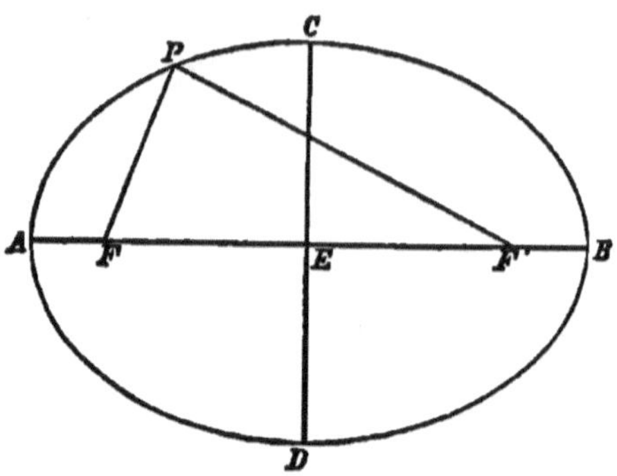

FIG. 12.

The middle point, E, of the transverse axis, is the *centre* of the curve, and bisects every line drawn through it and limited by the curve. Every such line is a *diameter* of the curve. The shortest diameter, CD, is perpendicular to AB, and is called the *conjugate axis.*

25. The above definition affords a familiar mechanical construction of the ellipse by string and pencil, the string being equal in length to AB, and fastened at F and F′. Also a construction by points, with dividers, by drawing pairs of arcs from F and F′ as centres, and whose radii, taken together, shall equal AB, the lesser one being always greater than AF. Such arcs will intersect, so as to give four points of the ellipse, for each pair of lines taken as radii. This construction is not figured, either of the two following being better for the draftsman's use.

26. *To construct an ellipse by radials from the extremities of the axes.*

Let AO and CO, Fig.13, be the given semi-axes of an ellipse. Make AE = OC, and parallel to it. Divide AO and AE into the same number of equal parts, and number them as in the figure. Then lines from C and D, and through corresponding points of division, will intersect at points, as *a*, *b*, and *c* of a true ellipse. The other three quarters of the ellipse can be similarly constructed.

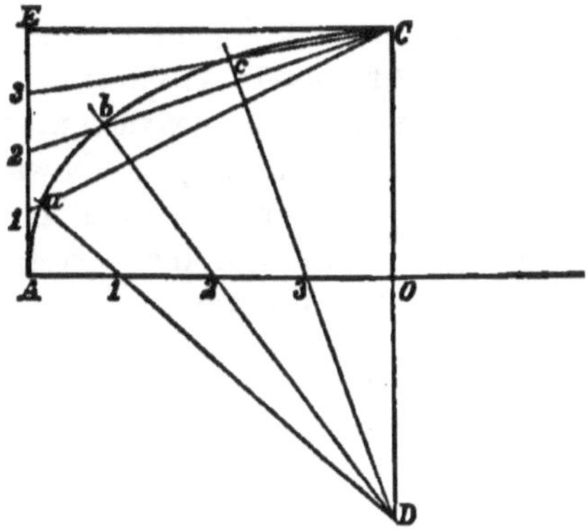

FIG. 13.

It would have been equally correct to have equally divided CE and CO, and to have drawn the radial lines through the points of division and from the opposite ends of the transverse axis.

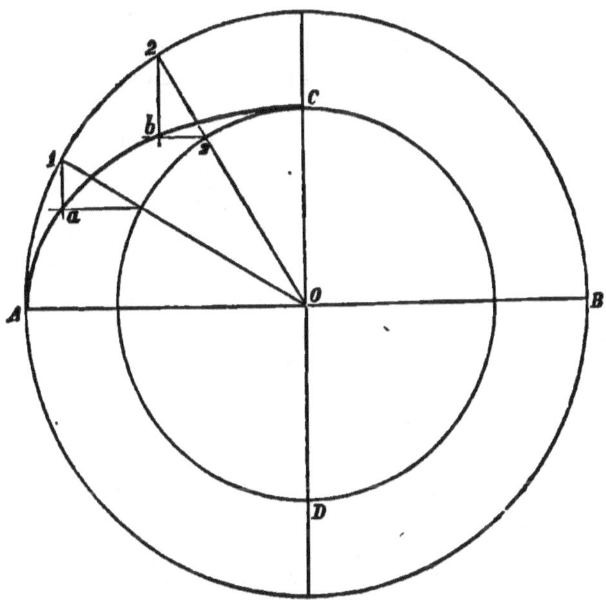

FIG. 14.

If AE were made equal to AO, the same construction would

have given the circle of which the ellipse AC is really the oblique projection.

27. *To construct an ellipse, by concentric circles on its two axes.*

Let AB and CD, Fig. 14, be the given axes. Describe circles on them as diameters, as shown. Divide the circumferences of these circles into any convenient number of equal parts, and number them similarly, as shown. Then, parallels to AB, through the points on the inner circle, will intersect perpendiculars to AB, through the corresponding points of the outer circle, at points, as *a* and *b*, of a true ellipse, whose axes are AB and CD.

Special Definitions.

28. A MACHINE is an assemblage of pieces, attached to a common support, and acting upon each other to produce a certain result; and so that *a given position of one will determine that of all the rest.*

29. It is here convenient to distinguish the terms: "Engine," "Machine," "Tool," and "Instrument."

An engine, and a machine, are not essentially, or necessarily, different things; but different names for things essentially alike, and expressive of different ways in which the latter may be regarded.

Thus, from the etymology of the terms, *engine*—an invention—is a product of intelligence; and machine—a means—is something adapted, as a cause, to a certain end. Hence, therefore, when a given combination of working parts is generally thought of, more as a product of intelligence than otherwise, it is called an *engine*, as a steam engine, or a dividing engine. But when the same is thought of chiefly in regard to the end for which it is made, it is called a *machine*, as a spinning machine. Thus any piece of mechanism *may* be called indifferently an engine or a machine, and many *are* thus indifferently termed, as locomotives and steam fire engines.

30. *Instruments* are distinguished from machines in being more intimately and continuously controlled by life in all their movements. Thus an organ acts only when, and just as, it is played on; and the like is true of engineering instruments, and to a great extent, of mounted telescopes, since so many of their

parts are separately adjustable by the operator. A distinguishing feature of *instruments*, then, is that their parts are *separately* movable.

31. Again: *tools* are mainly the servants of manual skill, or training in processes; *instruments* are servants of a higher order of intelligence, such as results from a training in principles. Thus we say "the *tools* of a *trade*," and "the *instruments* of a *profession*;" a carpenter's *tools*, but an engineer's or a surgeon's *instruments*.

In reference to their material uses, *tools* are used in making the machines by which in turn consumable products, used in common life, are fabricated. Thus machine shop machines, are often called "*machine tools*," or "*machinist's tools*," while those used by hand are called *hand tools* or *bench tools*. In these the *train of pieces* forming a machine, is wanting; while in machine tools, it is the final piece acting immediately on the work, and driven by a machine, rather than directly by hand, that is strictly called the *tool*.

Classification of Machines.

32. The world of machinery is too vast and varied to yield readily to attempts to classify its members. Moreover, the components of mechanism need to be differently classified for mathematical and for descriptive treatment.

The following articles present an outline of classification suited to the subject of Descriptive Mechanism.

33. The *immediate* source of the power which moves any one or more machines, is some form of prime mover in which some force of Nature, as muscular power; the weight or impact of water; the elasticity of springs, or of expansive vapor; or electrical attraction, is made available for producing motion.

The first grand division of machines, is, therefore, into *Motors*, or *Motive Machines*, in which a force of Nature is made an available power; and *Workers* or *Operative Machines*, which perform some special duty, as with or upon raw material, by virtue of their design.

34. Again: some machines give only numerical or abstract or intangible results; signs or data, rather than substantial products; while others do produce such products.

Hence Operative Machines may be grouped into the two divisions of *Registrative*, and *Productive* Machines.

Here it is important to explain that a productive machine does not necessarily produce a finished result, but if, in connection with others, as in case of the cotton gin, or a dredging machine, it contributes towards such a result, it is entitled to its name.

35. Registrative Machines may be enumerated as follows:

1°.—*Counting Machines;* such as those sometimes attached to steam engines,* or turbines, to indicate their revolutions; or to Burden's horseshoe machine, or Hoe's power presses, to register their production.

2°.—*Measuring Machines;* of time, space, motion, magnitude, and force, as timekeepers; "Atwood's machine" for determining the laws of falling bodies; water and gas meters; dynamometers and pressure gauges; weighing machines, etc.

3°.—*Copying and Drawing Machines;* such as pantographs, elliptographs, Olivier's instruments for drawing certain curves,† and ruling machines.

4°. *Calculating Machines.*

5°.—*Recording Machines;* as telegraphic machines and steam engine indicators.

36. Productive Machines. These modify matter only in respect to its *position*, its *form*, and its *dimensions;* that is, *geometrically* and *physically;* and not in its atomic constitution, or chemically. We have then—

I.—Machines for changing the position of matter.

1°.—*By simple removal by stationary machines*, as by capstans, windlasses, cranes, hoisting machines, derricks, and suction pumps of all kinds.

2°.—By *conveyance*, whatever the direction or distance, as in "rolling stock" generally; and the moving mechanism of "atmospheric despatch" apparatus, common road engines, etc.

3°.—By *projection*, as in ancient, or mechanical, and modern, or explosive artillery; also in all kinds of forcing pumps, the hydraulic ram, etc.

* "Engineering," vol. iv., p. 371.

† Olivier's Des. Geom. and Applications.

4°.—By *separation*, as in reaping, ploughing, digging, dredging, and stumping machines; in fruit paring, fulling, washing, and churning machines; in machines for expressing or dispersing fluids from fruits or mixtures, and in ginning, threshing, and smut machines.

5°.—By *distribution*, 1*st*, of determinate *bodies*, as in pin-sticking, wire-card making, type-setting, pile-driving, and seed-planting machines.

2*d*, of matter indefinitely, as in elevators and blowing engines.

3*d*, of *films* or *material impressions*, as in printing machines of all kinds, upon all sorts of materials.

6°.—By *uniting*, 1*st*, by *interlacing*—*first*, of *fibres*, as in felting, paper-making, carding, roving, and spinning machines; *second*, of *threads*, as in weaving and knitting machines.

2*d*. By *union of particles*, as in mixing machines.

3*d*. By *union of pieces*, as in sewing, pegging, and rivetting machines.

II.—Machines for changing or perfecting the FORM of matter:

1*st*.—By *definite division*, *i. e.*, into definite parts, as in sawing, cutting, shearing, and punching machines.

2*d*.—By *surface abstraction* of portions of indefinite form, as in planing, turning, shaping, milling, boring, polishing, mortising, drilling, slotting, paring, carving, and screw-cutting machines.

3*d*.—By *moulding pressure*, and often, or always, without loss of material; as in rolling, forging, squeezing, wire-drawing, coining, brick-making, moulding, bending, folding, and swaging machines.

III.—Machines for changing the DIMENSIONS of matter:

1*st*.—By *condensation*, as in road-rolling machines and presses for compressing matter.

2*d*.—By *indefinite division*, as in chopping, tearing, grinding, crushing, and stamping machines.

Functional Classification of Mechanical Organs.

37. The foregoing reconnoissance, so to speak, of the field of mechanism may give an idea of its extent, and may direct the

student's reading and practice. But, for present purposes, it is to be observed, that the vast range of mechanism here opened to view is composed mostly of endlessly varied combinations and proportions of a few mechanical elements or organs.

The drawing of the separate elements or organs of mechanism will be the chief subject of the following pages, in connection with so much description of their successive forms in progressive practice, and of their action and use, as will lend additional interest and value to the drawing of them.

Some instructions upon the drawing of connected trains of mechanism will be given afterwards.

38. Mechanical organs may be conveniently classified as follows, according to their functions, into—

Supporters,	*Regulators,*
Receivers,	*Modulators,*
Communicators,	*Operators.*

39. Supporters, as their name implies, are the frames or other fixed supporting parts of machines, whether *general* supports of the whole machine or *local* ones of particular parts.

Receivers are those parts to which the motive power is first applied in any machine, as in the piston of a steam engine or the endless platform of a horse-power machine.

Communicators are the pieces which communicate the motion of the receiver to that of the parts which act on the material presented to the machine.

Regulators are those organs which determine the effort exerted, the equalizing of its expenditure, or the supply or discharge of engines, etc.

Modulators are organs for the purpose of changing the relations of motion, as by reversing, disengaging, intermitting, etc., or by changing the ratio of the velocities of connected pieces.

Operators are the organs which act directly on the raw material, or to immediately accomplish the object of the machine.

The following table presents a more detailed view of these organs.

TABLE.

FUNCTIONAL CLASSIFICATION OF MECHANICAL ORGANS.

2

Functional Classification.

- I. Supporters
  - *General*
    - Beds.
    - Housings.
    - Standards.
    - Frames.
  - *Local*
    - Local bed plates.
    - Brackets and arms.
    - Pillow blocks and axle boxes.
    - Bracket and suspension hangers.
    - Footsteps and bolsters.
    - Face plates.
    - Travelling tables.
    - Guides and stuffing-boxes.
    - Steam cylinders.
    - Pump barrels.
    - Cases or chambers.
- II. Receivers
  - *In circuit motion*
    - Winches.
    - Levers of horse mills.
    - Endless platforms of horse powers.
    - Driving pulleys.
    - Vertical water-wheel buckets.
    - Turbine buckets.
    - Windmill vanes.
  - *In reciprocating motion*
    - Pistons.
    - Treadles.
    - Beam levers, as of hand fire-engines.
- III. Communicators
  - *Principal*
    - Bandwheels.
    - Excentrics.
    - Screws.
    - Gears.
      - Spur wheels.
      - Bevel wheels.
        - Racks.
        - Circular, elliptic, etc.
      - Hyperboloidal wheels.
      - Spiral wheels.
  - *Auxiliary*
    - Bands.
    - Cords.
    - Chains.
    - Articulations.
      - Fixed couplings.
      - Cranks.
      - Rockers.
      - Links.
      - Working beams.
      - Jointed rods, etc.
      - Hooke's joint.

IV. Regulators.

- *Of supply.*
  - Low water detectors.
  - Giffard's injector.
  - Cocks.
  - Valves....
    - Slide.—Greene.
    - Puppet.—Putnam.
    - Rotary.—Corliss.
- *Of effort.*
  - Governors.
    - Ball.
    - Fan.
    - Oil.
      - Of st'm pipe opening.
        - Huntoon.
        - Snow.
        - Judson.
      - Of st'm v'lve opening
        - Putnam.
        - Greene.
        - Corliss.
  - Fly wheels.

V. Modulators.....

- *Reversing actions.* Band shifters, etc.
- *Intermitting actions.* Escapements, etc.
- *Disengaging actions.* Couplings, etc.
- *Speed changers.* Cone pulleys, etc.
- *Tool-holders*, and slide rests.

VI. Operators......

- Saws.
- Drills.
- Cutters.
- Paddle wheels.
- Screw propellers.

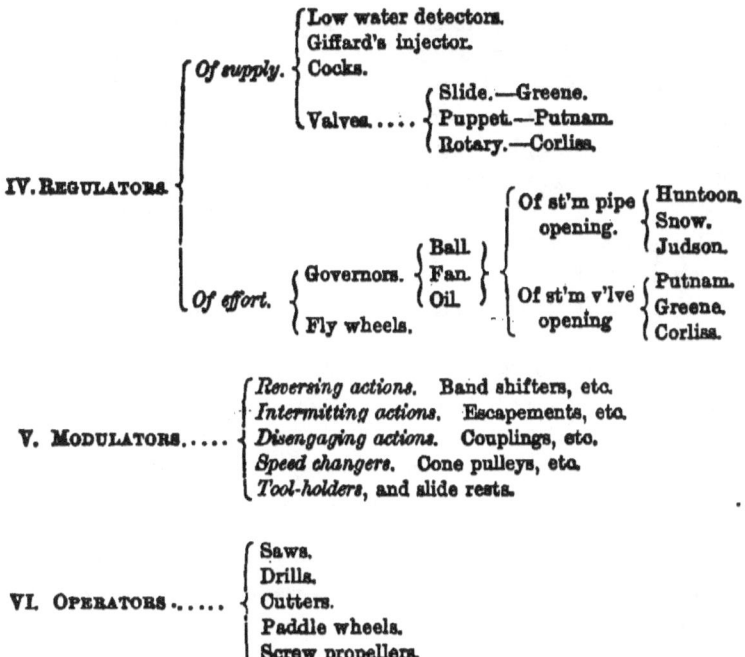

40. It should here be noted that many of these elements, especially among *modulators* and *regulators*, are *compound organs*, or *sub-machines*, consisting of a train of parts, as in Reversing actions, Slide rests, Escapements, and Governors.

We therefore define a *sub-machine* to be a series of connected pieces, designed to perform a part subservient to the main object of the machine to which it is attached.

Geometrical Classification.

41. Though the foregoing may seem an elegant classification of the elements of mechanism, yet it is partly obscure; for a given element does not, inherently and always, belong only to one and the same class. Thus a *piston* as a *receiver* in a steam engine is not conspicuously, if always at all, different from a *piston* as an *operator* in a pump. Also, a *spur wheel*, which is usually a *communicator*, becomes an *operator* and *regulator* combined, in the geared fly wheel of an engine.

Still the foregoing classification is generally useful.

42. The following, which is new and entirely different, is combined with the former in the "general table." It consists in arranging mechanical elements according to the geometrical magnitudes which express their essential or ideal character, and to which they may therefore be reduced.

For example, a *shaft* revolving in *two pillow blocks*, is essentially a *line* supported at *two fixed points.* Hence a pillow block is classified as a *mechanical point.* Likewise, material *governor balls* reduce, in thought, to *heavy points* at which their whole action is concentrated.

Other elements, which act by virtue of their surfaces, or are equivalent to certain mutually acting surfaces, are classified according to their surfaces. Thus, *spur wheels* are equivalent to *rolling cylinders*, etc. By following out this idea, the classification expressed in the horizontal columns of the general table, with the headings at the left, will be intelligible.

43. We now proceed to develope the foregoing double scheme in a series of illustrations of the several classes of mechanical organs there named.

These illustrations embrace, *primarily*, practical examples, drawn to scale, in a series of plates of a size convenient for adoption in actual class practice; and, *secondarily*, brief notices and simpler illustrations, of various other forms of each of the organs selected as examples.

The course, thus composed, is based upon the idea of *at least one* good representative plate, of each of the six classes of organs, as a minimum for the student's practice under instruction. But, to ensure agreeable variety, as well as material to supply the wants of more rapid workers, several examples, drawn to scale, are given in each class of organs, so that all the members of a class need not necessarily draw the same objects.

Reduction of Scales.

44. Where the admirable French colored mechanical lithographs, or suitable actual objects, are at hand, students, so far as qualified, may profitably make drawings from them. But in drawing from the French plates, those should principally be

chosen which give a *scale* and *measurements*, and the copy should be drawn from the measurements to a new scale.

The scales given on the originals, being in French measures, the following examples will illustrate their transformation into suitable English measures.

First, scales are expressed by one or more units of *lower* denomination to one of the same or of a higher, as a scale of two and a half inches to one inch or to a foot. Suppose, then, that we have a French drawing of some small machine, on a scale of 24 decimetres to 1 metre.

1 metre=39.4 inches, very nearly, and as 100 decimetres=1 metre,

1 decimetre=.394 inches, very nearly, and

24 decimetres=9.456 inches, or a scale of $\frac{1}{4}$, very nearly. Suppose, then, that we wish a scale of about $\frac{1}{5}$. We see that one decimetre=$\frac{2}{5}$ of an inch, very nearly, then take $\frac{1}{3}$ of an inch for a decimetre, and 24 decimetres=8 inches; that is, 8 inches to 1 metre=a scale of $\frac{1}{5}$ very nearly.

Second, scales are expressed in terms of one or more units of *higher* denomination to one of the same or a lower, as a scale of four inches to an inch=a scale of $\frac{1}{4}$; or of five feet to one inch $=\frac{1}{60}$. Suppose, then, a French drawing, on a scale of 3 metres to 1 centimetre, is to be copied. 1 centimetre=3.94 inches, nearly; and one-third of this on the scale=1.31 inches, nearly, will be a metre of a scale, giving a scale of about $\frac{1}{30}$, since a metre=39.4 inches, which would be adapted to a very large machine, with the smaller parts omitted. To enlarge the scale to about $\frac{1}{20}$, construct a new scale in which 2 inches shall be called a metre; then, 2 inches to 39.4 inches is a scale of $\frac{1}{20}$ very nearly.

45. We will now proceed directly with the explanation of the plates, which are of about the size recommended for student practice, viz.: $8\frac{1}{2}\times12\frac{1}{2}$ inches, that is, such as may be made by dividing a sheet of super-royal drawing paper, stretched upon the board, into four equal plates.

If other sizes be preferred, we would recommend *imperial* paper in four plates of $9\frac{1}{2}\times13\frac{1}{2}$ inches, or *medium* paper in two plates of 10×15 inches; but that all the plates of each one's set should be of uniform size.

PART II.

THEOREMS, PROBLEMS, AND EXAMPLES ON ELEMENTS OF MACHINES.

CLASS I.—SUPPORTERS.

SECTION I.—LOCAL SUPPORTERS.

46. Local supporters are very various, and difficult to classify. The following partial catalogue may therefore serve to suggest other kinds and forms of special supports.

Local beds; as those of especially large and heavy parts.

Brackets and *arms*, *pillow-blocks*, *axle-boxes*, *bracket* and *suspended hangers;* supporters of horizontal revolving shafts.

Footsteps and *bolsters;* supporters of vertical revolving shafts.

Simple tool-rests or *holders;* supporters of operating tools.

Simple chucks and *face-plates* to support revolving material, as in common and wheel-turning lathes.

Travelling-tables; as in planing, milling, drilling, and shaping machines.

Guides and *stuffing-boxes;* as in steam engines.

Cylinders, *barrels*, *chambers*, *chests*, etc., for water, steam, air, etc.

A—Point Supporters.

EXAMPLE I.

A Heavy Pillow-block.

Definitions and description.—The general term, "*bearing*," is applied to the supporting surface on which any piece, as a revolving shaft, rests; whatever may be its position. The piece which supports a horizontal revolving shaft is called a *pillow-block*, or plumber-block, when itself is supported from below, and open at both ends or sides, as in Pl. I., Figs. 1 and 2.

In the pillow-block, Pl. I., Fig. 1, there is the *body*, B, and the *cover*, C. The part, S, of the body is the sole, through which, as at *ac*, *holding-down-bolts* pass, to confine the block. *bb′* are *brasses*, whose inner surfaces are cylindrical, and form the bearings for a shaft. They are flanged so as to prevent lateral displacement, and are therefore, as at *b*, dropped into place before putting on the cover C. That part of the shaft which is within the pillow-block is the *journal*. *s* is one of four *set-screws* on each side of the body, to set up the side brasses, *e*, against the shaft. Each has a check-nut, *n*. At A are the nuts of the cover-bolts, whose heads, not shown, are in recesses in the under side of the sole, as at *g*, Pl. II., Fig. 1. The bolt-holes, *ac*, are longer one way than the other, and are hence said to be *slotted*. They are thus made to allow the position of the block to be adjusted between the lugs, as *dd′*, Fig. 5, so as to bring the two or more pillow-blocks on the same shaft into line; or, in case of a steam engine, to adjust the distance from the centre of the shaft to the centre of the cylinder.

The pillow-block shown in the figure, being for a 14″ horizontal shaft, the bearings, *b*, are continuous only in the lower half of the block. The bolts, as A, are relieved from the lateral pressure of the shaft upon the cover by forming the latter, as shown, to be embraced between the walls, B, B, of the body.

This pillow-block was designed for a *vertical* engine, used in driving the rolls of a steel-rolling mill. At the beginning therefore of the *down*-stroke of the piston, the cylinder being overhead, the thrust of the connecting-rod and crank, and the weight of the 14-inch shaft, come upon the bottom of the bearing at D. When the up-stroke begins, the weight of these three pieces relieves the upward thrust, and bars, as *b′*, afford sufficient bearing on the upper side of the shaft. The wear being mostly at these points, provision for sufficient adjustment is made by the space between the cover and the body of the box.

Construction.—From the above description, with the given scale and measurements, the drawing can be made. Observe that each elevation has a centre line, and that the plan, which the student should make, with, or instead of one of the elevations, would have two centre-lines. A section should also be made.

The scale might well be increased to *an inch and a half to one foot.*

Example II.

A Putnam Pillow-block.

Description.—This beautiful pillow-block, Pl. I., Fig. 2, is not shown in finished drawings, like the previous figures, but only in sketches, with measurements, from which the student can make finished drawings.

This design is from the Putnam Machine Co., at Fitchburg, Mass., and is adapted to a horizontal engine of about 24-horse power.

In a horizontal engine, where the piston is at either end of its stroke, the connecting-rod from the piston-rod to the crank, and the crank, a short, stout arm attached to the shaft by which the latter is revolved, are both horizontal.

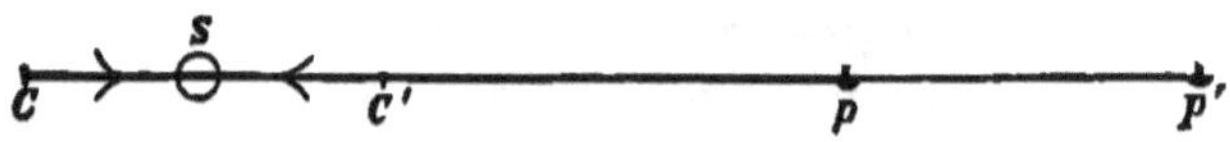

Fig. 15.

Hence when the connecting-rod is at its extreme back position, *pc*, Fig. 15, and about to turn forward, it acts to *pull* the crank CS against the shaft S, and the latter against the front of its bearing in the pillow-block. Likewise, when the connecting-rod is at its extreme front position, *p'c'*, and about to move backward, it acts for a moment to push the crank C'S, and thence the shaft S, backward against the pillow-block bearing.

Thus the pillow-block of a horizontal engine is mostly worn at the points A and B, Pl. I., Fig. 2. Separate adjustable brasses are therefore provided at those points, in the design here shown. A recess, *abcd*, contains the brass *ad*, and wedge *bc*, each ten inches long, see also Fig. 3, whose tapering faces lie together as at *ef*. Hooked bolts, as C, enter the holes *gg'* in the wedge. By turning on the nuts, *h*, of these bolts—of which there are four in all—the bolts and wedges are drawn up and the brasses crowded in against the shaft. Shallow recesses, as FG—F'G', in other parts of the bearing, are filled with an anti-friction alloy called Babbitt metal, and the wear on these parts being very small, the cover, H, is closely fitted to the body, I, of the box. The collar LL—L'L' affords a long bearing for the shaft.

The sole, holding-down-bolts, and cover-bolts will be recognized on comparison with Fig. 1.

Fig. 4 shows a sketch of the manner of fastening the block to the top flange, AA, of the bed-plate, or general support of the engine, by means of a key, *kk'*, through the holding-down bolt *bb*, and under the flange AA. In Fig. 2, MNN shows the plan of the wider part of this flange, on which the pillow-block rests.

Construction.—Since this pillow-block has two vertical planes of symmetry through its centre, O, it is sufficient to show the exterior of one half of it. In the elevation, therefore, all to the right of the line O′O″, is a section on the plane OR; and in the plan, the right-hand half shows the cover removed. The arcs, as *nk* and *om*, are drawn from O as a centre. The points *rr'*, *ss'*, and *tt'* are fully lettered, as the large and small curved outlines of the block in their neighborhood sometimes perplex learners.

The parts on which shade lines are scattered in the elevation, are sections of the solid portions of the block, which is hollow. In the finished drawing, these portions should, of course, be filled with fine shade lines, omitting the bolt holes, as K′, and the bolt C. The figure—2—is a sketch from a model, having quite different measurements from those given. It will therefore be sufficient for the student to give the same views as in Figs. 2, 3, and 4, though an end view, partly shown in the right hand one of Fig. 4, might usefully be completed from the given measurements.

47. *The heavy lines*—or lines of shade—are shown on the principal figures of this plate. Taken in connection with the *fact* that light is usually so taken in practical examples, that its *projections* make angles of 45 degrees with the ground line, and the *principle* that they divide surfaces in the light from those in the dark, they will assist the student in adding the lines of shade to other figures where they are not shown.

Example III.

A French Pillow-Block.

Description.—This example, Pl. II., Fig. 1, is given on

account of its beauty of design, rather than for its mechanical merits. The spherical ends of the nuts, their raised seats, *f* and *a*; the tapered collar, FF′, the three-centred tops of the cover, two of whose centres are O and *p*, and the slight slope of the surfaces which meet at *bc*—*b′c′*, all give a fineness of figure which pleases the eye. But, designed for a horizontal engine, as it is, there being but two bolts to the covers, there is no provision for the extra wear of the brasses G, at the lateral points as I; and as the cover does not slide *within* the body of the block as in Ex. I., it is less capable of resisting the horizontal thrust upon it.

Construction.—This block, having two axes of symmetry, only one-fourth of the plan is shown. The scale should be increased to *one-half*, or *three-fourths*, for the best effect, and *half* of the plan should be shown. The measurements are left to be found by the given scale, or assumed.

The end elevation, which is very neat, can easily be made from the projections here shown; and the whole, owing to its numerous curved and oblique surfaces, would be a particularly *good example to shade with graded tints.*

Another pillow-block, which may, if desired, be taken as a separate example, is shown at MP—MP, Figs. 1 and 2, Pl. VII.

Example IV.

A Locomotive Main Axle Box.

Description.—Pl. II., Fig. 2. In the pillow-block, the shaft rests in the block. In a locomotive, the shaft or axle is supported by the wheels, which, in turn, rest upon the rails of the track. The weight of the engine then bears upon the tops of the axle boxes, which, again, bear down upon the uppermost part of the axles. Hence the main provisions for wear and support are made in the upper part of the box, which, indeed, is essentially a pillow block inverted, and modified to suit the frame of the engine.

The example shown is from recent practice on the New York Central R. R. By comparison of measurements with those of the engine frame, Pl. VI., Figs. 3 and 4, it will be seen that the wedge key, W, Fig. 3; *a′p*, Fig. 4, is between the flanges B, B—B′B″ of the box. G G, G is the body of the box, level on the top surfaces, AA, and depressed at HH,—H' *m* H'. The depressions SS—S'—S" are the seats of "the stirrup blocks," on which, through the medium of very stout springs, the engine rests.

C is the brass lining, made thickest at E, by centring its outer curve at *o*, $\frac{1}{2}$ an inch above O, the centre of the axle. *e—e—e* are recesses for Babbitt metal. KK are the front and rear walls of the oil cellar, which is packed with cotton waste and oil, and whose lateral walls, L, are $\frac{9}{16}$ of an inch thick. The outside recess, FF', in the cellar, keeps the bolt *cd*, which passes through the front and rear walls to hold the cellar in place, from passing through the oil.

Construction.—The titles of the separate views, and the given traces of the planes of section used, and the given measurements, leave little need for minute directions here. The scale may be changed to $\frac{1}{4}$ or $\frac{1}{8}$, and a horizontal section through O might well be made. It is left for the student to determine, by comparison of the different views, which parts of the sections should be filled with lines of shading, as being in the planes of section.

Shaft-Hangers.

48. On entering any mechanical establishment, a noticeable feature consists in the many band wheels, revolving on a common axis, or "*line of shafting*," supported from the walls or ceiling; or from posts. The band wheels go by the name of "*overhead pulleys*," and their supports by the general name of *hangers*, though this name may be more strictly applied to supports from the ceiling timbers.

Now a wall, or row of posts, or ceiling timbers, are liable to warp, or spring, lean or settle, and thence to throw any bearings attached to them out of line with each other, and thus to produce an injurious binding of the shafting in its bearings. Again, overhead shafting is less accessible for oiling than that which is near the floor; and, if unprotected, may drip blackened oil disagreeably upon persons and things below.

Hence, the main points of a good hanger are, first: that it shall be adjustable both vertically and horizontally, so that its bearing shall be in line with all the others of the same row;

second, that it shall seldom require oiling; and third, that it shall not drip.

Pl. III., Figs. 1 and 2 represent two very good hangers, the second of which fulfils all the conditions just mentioned, while both are good exercises in the construction and reading of drawings.

Example V.

A Bracket Hanger.

Description.—Pl. III., Fig. 1. This design is from the Industrial Works at Philadelphia, and as made in 1858 and subsequently.

It is shown in two complete elevations and a plan, from which the bearing is removed; and a horizontal section through the case F,F′,F″ is shown.

A,A′,A″ is the bracket, fastened by bolts, at *b*,*b*′,*b*″, to the wall. A two-inch shaft is supported at C′,C″D″, and the box C″D″ is self-adjustable by its spherical curvature shown in dotted lines, where it passes through the close-fitting ring, B′,B″; the upper and lower parts of which are held together by bolts as at *c*′,*c*″. At *d*′*d*″ is the oil hole; *e*″,*e*″ are dripping cups to catch any oil that may work out at the ends of the bearing. They rest in the ears or recesses at *e*,*e*′. The ring B′,B″, is attached to the screw S,S′,S″, which affords a *vertical* adjustment to the bearing. The latter is adjusted horizontally by the three screws *m*, which bear against the hollow cylinder *p*, within which the screw S works. Just above the letter S is seen half a thread of this screw. The shape of the chamber, F, allows for horizontal adjustment, principally in a direction perpendicular to the wall, to which the bracket is fastened, as is obviously most necessary. The whole being adjusted, the check nuts *n*,*n*′ and N,N′ hold the bearing fast in the desired position.

Construction.—The measurements not given, may be determined by the scale, or may be assumed. The student may advantageously increase the scale to one *fourth*, one *fifth*, or one *sixth*; and may make out a vertical section through the axis of the shaft.

The heavy lines are indicated by small double marks across them. The student should, however, always note the heavy lines for himself before calling for assistance; guiding himself by the principle that the light is taken so that its *projections* make angles of 45° with the ground line; and then that surfaces illuminated are separated from those in the dark by heavy lines. (47, Ex. II.)

EXAMPLE VI.

A Self-Oiling Drop-Hanger.

Description.—Pl. III., Figs. 2, 3. This design is from Messrs. Bullard and Parsons, Hartford, Conn. It is claimed for it that it requires oiling but twice a year, thus saving more than half of the oil, and nearly all the labor required by a plain box.

A, A′ is the top plate, solid with the drop, which extends in one piece to the line M*m*′. The moulded cylindrical part MM*f*′*e*′, is hollow, and receives the swivel *h*′MM*u*′, whose bearings are as indicated at M and M, the space between the lines *b*′ and *c*′ being hollow all around. The swivel also is hollow between *a*′ and *a*‴, from MR to *k*′; the ring QQ′, being in one piece with the swivel. Fig. 3 shows an end view of the ring and swivel, where O″O‴ is the form of a section of the ring at the top, and the similar small figure at *v*″ is a section at the bottom. By means of the nut N, working through the head of the swivel, the latter may be raised or lowered, and turned in any direction. The journal box itself, *gs*—*q*′*t*′ is held in position within the ring by the opposite set screws *n*,*n*′,*n*″, which adjust it laterally, and work through bearings *t*,*t*, not shown in plan. The box being thus held at two points is self-adjusting to imperfections in the straight line of the shafting. L′ is the oil cellar, the spiral grooves, L,L, in which, hold oiled packing, which draws up the oil from L by capillary attraction; and the circular channels at *q* and *g* catch any oil that might otherwise drip out.

In *reading this drawing*, we notice a set of circles with Y as a centre in the plan, and another, with X as a centre in the elevation. Of the former, *a*,*b*,*h*,*c* and *d*, with *a*′,*b*′,*h*′,*c*′ and *d*′, represent the several vertical inner and outer cylindrical surfaces of the drop and swivel, and *e*—*f*, is the plan of the extreme circumferences at *e*′ and *f*′. Of the latter, the letters of reference show the position, being the same on the two projections of the same circle, as *r*—*r*′, the end circle of the box; or the same cylindrical surface, as *pp*′, the inner surface of the box.

The letters CC′, etc., and HH′, etc., clearly indicate the form of horizontal sections of the drop and the ring.

To avoid the indistinctness of too many dotted lines, the plan, HQ, of the ring is made in full lines.

Construction.—The measurements not given may be ascertained by a scale, or suitably assumed. By placing the figure lengthwise on the plate, the scale may properly be enlarged to one-*third*, or one-*half*.

Example VII.

Turbine and Spindle Foot-Steps.

Description.—A foot-step is the support of a vertical revolving shaft at its lower extremity. Pl. II., Fig. 3, represents the footstep for a Jonval Turbine, substantially as made by Collins and Co., of Norwich, Conn. SS—S′ is the *bridge tree*, extending across the wheel case at the bottom, and stiffened by the rib RR′. The socket, *dd*—*d′d′*, is solid with the bridge tree, and surrounds the *cup*, *ab*—*a′b′*, whose position is adjusted by set screws, as *pp′*, roughly shown. The remaining parts are not shown in the plan. B, the step itself, is of lignum vitæ, immovable in the cup *ab*—*a′b′*. On it rests the *step bowl*, CD, of iron, which is keyed to the shaft, E, of the wheel, as seen at *k*, and is solid with the lower plate of the wheel.

Construction.—The measurements may be determined from the scale, and recorded. The cup, and parts above it, are shown in section, and may be shaded accordingly. Also the elevation is shown in section, as cut by a plane a little in front of RR.

The *spindle foot-step*, Pl. II., Fig. 4, gives a very simple drawing exercise, but is noticed on account of its utility. Where thousands of spindles are running in the same mill, any device which lessens the frequency of oiling is valuable. In this foot-step, any convenient fibrous packing is placed in the annular space, DD′, and well saturated with oil. Openings, *aa′* and *bb′*, conduct the oil from this space to the vertical revolving spindle, which rests in the step CC′. Near the upper end, the spindle is supported by another bearing, similar to A″B″, but open at both ends, and called a bolster.

A spindle making 4,500 revolutions per minute needs oiling

not oftener than twice a week with this foot-step and bolster, instead of once or more every day.

There is a somewhat similar device, but without the fibrous packing, known as Gilman's spindle step for "Roving Frames." These machines act in an earlier stage of the formation of the thread, and their spindles revolve more slowly, or at 500 revolutions per minute. These steps require oiling but three or four times a year.

49. In leaving the subject of shaft supports, an improvement in the shafting itself may be mentioned. This is what is known as *cold rolled shafting*. Merchant, and other manufactured iron is generally rolled hot; but, by a patent process, bars, rods, axles, also plates and sheets, are now rolled cold. This, as experiments show, compresses, hardens, and strengthens the iron; and also leaves it highly polished, and perfectly true in straightness and roundness, and firmest in its outer surface or "*skin*," which is cut away in other shafting, by the process of turning it true in the lathe.

B—Line Supporters.

Example VIII.

Locomotive Guide Bars and Cross-head.

Description.—The outer end of the piston-rod of most engines is attached to a block, or transverse piece, which slides back and forth as constrained by fixed guides, upon which it moves. The block or transverse piece is called a *cross-head.*

When the guide bars are separated only by the cross-head they are, ideally, one to four straight and parallel lines, on or between which the cross-head, reduced to a *point*, moves. When, as in side-lever, and some other engines, they are necessarily separated by the diameter of the cylinder, the cross-head becomes extended into a transverse *line*, attached to the piston-rod at its middle point, and having its rectilinear movement determined, at its extreme points, by the guides.

On some accounts, the cross-head might be classed with communicators, but it is so convenient to represent it in place, as working between its guides, that it is here accounted a *supporter*, which indeed it is, to one end of the *connecting rod*, which actuates the *crank*, and thence the *main shaft* of the engine.

Pl. II., Fig. 5. T,T′ is a collar on the back end of the cylinder, from which project the pieces, one of which is E, to which the front ends of the guide bars are bolted. DD′ is an arm, open like a loop or ring, or like an ox-bow, at the part to which the guides are fastened, so as to allow the vertical play of the connecting rod. Here the pieces E′, to which the guides are bolted, are themselves bolted to DD′ by the nuts and bolts at NN′ and *n*N′. Now, BB′ is the front upper guide, B,C′ the front lower one; A,B′ the back upper one, and A,C′ the back lower one. That is B, for example, is the horizontal projection of two bars, one vertically under the other; and C′, for example, is the vertical projection of two, one of which is exactly behind the other. RR′R″ is a portion of the piston-rod, whose full diameter is shown in the elevation, by nicking out a little of the guide bars, as shown.

The cross-head, which is quite an irregular solid, is shown in plan and elevation, partly hidden by the guides; alone, in rear elevation, in Fig. 6; and with a cross section of the guides and brasses in Fig. 7. M,M′,M″ is the body of the cross-head, flush with the tops of the upper guides, and the under surfaces of the lower guides, but entirely hidden in the side elevation. VV, V′V′ are the vertical wings of the cross-head, giving it a longer bearing on the inner faces of the bars. H,H′,H″,H‴ are the horizontal wings, which in some engines are as thick as the space between the upper and lower bars. In this design, brasses *b*,*b*′*b*′,*b*″*b*″, shown also in section above and below H‴, intervene between the wings H,H′, and the bars. They cannot slip out to right or left, being hooked at both ends, as shown at *b*′ on the lower one. They are otherwise confined by the plate FF′, which is bolted to the wing H,H′. The back end of the piston-rod is conical, and goes through the body of the cross-head, as shown by dotted lines. It is fastened by the key *kk*′*k*″*k*‴. The pin PP′ is cylindrical, and forms a point of attachment for the connecting rod. K,K″ is an arm, projecting from the back plate G, to carry the pump-rod L.

Fig. 7 shows a section of the back bars, brasses and wing in the plane Y*y*; and a section of the front bars in any plane, as X*x*, to the right of the brasses.

Construction.—With this description, and with the full measurements and lettering of parts broken away, the construction can readily be made. The scale may well be increased to *one-sixth*, or even *one-fifth;* in the latter case by breaking out a part of the length of the guides.

50. As an example of the gradual development of a mechanical idea, it is interesting to note the successive forms of locomotive cross-heads that have appeared. Fig. 8 represents, roughly, the general form of a cross-head often seen from about 1845 and onward. Here the single guide bar, B, running through the cross-head, the latter has the greatest leverage for working itself in a rotary direction around B. The piston-rod was inserted at R, and P is the end of the pin to which the connecting rod was attached between two ears, one of which is Q.

Fig. 9 represents an improvement relative to steadiness of motion in the cross-head H, by making it move on two guide bars, B,B. Here, too, we have an elementary illustration of unessential variations of one idea, for the guides were sometimes of circular section instead of a square one; and square section guides were sometimes set diagonally, or so that opposite edges as *aa* should be in a vertical plane. This form was common between 1850 and 1860.

Finally, the last example, Figs. 5–7, represents the fully developed idea of steadying the cross-head to the utmost, by confining it between four exterior guide bars; which is the extreme opposite in effect of the form shown in Fig. 8. This form has prevailed in the United States since about 1860, especially on "outside cylinder" engines.

Where there is a greater tendency to a *vertical* than a *horizontal* displacement of the cross-head, as in the common four-driver switching engines, without trucks, which rock vertically a good deal, the guides now often consist of *two* bars in a *vertical* plane, with a cross-head of greatest width *vertically;* as if the *plan* in Fig. 5 was an elevation of guides consisting of only two bars.

C—SURFACE SUPPORTERS.

a—Plane Supporters.

51. Passing these without figured illustration, we merely define iron-planer tables and face plates of lathes as *movable* supporters. Each is pierced with many cross-shaped openings to allow large or small work to be conveniently fastened at any point of it.

b—Developable Supporters.

52. Associating *prismatic* and *pyramidal* forms with *cylindrical* and *conical* ones, we distinguish *surface* from volume elements, when it is only the *surface* of the supporter, and not its *interior capacity*, which we have to consider.

EXAMPLE IX.

A Local Bed Plate.

Description.—Pl. I., Fig. 5, represents the bed for a 60-ton fly wheel, at the Bessemer steel rolling mill, in Troy, N. Y. Its principal parts are the sole, AA′; the vertical web, B—B′B′; the top plate, C,C′; the gussets, DD′, and EE′; and the transverse supports, as FF′, through which the holding-down bolts pass into the masonry below.

Both projections have a transverse centre line, OO′. The part of the plan to the right of the broken edge, *ab*, shows a horizontal section in the plane, MN; *dd′* is one of two lugs to confine the pillow block which rests on the plate CC′.

Construction.—With the given measurements, sufficient data are afforded for drawing this bed, as shown, or with the substitution of an end elevation, and longitudinal and transverse sections; some one or more of which variations from the given figure should be made by the student.

D—VOLUME SUPPORTERS.

EXAMPLE X.

A Locomotive Cylinder.

Description.—Pl. IV., Fig. 1. This example is from a first-class engine of the New York Central R.R., taken from working drawings of an engine not then built.

The drawing shows an end elevation, with the cylinder head removed, and a vertical longitudinal section.

The end elevation also shows a part of the *saddle*, EFLH, ex tending across the engine, and bolted to the smoke-box, *uv*, while the cylinder flanges, SJD, and MN, are bolted, the former to the smoke-box, and the latter through the main frame, whose section is I, to the end, LH, of the saddle.

The drawing further shows, incidentally, for convenience, a bottom plan, T; a transverse section, T′; and longitudinal section, T″, of the *steam valve;* the *piston,* P, with its rod, R; and the stuffing-box, UV*w;* consisting of the collar, U, of the back cylinder head; the gland V, bolted to U; and the annular space, *w*, in which the steam-tight packing is confined by the gland, V, and ring or lining, *x*.

For the rest, the correspondence of the letters well shows the different projections of the same parts. Thus, *g′g″—gt*, is the valve seat; *h′h*, indicating lines by one point, is the floor of the steam-chest, whose sides and top are removed, and into which steam enters through a pipe behind K′K at D′, and the port *d″*, which may be 12″ to 14″ long. The annular surface, of the width, *j′l′—j″l*, is on the cylinder head, and is set a little back from *l′m′—ln″* to allow a ring of packing to be inserted. Y*y′y″n′—nmky* is a steam-port, extending, as the end view thus shows, through nearly a third of the circumference of the cylinder, at each end; C′—CC″, is the axis of the cylinder, C′*o′* the radius of its *bore*, and the minutely greater distance, C′*p′*, is that of the *counter-bore* for a short distance at each end, as shown at *op*, and intended to facilitate the discharge of water of condensation. BW is the front cylinder cover which, like the rear one, is a little concave, so as to conform to the piston, and thus reduce the volume of old steam left in the cylinder at the beginning of a new stroke. XXX is the front cylinder-jacket of brass, the confined air within which keeps the heat of the cylinder from escaping, and is ornamental.

In view of a prevailing disposition in some quarters to strip the locomotive of all its ornaments, it is not an improper digression to say, here, that it is probably all a mistake to do so. It is not for the sake of the engine, though *that*, as a thing quite analogous to life, deserves ornament, nor for the sake of the public only, nor in regard to the character of the train, as *express*, or freight, that an engine is to be ornamented; but it is chiefly for the sake of the *men* who operate it. If $300 to $500 apiece, spent in beautifying the passenger engines, and $200 to $300, each, on the freight engines, interests and rationally gratifies their operators, and so raises the morals of the entire force of a road; it is money well spent. It may be true, however, that brass and scarlet are not the chief means of locomotive decoration. An abundance of smoothly rounded and finished surfaces of iron and steel may have a greater as well as more quiet elegance.

Construction.—With the full measurements given, this needs no special explanation. By turning the figures crosswise of the plate, and substituting a mixed plan and horizontal section for the end view, the scale could well be increased to one-*sixth.*

EXAMPLE XI.

A Jet Condenser.

General explanations.—Steam engines are distinguished, in one of many ways, as *condensing*, or *non-condensing;* popularly called, *low* and *high* pressure, respectively.

The latter terms are quite loose, since there is no particular point at which pressure may be said to cease to be low and become high.

High pressure engines work *against* the pressure of the air, since their passages for the escape of steam from the cylinder open into the atmosphere; while the cylinder, acting as an air-pump, tends to exhaust all the air from the boiler, so that there shall be only steam on the side of the piston which is at the moment open to the boiler.

Low pressure engines, on the contrary, have a vacuum more or less perfect on the opposite side of the piston from the steam. Hence, with any given pressure, they have an advantage of about 14 pounds per square inch over high pressure engines.

In short, each has steam, only, on one side of the piston; while the *high* pressure engine has an opposing atmosphere, but the low pressure one a vacuum, on the opposite side.

The vacuum, maintained in the low pressure engine, exists primarily in the condenser; a vessel immediately communicating with the steam-cylinder, and into which the steam passes after effecting a stroke of the steam-piston, and is condensed.

This vacuum is produced at first by the action of an air-pump, which is a part of the engine, and which removes not only the air at first found in the condenser, but the water of condensation also. It is maintained by the air-pump and by the process of condensation itself.

There are two classes of condensers, according as the escaped steam is brought into *direct*, or *indirect* contact with cold water as a means of condensation. The former are called *jet*-condensers, the latter, surface-condensers.

Description.—Pl. V., Figs. 1, 2, 3, represents a jet-condenser of the form frequently found on American lake and river boats. Fig. 1 is a partial plan; Fig. 2 a vertical section on the vertical plane M*n*, Fig. 1; and Fig. 3 is a partial elevation, looking in the direction indicated on Fig. 1 by *n*M.

MM—M′N′—M″N″ is the wall of the condenser, which is vertical and cylindrical. AB—A′B—′B″ is the *injection-pipe*, conducting cold water to the upper part of the condenser, whence it falls through the *strainer*, CC′, and meets and condenses steam which enters from the cylinder through the nozzle D,D′,D″. A few, only, of the numerous holes in the strainer are shown in the plan.

E′E″ is a manhole, covered by a plate, for affording access to the interior of the condenser. F,F′,F″ is a slanting flange by which the condenser is kept in place relatively to the *gallows frame* which supports the working beam. The lugs, *bb*, and the lower brackets H,H′,H″, afford bearings for bolts which fasten the condenser to the *bed-plate*, Pl. VIII., and parts adjacent. The upper brackets, G,G′,G″, give bearing for tie rods which bind the beam-pillow block and the condenser to a fixed relative position. The strainer rests on lugs, as *aa′*. The lower surface N′N′ of the condenser, rests on the bed-plate, and its top rim *c′d′*—*c″d″* is the bearing for the cylinder.

Construction.—The small scale of $\frac{1}{2}$ inch to 1 foot may well be enlarged to not more than 1 inch to 1 foot. The curve KL—K′L′—K″L″ is the elliptical intersection of the oblique front plane of the flange, F,F′,F″, with the vertical cylindrical outside of the condenser. It is readily constructed by points, by simply considering that any ordinate, as *hf*, upon the centre line M*n* of the plan, will be vertically projected at F′, and on Fig. 3 at $h''f''=hf$, and laid off from the centre line $n''u''$.

Example XII.

A Surface Condenser.

Description.—Any surface condenser is an arrangement of parts such as to bring confined steam into contact with a large area of surface which is kept cold. Pl. V., Figs. 4, 5, 6, 7, and 8, gives sufficient, though not entirely complete, views of a surface condenser, as built by the Novelty works at New York for the recent Pacific Mail steamers.

In this condenser are 4,224 tubes of galvanized brass, each about 9 feet long, $\frac{5}{8}$ inch outside and $\frac{7}{16}$ inside diameter, inserted at the ends in tube plates, I., Fig. 7, where a tight joint is made by a collar or packing of compressed wood, *pp*, around the tube.

A,A″ is the bed-plate, see Pl. VIII., through a passage in which, water is forced, entering the condenser at L, and passing through the tubes and out through the out-board delivery, O, as indicated by the arrows W,W,W, Fig. 5. B,BBB is the condenser proper, into which steam enters from the cylinder, C, through the exhaust valve at *ee*, and as shown by the arrows *s,s,s*. It is condensed by contact with the cold tubes; and as it is not mixed with the cold water of the tubes, it forms fresh hot water for the supply of the boilers. This water flows into the lower part of the bed-plate A, whence it is lifted by the air-pump, P, into the hot well, and thence pumped into the boiler.

The covers, FF, Fig. 6, of the separate openings in the skeleton frame of the condenser, are called *bonnets*. F′, at the left, is an edge view of one of them. At HH, one of the bonnets is removed, showing some of the tubes. D is the rounded condenser cover. KK″, not shown in the plan, Fig. 4, is the flange resting against the gallows frame GG. Fig. 8 is its horizontal projection, corresponding in position with K. S″,SS is the steam-chest; N, the steam-pipe nozzle; Q and Q′, the steam and exhaust pipe.

Fig. 7 is a detail, enlarged, of an edge view of part of a tube plate, showing the wood packing *pp*.

Construction.—The scale, except in Fig. 7, is very small. A scale of from $\frac{1}{8}$ to $\frac{1}{16}$ would be much better.

SECTION II.—GENERAL SUPPORTERS.

A—Point Supporters.

B—Line Supporters.

Standards.

53. Standards, otherwise called, without much distinction *posts* or *columns*, are those upright supporters *around* which the working parts are mostly arranged. It may be said that standards and posts are fastened only at bottom, but columns at both ends.

This class of supporters is found in connection with upright drilling machines, power hammers, etc.

The two following examples are chosen for their excellence in affording practice in drawing compound curves, and, in part, the intersections of surfaces.

Example XIII.

The Standard of a Power Hammer.

Description.—This example, Pl. VI., Fig. 1, represents the standard for one of Shaw and Justice's patent dead-stroke power hammers, with a 100 lb. hammer. It presents some points of such novelty and interest and value, as indicated by extensive use, that the following general description precedes that of the standard separately.

Fig. 16 gives a general view of the whole machine, in two elevations.

A is the hammer, working vertically in guides B. It is attached by the belt and links, CC, to the heavy bow spring, DD, which in turn is actuated by the connecting rod, E, from a crank pin on the wheel J.

A band wheel, L, on the same axis, actuates the whole machine. Its band, however, is loose, and is made to act by pressing down the treadle, HH, which draws up the "idler wheel," GG, against the band (not shown), and makes it bind.

The same operation also slacks the leather brake, MM, and leaves the machine free to act. When the treadle is let go, the *band* is slacked and the brake tightened, by the falling back of G, and the hammer is instantly stopped.

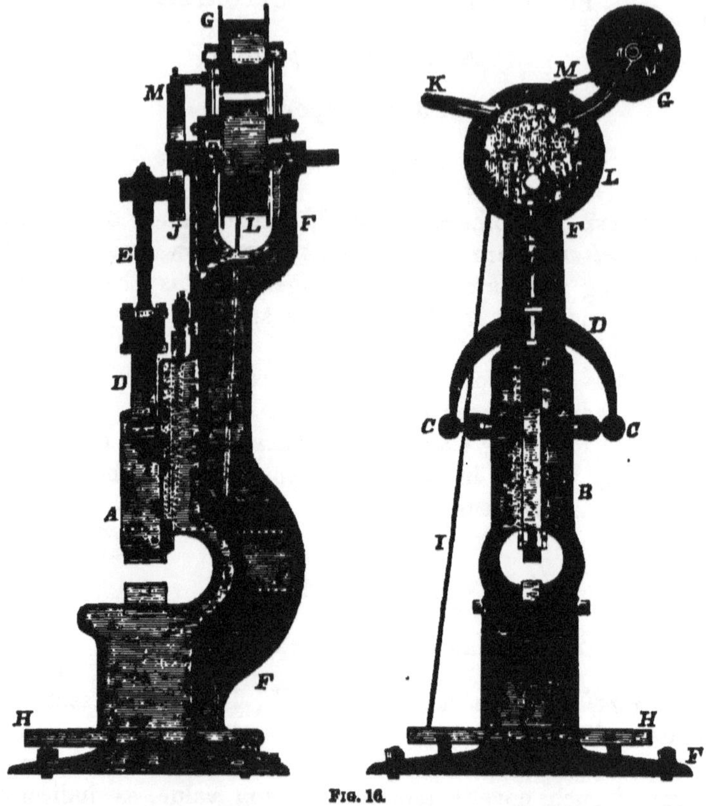

FIG. 16.

To understand the action of the spring, it must be understood that the hammer acts with great velocity, making, for a 100 lb. hammer, about 250 strokes per minute. At the instant, then, that the *spring* begins to *ascend*, the resistance of the hammer compresses it somewhat, and when it begins to *descend*, the momentum of the hammer carries it upward, still, a short distance, which causes a strong compression of the spring, while the belt CC will be slightly curved *upwards*. Then, in the remainder of the descent, the recoil of the spring acts with great force to straighten the belt and draw down the hammer much more powerfully than its mere free descent, through so short a space, could do. The spring further acts to pick up the hammer instantly after its blow has been given, so that the foundations are less beaten than by a drop hammer.

The connecting rod is in several pieces, coupled with right and left screws, so that its length can be adjusted by turning its parts by pins inserted in its holes, so as to give any desired distance between the hammer and the anvil.

Construction.—This is shown in Pl. VI., Fig. 1, by two elevations and a horizontal section through the guides. The scale used was that of half an inch to the foot. A scale of double that size, as shown in the section, is recommended.

As any two projections of an object often reveal all its dimensions with sufficient clearness, the student can often exercise himself to great advantage by constructing other views than the ones shown, from the measurements given. Thus, in the present case, a plan could be constructed, and a vertical section.

With the full measurements given in this example, no further directions for its construction seem necessary.

Execution.—Under this head but few remarks have hitherto been made, as the plates have been supposed to be executed simply in lines. Some of the figures, and this one among others, might be fully shaded with excellent effect; in this case, on account of the numerous bluntly rounded edges. When thus shaded, a figure need have no ink lines at all upon it, all the sharp edges, if there be any, being well shown by the contrast of shades between the adjoining surfaces, or by leaving a narrow line of lighter shade on edges exposed to the light, and of darker tint on edges which are lines of shade.

54. Having in view the advantage of comparing different means of attaining the same object, the following figures and description of a very elegant species of spring power hammer are added. The description is nearly in the words of the manufacturer's circular. The machine is known as Hotchkiss's patent Forge Hammer.

Description.—The Hammer represented by Figs. 17 and 18 claims

1st. Simplicity and Durability.

2d. Economy of Power and Space.

3d. Striking square with a sharp and elastic blow.

It runs with little noise, and by the peculiar arrangement of the cylinder and piston, the hammer is driven by air-springs, which saves the machine from jar, other than the blow on the anvil or work; and can be used in any building without injuring the foundation or walls.

The cylinder and hammer moving in vertical slides, each blow is square, exactly in the same place, and some kinds of die work can be forged as exact as under a drop, with greater rapidity. It is under the perfect control of the operator, can strike light or

FIG. 17.

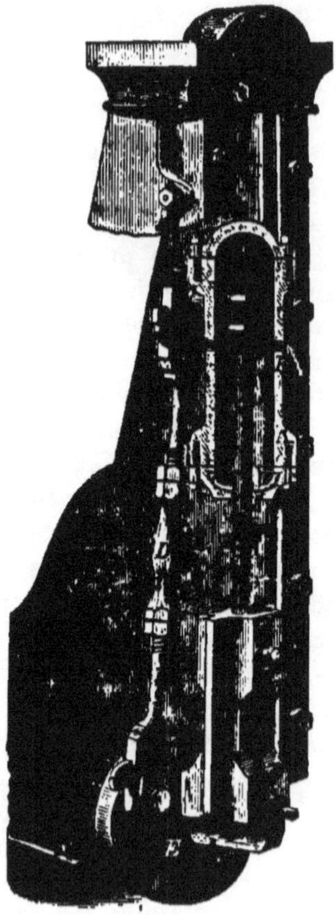

FIG. 18.

heavy, slow or fast, as desired, and will draw, weld, or swage.

The hammer derives the increased force of its blow from the *reaction of compressed air upon the piston.* The air is compressed within the cylinder A by the piston B which fits the cylinder, air-tight. (See Fig. 18.) The cylinder moves in the slides C by the action of the connecting-rod D, driven by the wrist pin in the face plate E by belting, in the usual manner. The cylinder is air-tight at each end; there are two small holes F in the cylinder, through which the air passes freely in and out, which supplies any leakage, and prevents a vacuum behind the piston as it passes beyond the air-ports either way.

The *piston, piston-rod, and hammer* are *entirely independent* of the cylinder, and can be moved up and down *without moving the cylinder;* when the cylinder is moved either way, the piston passes the air-ports F, confining the air in either end of the cylinder, which prevents the heads striking the piston, and acts as a spring, lifting the hammer up or accelerating its downward movement. The cylinder has a definite motion, governed by the travel of the crank, but the hammer has more lift, according to the compression of the air.

If the cylinder is moved up and down slowly, there is no blow given, as the weight of the hammer hangs on the cushion of air, *under the piston*, in the *bottom* of the cylinder. If the cylinder is moved up quickly, the air under the piston lifts the piston and hammer as quickly to the highest point the crank will allow the cylinder to go; then, the momentum the hammer has acquired, causes the hammer to go higher, which pushes the piston up in the cylinder and compresses the air in the upper portion, which acts as a spring to accelerate the downward motion of the hammer.

In addition to the weight of the hammer, and the reaction of the upper air spring upon the piston, the upper head of the cylinder as it comes down drives the piston and hammer down with the same rapidity with which it was raised; thus, by a rapid reciprocating motion of the cylinder, quick and sharp blows are given.

The blow is according to the speed at which the hammer is run, for when running at high speed the upper air spring is more compressed.

The speed is regulated by the idler pulley operated by the treadle. But if it is wished to run rapidly and lightly, raise the cylinder by lengthening the rod by the double nut on it, which allows the lower air-cushion to take the bulk of the blow. The hammer, after being driven down, is instantly picked up by the ascending air-cushion, without any shock or jar; and so long as the packings are tight, it can be run for years with little wear. These packings are as simple as a pump-packing, durable and easily renewed.

C—SURFACE SUPPORTERS.

a—Plane Supporters.

Frames.

55. Frames are those general supporting members of machines which, according to the usual meaning of the term, enclose certain open areas, one or more. They are also the more immediate supports of moving pieces, whose *centres or paths of motion they hold in fixed relative positions.*

So various are the forms and uses of machines, and so dependent is the form of the frame, in each separate case, upon the intended use of the machine, that it may not be possible to present a complete or well-defined classification of frames.

Still, the mind may be guided, in ranging through multiplied examples of frames, by the following view of the more conspicuous varieties of familiar or novel designs.

Frames, then, are

Beam frames; as in the general frames of locomotives, car trucks, etc.

Webbed; when thin, and embracing many regular and irregular openings. Webbed frames appear in the two principal forms of *plane*, as in the end frames of spinning machines; and *closed*, as in case of some upright engines, whose vertical prismatic, or more commonly *pyramidal* frames, consist of perforated plates joined at the corners to form a frustum of a hollow pyramid, upon and within which the working parts are supported.

Trunk frames; as that of a Corliss' upright engine, which is a frustum of a cone with a circular or oval base, as may be most convenient, and whose convex surface is continuous, except as broken by one or more large openings, to allow access to the interior.

Jointed; as those of many beam engines, and vertical direct-acting engines. The columns found in such frames are sometimes inclined, forming an open pyramidal frame. Jointed frames are also sometimes composed in part of rods united by joints, as in the frame of Wheeler's "tumbling beam" engine, etc.

Consolidated; where, for the greatest rigidity, and security from displacement by shocks, the local and general supporters, and other fixed parts, are, as far as possible, consolidated into one piece, as. in Reynolds' three-cylinder engine for reversing at full speed under full steam. The same idea is also illustrated in those steam hammers in which the steam cylinder and frame are one piece, and in the Corliss, and the Babcock and Wilcox horizontal engines.

In the *study of frames*, the principal things to be sought are, first, the combination of lightness with strength; second, easy access to all the attached working parts; third, whatever grace of outline can be had; fourth, and as a means thereto, the solution of the numerous problems of tangent curved and straight lines which occur in designing open frames.

In illustrating a few specimens only of the above descriptive list of frames, which is all that seems to be necessary, we have taken examples differing from each other as much as possible, and presenting, otherwise, useful exercises for practice.

Example XIV.

Locomotive Frames.

General Principles.—In the construction of the modern American locomotive, the objects sought are unity and firmness in the assemblage of principal parts, and ready self-adjustment, with durability, in the running parts.

To secure the first object, the boiler, with its enclosed fire-box; the frame, and the heavy castings which embrace the cylinders, are all strongly united so as to act substantially as one piece.

To secure the requisite flexibility, with steadiness, the springs over the two driving axles are linked to a balance beam, the flange on the forward driving wheel is omitted, and the front end of the engine is supported at points on the transverse centre line of the truck, whose wheels are far apart, so that a slight vertical displacement of any of them occasions but a slight movement of the central points of support.

The frames are of wrought iron, in a few heavy forgings, and are next to the inner sides of the driving wheels.

The figures 2, 3, and 4 on Pl. VI. show two different styles of frame.

Description.—Pl. VI., Fig. 2, shows the essential parts of the frame of locomotive 21, on the Boston, Hartford, and Erie R. R.; now (1868) nearly new.

GG—G′G′ is the main portion, and is solid with the jaws, EE, which embrace the axles, K, and the axle boxes not shown.

The portion at *cd* is reduced to the thickness shown at *e* in the plan. HH—H′H′, the forward section of the frame, is fastened to the rear section GG′ at *e*—*cd* and to the forward jaw E by bolts, as shown. The stirrups, FF, are bolted to the frame by four bolts each. The lugs, LL, mark the points of attachment of the cylinder to the frame.

The plain portions of the frame are broken out, that the more important parts may be shown on a larger scale.

Figs. 3 and 4 show a slightly different style of frame, and also one of the adjusting wedges, not shown in Fig. 2, showing the manner of setting the axle boxes, so as to secure the correct distance between the axles of the two driving wheels, on the same side of the engine.

In this frame, the stirrups are all alike, with horizontal bearings, and each is held by two bolts.

The inner sides, as *ac*, of the jaws, converge upwards, as seen in both frames.

One of the adjusting wedges is separately shown in Fig. 4. Its flanges *ha*—*h′a′*, and *pq*, clasp the jaw E, its interior width being seen to be the same as the thickness of the frame. The surface *ac* bears against the jaw, and the vertical surface, *ed*, against the axle box. By the nuts, N,N′, the wedge is raised and lowered. When raised its ascent along the taper of the jaw crowds the axle box to the right, or forward. D is a clamp screw, which holds the wedge tight against the box. A similar one, bearing against the opposite wedge, not shown, each acts as a check nut to the other.

The action of a check or jam nut is thus explained: When a single nut, as N, is screwed up to its bearing, any jar which turns it free from the bearing leaves it loose on the screw, and liable to work off. But when a second nut, N′, on the same screw is brought home, each clamps the other *to the threads of the screw*, as well as to the bearing, so that neither is so apt to get loose.

The nuts may bear directly against each other as well as on opposite sides of an intervening piece S, as in Fig. 3.

The bolt heads in the main joint MM′ are countersunk, that is, let into the iron, nearly their whole depth.

In the West Albany Machine Shops of the N. Y. Central R. R., where the data for Fig. 3 and several other examples were obtained, locomotive building, as well as repair, is carried on to such an extent that an admirable uniformity of like parts in all engines of the same class now built there, has been secured by means of a system of steel gauges for internal and external turned work, and of nuts and bolts for all parts of an engine.

Construction.—From the full measurements given, either or both of these frames can now be drawn without further explanation; simply observing that always, where a large number of equal bolts are used, it is sufficient to show one or two of them and indicate the positions of the rest by their centre lines, as shown in figures 2 and 3.

Pl. VII., Fig. 6, gives a sketch, merely, with measurements of a web-frame for the support of a lathe bed. It affords very good practice in compound curves, suitable centres for which can be adjusted to the given measurements, by the student.

b—Developable Supporters.

Beds.

56. Bed plates, or, simply, beds, are the general supporters of engines and other heavy machines, consisting of several parts resting on a common foundation.

They are usually of cast iron and made in one piece, or in sections, firmly bolted together, according to their size. They rest immediately upon the masonry foundation beneath them; and, being in one piece, accord with the important structural principle of *continuous bearings*, for what they serve to support.

Bed plates may be described, as to their varieties, as *flat* bed plates, when consisting simply of a flat plate on which the supporting frame-work of an engine rests; *box* bed plates; open, when consisting of four sides in one piece surrounding an interior space, open at top and bottom, as seen in many horizontal engines. These bed plates are, moreover, usually symmetrically divided by a vertical plane through the axis of the cylinder. See Fig. 19.

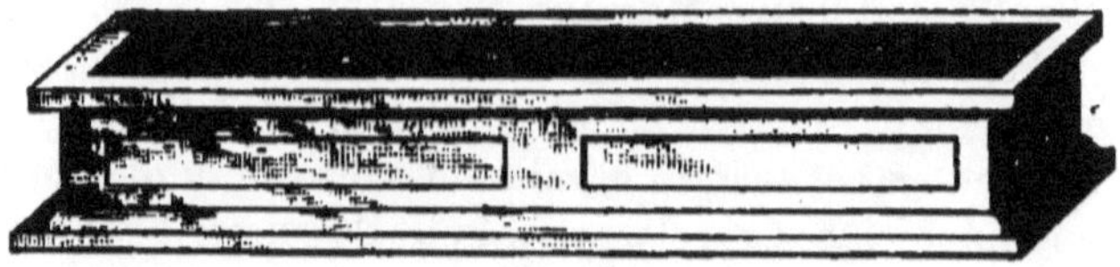

FIG. 19.

There are also *covered box* beds, that is like the last, only closed on top, narrow, and having the guide, cylinder, etc., bolted to them on one side; as in N. T. Green's horizontal engines; and *tank beds*, which are hollow and answer for other purposes than that of mere support. Among these we distinguish, marine engine tank or hollow bed plates—see Pl. VIII.—and pier tank beds which are quite massive, and proportioned with such breadth of base as to preclude the necessity in many cases of a separate masonry foundation. These are used more for portable horizontal engines, though a very common practice is to make the boiler strong enough to serve also as a bed for such engines.

EXAMPLE XV.

A Prismatic Beam-bed and Pedestal.

Description.—Pl. VII., Figs. 1–5. This beam-bed is from a Babcock & Wilcox horizontal engine of 10″ bore and 30″ stroke of piston. It is shown in elevation, and three vertical sections, at AB, CD, and EF. The two sections at CD and EF are taken looking towards the cylinder, and show all *in* and *beyond* their planes.

By pointing out the different projections of numerous points, and by comparison of measurements, and distances with the dividers, the student will be able to apprehend the form of this somewhat irregular, but very neat and substantial frame or bed. The entire frame, or engine support, embraces the standard pillow-block or pedestal, MP, the bed-piece, HH, and the cylinder with its pedestal, not shown, which are all bolted together, as at *gg* and *mm*, so as to act as one piece.

In the bed-piece, on all the figures, L is its vertical web, KK its ribbed lateral wings, on the back, and GG those on the front, which are shaped to act as guides. I,R, is the back cylinder head, and the parts within IIIH′, form the stuffing-box. The inner circle, *c,c,c,* is that around the piston-rod, and *d, e,* and *f* are circles of the cylinder-head, as seen by comparing Figs. 1, 4, and 5. *m,m,m,m,* are some of the bolt-holes through which the bed is bolted to the cylinder-shell. Q is the back end of the steam-chest, and *h* the collar around the valve-stem. The hollow square guide *k* and parts adjacent, Fig. 4, are for the support of some of the moving parts. The disappearance of edges into a plane surface is shown, as at *r,r.*

The ribs, K,K, are partly shown, dotted, in Fig. 1. The long curves of the bed and pillow-block are constructed by ordinates; and in part by circular arcs, as shown.

The pillow-block, MP, Figs. 1 and 2, may, as said before in Ex. III., be drawn separately; and on a scale of one-*sixth*, one-*fifth*, or one-*fourth*. It is largely hollow, as indicated by the dotted lines, next to the outer ones. *ss* is the diameter of the shaft. Only the larger measurements are given. The rest can be found from the scale. The boxes are adjustable laterally, as is proper in all blocks for horizontal engines, by set-screws and jamb-nut, as at *p, q.* N and O are moulded edges.

A plan may be substituted for Fig. 2, or a vertical cross section through its centre.

Figs. 1 and 2 should stand on the same line, so as to favor the projecting of points from one to the other.

Construction.—As in all such cases, lay out the longer and outer lines first, and fill in the smaller parts afterwards. The proper heavy lines are nearly, if not all, indicated by two dashes across them.

D—VOLUME SUPPORTERS.

Example XVI.

A Tank Bed-plate.

Description.—This example, Pl. VIII., is an excellent one, representing the bed-plate for a beam-engine, as built by the Novelty Iron Works of New York for the Pacific Mail Steamship Co.'s steamers, of the class having cylinders of 105″ di ameter and 12 stroke of piston.

Fig. 1 is a plan of the bed-plate. Fig. 2 shows a longitudinal section made by the vertical plane MN. Fig. 3 shows a transverse section made by the vertical plane PQ. Fig. 4 is a fragment of an elevation, as seen in looking into the opening at BB′.

Over the chamber EE′, on the surface CCC—C′C′—C″C″—C‴—see the several figures—the condenser, square in plan and fastened by bolts through *bb*, etc., is set. Over the condenser is the steam cylinder, which is vertical. The condenser here supposed is of the kind called surface-condensers, Ex. XII., consisting of a tight central chamber over E*e*, *through* which many tubes pass, *from* a side-chamber over DD—*d*′, *to* an opposite side-chamber over II,I′. Water is delivered to the condenser *from* a steam-pump, *through* the passages DD—D′D′—D″D″, which begin at BB″—B′—B‴. These passages lead over the arch, *ff*, Fig. 3, which is at DD in the bed-plate, and the water enters, through DD—*d*′, the side-chamber over that opening.

Exhaust-steam from the cylinder, entering the condenser, is liquefied by contact with the multitude of cold tubes which traverse its central chamber, and the water of condensation flows to the right-hand part, KK′, of the hollow bed, where it is removed by the air-pump, which also maintains the vacuum in the condenser.

The air-pump stands at AA—A′A′—A″A″—A‴. Valves at FF—F′, called the foot valves, prevent the re-flow of air and water to the condenser, when the air-pump bucket or piston descends. FF—C′G′ are openings to give access to the foot-valves, and are covered by a bonnet. H is the opening for a pipe leading from the bilge to the condenser, and used in case the vessel springs aleak. The vacuum in the condenser causes water to flow into it from the bilge, which is then removed by the air-pump. *mm*′ is a manhole for entrance to the bed-plate.

The surfaces to which the word "faced" is attached are planed, to secure accurate bearings and tight joints.

Construction.—The plan being entirely symmetrical, except the difference between the holes, *mm*′ and H, it would be sufficient to draw just enough more than half of it to include both of those openings.

The completeness of the measurements, and the repetition of many of them, will aid in understanding the foregoing description; and as no directions, besides those already given in previous problems, are necessary, the drawing is here left to the student, with the correct location of the heavy lines on Figs. 1 and 2, as a study. These lines can generally be found by inspection, by careful attention to the principles of (47).

Example XVII.

Housing or Chambered Frame for a Reversible Rolling Mill Engine.

Description.—Pls. IX., X., and XI. [Let Plate X. be cut out and pasted at the top edge to Plate IX., so that their centre lines AB and AB shall coincide in direction. Then paste Plate XI. to the right hand of Plates IX. and X., so that the lines CD shall be in the same horizontal direction. The three projections will thus, if it be desired, be brought into proper relative position for reference.] This remarkable design is from an engine designed by G. H. Reynolds, of New York, in 1866, for a steel rolling mill at Troy, N. Y.

A very similar arrangement is described* as an English invention patented by a Mr. M'Naught, viz., three radially equidistant cylinders, fixed in a common frame, with the connecting rods jointed to a common crank pin, and the valves worked by one eccentric.

The rolls of a rolling mill usually consist, as described in the next example, of three lines of horizontal cylinders, one above another, and with circular grooves around them, shaped to the section of the bar to be rolled. If, then, a piece shoot through between the upper rolls as shown at *a*, Fig. 20, one rolled through the lower rolls will return as at *b*, from the workman at B, to the one at A. This is re-

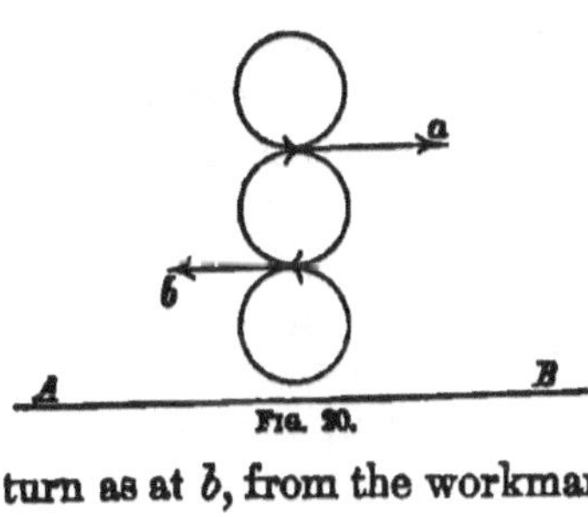

Fig. 20.

* Imp. Cyc. Machinery.

peated, through the different approximately shaped pairs of grooves which lie together along the same rolls, until the final form is given to the manufactured bar.

To accomplish this alternate passage of the bar with only two rolls, their motion must be reversed every time the bar passes them.

The rolls revolve very rapidly; they require an immense power to actuate them when numerous; and to save time, loss of which would fatally cool the bar, they must be reversed instantly. To fulfil these cardinal conditions by direct action, that is, without gaining the necessary speed by gearing from the engine shaft, is the object of the novel engine here partly illustrated.

This engine was to be of 3,000 horse power, to make some 300 revolutions per minute, and to be instantly reversible at that speed, with steam on, many times per minute. One of the lines of rolls was to be on the same shaft, S, Pl. XI., with the engines, and would be geared to the other.

To secure this high *shaft-velocity*, without excessive piston speed, the cylinders, of which there were to be three, placed 120° apart, as shown on Pls. IX. and X., were to be of three feet diameter, and only one foot stroke. To secure the great *power* required for many rolls, without cylinders too large for the required velocity, there were to be, as stated, three cylinders, of which one was to be vertical. This arrangement, also, would apply the power more equally to the shaft, and with never but one "dead point" at a time. Finally, to provide against the great strain, and *dislocation of parts*, arising from the many quick reversals, the three cylinders and the frame, enclosing all the steam passages, were to be in one huge casting.

Pl. IX., Fig. 1, shows the plan of the engine, with a partial section of the vertical steam-chest, OO, and slide-valve, DD. Here EEE is the branching steam-pipe, and F,F,F are the vertical central planes of the three cylinders, placed far enough apart, laterally, to allow their respective connecting rods to act side by side on the same long crank pin, P, Pl. XI. Q*h* is the top cover of the vertical cylinder. L, O, and L′ are the three steam-chests, and the oblique cylinders are shown in dotted lines. Steam enters through the pipe, E,E, and distributes itself, as shown by the arrows, through the general steam-passage, T, to the three cylinders. Escaping through whichever steam-port is at the moment under the valve, it flows, as at D, into the exhaust passage, K. The cylinders are thus steam-jacketed, both by the live and exhaust steam.

Pl. X. shows the front elevation. CD is the sole of the frame, which is of cast-iron, whole. Its upper part, FGB, is not exactly of uniform radial width, its three outer semicircles being described from a centre, O′, 2 ins. below the principal centre, O. The three cylinders are shown in dotted lines, with their steam and exhaust ports. E and E, at the upper cylinder, are sections of the steam passages, where they enter the steam-chest, at O, Pl. IX. KK, at the same place, are sections of the exhaust passages, as at *hk*, Pl. IX. The arrows will make the course of the steam intelligible. H is a three-armed brace, solid with *ab*, and alternating with the centre lines of the three cylinders. O is the shaft; *e*, Babbit metal lining to the boxes, *d*; and *cd* are adjusting keys; *gf* is a wrought-iron ring, carrying the guides to the piston cross-heads, whose outer ends are bolted to ears, R.

The pilasters, I, are fluted in the original design, and can easily be made so by the draftsman. Also, the small panels of the upper part were finished, as at *m*, with bevelled edges and round corners.

Pl. XI., Fig 1, shows in part a vertical transverse section through the axis of the upper cylinder; and below an interior view, with the end outer wall of the frame, between D and G, Pl. X., broken away. Thus the interior of the cylinder, V, is shown. M are holding-down bolts passing through masonry. K is the exhaust pipe; L and L′ the oblique steam-chests, shown before, and O the vertical one, shown in section. The spherical piston, R, being at the top of its stroke, steam is just entering from O, through the port, *p*′, and escaping from below the piston, through *q* and K. Q*h* is the cylinder-head, moulded to conform to the piston and the nut that secures the piston-rod. The shaded portion, at *gf*, is a section of one of the two wrought-iron rings which carry the guides, *n*.

The main shaft, S, is cranked, as shown at each end of the long crank pin, P, to the middle of which the connecting-rod of the upper cylinder is attached.

Construction.—As the plan is only partially symmetrical with respect to its two centre lines, the only considerable reduction which can be made in the drawing is, to make but half of the elevation, Pl. X. By exercising great care, these plates may be made on a scale of $\frac{1}{20}$ of an inch to an inch; or of half an inch to a foot. Only the more important measurements are recorded. The rest may be determined nearly enough from the given scale.

Example XVIII.

Housing for a Rolling Mill.

Description.—Pl. XII.—This example presents some fine features, both of construction, and for practice in execution.

It was reduced from designs by Mr. John Fritz, of Bethlehem, Pa., for the mill for rolling Bessemer steel, at Troy, N. Y. The whole is in one piece from H to I′. It is here shown in plan, two elevations, and a horizontal section of the uniform column.

The elevations have each a vertical centre line, and the plan has two. From these lines many of the measurements can be laid off.

Several housings or frames, like this one, are ranged in vertical positions, parallel to each other, and supported on cast-iron ways, W,W′,W″, to which they are bolted by four bolts, shown as at *mm′*. These ways are bolted through oak timbers, 20″ wide by 18″ deep, to masonry, on which, in turn, they rest.

The face of the upper part of the housing, from *aa′* to *bb′*, consists, as shown in the plan, of three vertical cylindrical segments, tangent to each other, and to the plane portions exterior to *aa′* and *bb′*.

The largest of the several mouldings on the outside of the housing is cylindrical from *c* to *d′*,*d″*, and double curved above *d′*,*d″*, and also of increasing width above *e*, and, therefore, less sharply rounded, as its thickness, seen in front elevation, is constant.

The several circles, having O for their centre, are the vertical projections of circles, or of cylindrical surfaces, whose axes are perpendicular to the vertical plane at O.

The recesses at *b′fg*, seen in the section at C, admit the long bolts, DD. These bolts are attached, as at D, to levers, suspended as at E, and from whose inner ends, as F, depends by links the pile of weights, G, which contains two cubic feet of iron.

This arrangement is for use in plate mills, where two rolls are used. The bearing of the upper roll, being clamped to the bolts D, would be drawn up strongly with the upper roll, against the "top rider" or covering of the bearing, over the roll. This cover is, in any case, held down by powerful screws working through the aperture I,I′. As, then, this screw bears down on the rider and thence on the roll, it acts to depress the bolts D and raise G.

In the rolling of rails, etc., three rolls are used, and their bearings are confined and their separation adjusted within the recesses R and R′. Owing to the immense strain to which the rolls are subjected, which can best be realized by observing how they are sometimes broken in two in spite of their great size, some yielding point must be provided which should prevent the breaking of the rolls by being a little weaker than they. Therefore, a piece, called a break piece, is inserted between the point of the screw which works through I,I′, and the top rider or upper bearing of the upper roll.

The wings, as K,K′, support rollers which aid in handling the bars to be rolled. The apertures, as *h*, enclose the bearings of similar rollers used in passing the bars through the upper rolls.

Long rods pass through the holes at *n*,*n*′,*n*″, to couple the several housings together.

Construction.—The scale of Pl. XII. is $\frac{1}{16}$. Besides the familiar directions to draw as much as possible from the centre lines, and to lay down first the main outlines, and then the smaller parts, the main features in the construction of this housing are the nice union of its numerous and crowded tangent arcs, and the construction of the curves as at *kp*.

To secure the first point, the pencilling should be constructed in the finest lines, with the utmost care, and without wearing holes at the centres of the groups of concentric arcs; and then they should be followed with the utmost exactness by the pen.

The scale of one-sixteenth or three-fourths of an inch to a foot is a very small one for such an object as is here described. Except as a test example for fine drawing, it might better be made on a double plate to a scale of one-eighth.

The construction of the curves at *kp* is an easy application of the problem of two intersecting right cylinders, as will next be explained. (See Fig. 2.)

Let ACD be the horizontal projection of the segments, AC and CD, of two vertical cylinders—analogous to MN and N*b*, in the plan, Fig. 1—tangent to each other along a vertical element at C. A″D″D′ is the vertical projection of the same surface. EFD—E′F′ is a cylinder whose axis, EO—O′, is perpendicular to the vertical plane, and which answers to a cylindrical surface having *qt′q*, Fig. 1, for its vertical projection. ACD and E′C′F′ are thus two projections of the intersection of these horizontal and vertical cylinders. To find the projection B″C″F″, answering to the curves at *ko″p* in the end view, Fig. 1, take the auxiliary vertical plane, PQP′, perpendicular to the ground line, then any point, as CC′, being projected upon it as at *cc′*, may be revolved in the arc *cc″*—*c′*C″ to C″.

In like manner, to find any points as *o″o″* in Fig. 1, assume *o*,*o*, in the plan, project them upon the vertical projection of the same curves at *o′*, project them across to the end elevation, and there make the points *o″* equidistant from the centre line *cd″*, by the distances of the same points *o* in the plan, from the centre line *nd* of the plan.

In the plate this construction is not followed, as is evident upon inspection, since it would be very difficult to draw properly so many parallel irregular curves. Besides, it is fortunately unnecessary, for the true width of the housing, at any point, is learned from the plan, and the height at which any given width occurs is found from the front elevation. These curves are therefore drawn in compound circular arcs, as shown, only the extreme points, as *p* and *k*, being correct.

Example XIX.

A Passenger Car Truck.

Description.—Pl. XIII., Figs. 1—4, represents a four-wheeled passenger car truck from the Pennsylvania R. R. As the description of it proceeds, it will be seen to be very complete and well designed, with abundant provisions for safety under all circumstances.

Premising that the same letters refer to like parts on all the figures, they will not here be repeated from all the figures for every part.

Fig. 1 is a partial plan. The right-hand half of Fig. 2 is a side elevation, and the left-hand half a longitudinal section through the centre. Fig. 3 is half of an end view, and Fig. 4 is a cross section showing the parts between the wheels.

AA′ is the outer side beam, and B′ the opposite one. CC′ is an inner side beam. DD′D″ is the end transverse beam. EE′E″ is the swinging bar, carrying the brake shoes W,W′W′, which are held from the wheels by the springs UU′. F,F′F‴ is one of two fixed transverse beams, framed into AA′, and bolted through C′C′. GG′G″G‴ is the swinging beam, resting, by the block *e′,e‴e‴*, on the springs JJ′J‴, which bear upon the stirrup board H′H‴, which is suspended by the short diverging links S′S′,S‴ from the hangers R,R′,R‴. These hangers are bolted to the cross beams FF.

I′I‴ is the equalizing bar, resting on the axle boxes M,M′,M″, which play up and down in the guides LL′L″. Between I′ and A′ are two rubber springs K′,K‴, kept in place by a cover *f′f‴* and step *g′g‴*, notched over the equalizing bar, as at I‴.

NN″ are the wheels, ribbèd on the back, and whose axles, OO′, rest in the boxes MM′. PP′P″ is the centre bearing, and T′T″ is one of the two opposite side bearings for the car body. Q,Q′ is a block, which supports the iron plate PP′. VV′V″ is a safety stirrup for catching the brake bar, EE′, in case of its breaking loose; X′X″ is a similar stirrup for holding up the axle in case it should break. It often consists of a simple bow of iron suspended as at *o′* and *p′*, but owing to the weight of the axle, it is here strengthened by the pillar bolts *r′r′*. YY′ is a guard strap, to keep the beam GG′ from being thrust up by any concussion. *h′h″h‴* is another stirrup for catching the swing board, H′H‴, in case of accident. ZZ′Z″ is part of the brake gear. Being drawn, as at *d*, by a chain from the brake stem on the car platform, it obviously tends to move the joints *s* and *t* as shown, and to draw the brake shoes W,W′ against the wheels.

a,a′a‴ are supports for the springs JJ′. *c,c′* is a saddle plate under which the truss rod *bb″* passes. *kk′* is the king bolt which holds the car body in place on the truck.

Ideal Conditions and Result.—If the plane of the upper surface of the rails of a railroad were a *perfect* and *perfectly unyielding* one, true as the ways of a lathe bed, if the track were *perfectly straight*, and if the *circularity and equality of the wheels* were perfect, the motion of every point of a car would be one of perfect translation, and would be as easy with no springs, or yielding parts of any kind in the running gear, as with them.

Actual Conditions.—None of the above conditions perfectly exist. The actual variations from perfect circularity and equality in the wheels are, however, insensible, compared with the defects in the other respects, and may be neglected. But the track, taking the centre line of each rail for reference, exhibits both vertical and lateral variations from a straight line. The former may, further, be simultaneous in the two rails, as in the depression at the joints of tracks whose joints fall on the same cross ties, and may be equally or unequally so; or they may be alternate, as in case of tracks, often seen at present, in which the joints fall on different ties, half the length of the rail apart.

Lateral deviations may also be simultaneous or irregular, equal or unequal; and in both kinds of variations there may or may not be, though there generally is, a return to the line departed from. Finally, owing to these irregularities, the surface determined by the tops of the rails is neither a perfect nor an unyielding plane.

General Idea of Rolling Gear.—This is, to secure in all the yielding points of the truck easy and brief oscillations about mean points, so that for a given vertical and lateral unevenness of track, every point of the line of support, T″P″, and hence the car body, shall move as nearly as possible with as pure motion of translation, when the track is straight, as it would do under the ideal conditions above stated.

Normal Condition of Springs.—Let a sixty-seat car be loaded with fifty people, for example, for a given journey, and be standing still. All the springs will then be compressed with what we will call their normal compression for that time. This degree of compression would remain constant under the *ideal* conditions already mentioned.

Action under Actual Conditions.—The car body rests on the swing beam GG′G″, at the three points, P″, and the two of which T″ is one. This beam rests on the springs, J, on both sides of the truck; and there, on the swing board, H′H‴, which hangs by the links, S′, and hangers, R′, from the cross bars, FF, of the rigid frame of the truck. This frame rests on the four rubber springs, of which K′ is one; these upon the equalizing bars as I′, and these, finally, on the axle boxes, MM′, which bear upon the tops of the axles OO′. The last are a final rigid support, being practically solid with the wheels which roll upon the track.

Now let *qn*, Fig. 2, be the top of a rail depressed at the joint *q* by the distance *mn*, exaggerated for illustration. When a train is at full speed, the point at *n* of the wheel is going horizontally, at from 40 to 70 feet per *second*. Hence the rise, *mn*, is practically instantaneous, and nearly the same as if the wheel were thrust up the same distance by the impact of a violent blow from below. Then, when *n′* thus instantly rises, a small distance; M′, and the end of the bar I′, rise the same distance, which compresses the spring K′. Now, if the force and quickness of the rise are no more than enough to compress the spring, the truck frame, AA′, &c., will be undisturbed; but if the upward action be not all taken up by the spring, the truck frame will be raised by the excess of that action, and will, through RR′ and S′, draw up the swing-board H′H‴, and compress the steel springs JJ′. If, also, this excess of upward action be not all spent on the springs JJ′, the swing-beam, GG′, and the car body will be raised a little.

We now see that with no springs the car body would sensibly be raised as far and as suddenly as the wheel at *n′*. But with springs, if lifted at all, the motion will be comparatively easy and gradual, through the gradual action of the springs, in their compression and expansion, till again in equilibrium.

Let us now consider the effect of lateral shocks. A sidewise lurch of the truck to the left, for example, will carry all its rigid parts, A‴R‴F‴, etc., Fig. 4, to the left. This will carry the points of suspension, *u*, of the links to the left, which will make the left-hand link more vertical, and the link S‴ more oblique than now. Equilibrium will thus be destroyed; and the links, in coming to equilibrium by returning to an equal inclination to a vertical direction, will carry H‴, and with it the springs J‴, and the swing-beam G‴, and car body.

Trucks are very often made with the top of G‴ no higher than the top of F‴, and with the links S‴ then swung from an axis resting on top of FF,F′F‴, and reaching from that to *o‴*. In such a case, the links are usually vertical, also; so that the oscillations would be slower, by reason of the length of the links, but longer, owing to the less upward component of motion in an arc of longer radius. The *shortness* and *convergence upwards* of the links in the present design would, we should think, make the truck quite stiff, though properly so, in respect to lateral oscillation; and not too much so for a smooth and firm track.

This descriptive statement of the theory of the four-wheeled car-truck is all that could well be given, for there could hardly be a clearer example of a piece of mechanism whose proportions must be determined experimentally, since the numberless variations in the intensity and direction of the forces applied would seem to make a mathematical, that is, an exact *quantitative* investigation impossible.

Six and Eight-wheeled Trucks.—The *idea* of both is the same, viz.: to secure a perfect rectilinear motion to the car-body, by having all the irregular wheel-motions occasioned by the track taken up within the truck itself. In the eight-wheeled truck a heavy intermediate timber frame, or carriage, rests on two common trucks, and the car-body on the middle transverse line of this carriage, where there is almost no jarring motion. In the six-wheeled truck the same end is not quite so fully accomplished, since there is one rigid frame as in the four-wheeled truck, and the intermediate carriage is shorter; but there are two swing-beams, one on each side of the middle pair of wheels, and a small carriage-frame rests on the two beams, while the car-body rests, as before, on the transverse centre line of this carriage.

In respect to these trucks, it may be said that, though a perfect track costs a good deal, yet the *difference* between that of a smooth and a rough one may be *small, as compared with the cost of either*, and the extra cost of compound trucks, at first, and for repairs, and of the *extra motive power* required to transport them, might go far to cover this difference.

Car Cross-sections.—While on this subject, it seems appropriate, from association, to mention two great points of car-moving economy, one of which seems to be more and more overlooked. These are, the steadily increasing size of the cross-section of the car, and the distance between the cars. A part of the needless extravagance of the times is the indiscriminate carrying of the luxury of high rooms to *live* in, into vehicles to travel in, as if travelling about were the normal condition of one's life; and forgetting the needless and great waste of means in dragging 100 square feet of car end, including truck surface, at 40 miles an hour through the air, when 70 would do just as well, and afford every essential comfort. Side-doors opening *outward* are dangerous; in one way if locked, and in another, if unlocked. If opening *inward* they waste room. Hence, end-doors and the centre-aisle seem best, as well as hardly to be dispensed with, as affording the comfort of a free inside passage through the train. But, observing that perhaps more travel singly or in threes than in pairs, cars might be better filled, as well as economically narrowed, by having a row of single seats on one side of the aisle. Then, as to *height* of car, who has not felt cold feet while the thermometer would be at 100° to 110°, at the ceiling of a car 7½ or 8 feet high? Let the car be heated by water-pipes at the passengers' feet, and coming from heaters under, or in open-screened compartments in the end of the car; and, with suitably screened ventilators, 6½ feet would be high enough for a car inside. Thus, the cross-section of the car-body might be reduced to only about 60 square feet; which is a great matter, when it is considered that the resistance of the atmosphere exceeds the sum of all other resistances on a level, at speeds of 30 miles an hour or more; and that it increases at least as the square of the velocity of the train, so as to be *four* times as great for *twice* the velocity.

Again, when the space between the cars is equal to the width of the car or more, the resistance of the atmosphere to each car after the first one, is from seven-tenths to nearly as great as it is to the first one. Hence, the platforms should be enclosed by painted canvas, or by permanent walls, with end and side doors, the latter to be opened just before arriving at stations by the train operatives. The car ends, having also suitably adapted couplings, might be thus not more than one foot apart instead of eight or ten as now, and with great economy of power.

The sums saved by these arrangements being spent upon the track, fencing and flagmen, the increased smoothness and safety, together with the diminished train-surface exposed to air-resistance, would allow increase of speed without increase of cost over present results; and this, by saving *time*, would be real economy, inasmuch as trains and travellers are not *producers*.

Reduction of Car Weight.—Once more, we must mention the enormous dead weight of train, including the engine, for each passenger, even of a fully filled train. This weight is not far from 1,500 pounds per passenger. To materially reduce this most power-wasting excess, it may be best to return to an essential feature of Nature's works; since the more truly they are imitated in man's constructions, the more perfect the latter will be. This feature, or principle, is, that Nature's great constructions are aggregations of many small ones, in the form of minute cells.

In other words: *component parts should be small and hollow.* Applying this principle to a car, its frame might well be composed of sheet-iron cell-work, prismatic or cylindrical. Applying it to the train as a whole, the superior strength of a small cell, to a large one with equally thick walls, would point to the adoption of small light cars, or of numerously partitioned large ones.

As the enormous dead weight of a train is bodily *lifted* through the vertical height of every gradient on the line traversed by it, while friction is the only car resistance, on a level, it may well repay railroad companies to make ample experiments, founded on these considerations, with a view to the *greatest possible reduction of car-weight per passenger.*

Construction.—By an oversight, the horizontal heavy lines of the plan were placed on the lower edges of the several pieces of the truck. This is the English fashion, but it supposes two different directions of light, one for the plan and another for the elevation.

The circles with cross-marks in them are washers, with nuts only indicated, not shown.

If desired, the scale could be enlarged to one inch to one foot, and the cross-section could be placed side of the end view to facilitate its construction.

CLASS II.—RECEIVERS.

57. In any machine there is a piece, which is the first in the order of *mechanical succession*, it being the one to which the motive-power is first applied. Such a piece is called a *receiver*.

In a vast number of instances, the receiver is a band-wheel.

When many machines, as the looms of a cotton-mill, or the lathes, planers, etc., of a machine-shop, receive their motion from an overhead line of shafting, which constantly revolves, it is important to stop any one or more of those machines at pleasure, without compelling the rest to stop. For this purpose, the band-wheel receiver is double, that is, two equal band-wheels play side by side upon the same shaft; but one is fast to the shaft and the other loose, so that when the band from the overhead pulley is shifted to the loose pulley, the machine stops. In some cases the fast and loose pulleys are on the line of shafting.

Another very common form of receiver is a *piston*, whether in steam, gas, air, or water engines.

Other forms of receivers are named in the *general table*.

A—Point Receivers.

B—Line Receivers.

Of these it has not seemed necessary to present any figures.

C—SURFACE RECEIVERS.

a—Plane Receivers.

EXAMPLE XX.

Locomotive Piston with Roth's Steam-piston Packing.

Description.—This piston-packing, Pl. III., Figs. 4—8, is made to pack the cylinder of any engine, or pump, uniformly, and no more tightly at any given pressure than is necessary; creating little or no friction. It embraces certain novel improvements in that class of packing for pistons, which is composed of metallic sections, forming, when put together, expansible rings which are applied within annular recesses, formed in the circumference of a piston, so that when acted upon by steam, the sections of a ring will be forced out against the inner surface of the cylinder and thereby pack the piston. In this class of pistons, the steam acts upon the packing from the inside outward, provision being made for the passage of the steam into chambers formed on the piston, and thence through an outer ring, which receives the packing-rings.

The main object of this invention is, to construct and apply sectional packing-rings to a piston, so that advantages shall be secured superior to those heretofore attained in that class of pistons which have their packing expanded by steam.

Another object of this invention is, to provide, in horizontally working pistons, against an unequal wearing away of the same, by the application of an elastic plate to that part of the piston which slides on the lowest point of the interior of the cylinder, which plate compensates for the extra wear caused by the weight of the piston.

There have been in use many patent piston packings, that have proved to be in some cases superior to the ordinary metallic packing, in so far as they have saved labor in packing the piston. In other cases they have proved to be inferior to the ordinary packing, particularly when an engine has been run to its full capacity. The reason of this is, that in all such pistons the whole inner surface of the piston is exposed to the pressure of steam in the cylinder, and this high pressure causes too much friction, consequently wearing away the cylinder and packing-rings unnecessarily fast.

Some kinds of packing-rings are in sections, which adjust themselves to the inner surface of the cylinder at a very small pressure, but at a pressure of from 100 to 150 pounds to the square inch cause so great friction as to waste, it is said, from 10 to 20 per cent. of the power of the engine.

In other cases, where the packing-rings are cut only once, those rings have to be tolerably thin, so as to spring out to the inner surface of the cylinder, at ordinary pressure, and are therefore easily broken in case there are any flaws in them.

In almost all such pistons there is a solid, or uncut guide ring, which has to be fitted as nearly as possible into the cylinder; but in horizontal-working pistons this ring wears more or less on its lower side, and, in course of time, throws the piston out of the centre of the cylinder, which causes an uneven wear to the piston-rod, stuffing-box, guides, and cross-head. All these evils are avoided in this piston, thus:—

1st. The rings are very strong, much broader, and have only a small part of surface of the packing rings exposed to the steam (water or air).

2d. By a peculiar and simple contrivance, when an engine is running with steam, the pressure on the lower part, between the piston and part of the packing, causes the piston to balance, or force it somewhat off the bottom of the cylinder. When steam is shut off, an elastic plate on the bottom of the piston, which is pressed out by a spring or set screw, or both combined, answers the same purpose.

The packing rings are cut in not less than three pieces, but this depends upon how true the cylinder is. When the packing is put into cylinders that are out of round and not parallel, it is better to cut the rings into more pieces, that they may adjust themselves better to the irregular form of the cylinder.

3d. In locomotives running with from 100 to 150 pounds of steam, and whose cylinders are true, one-eighth inch steam bearing under the packing rings is all-sufficient; on worn cylinders one-fourth to three-eighths of an inch may be necessary at first, afterwards it can be reduced to one-eighth and less. The surface of the rings bearing against the surface of the cylinder can be quite broad; the friction will be no more, and they last longer in proportion to their width. These rings must be fitted very closely, and must not have more play than will permit them to barely move between spider and follower.

Fig. 4 is a section through the axis of the piston rod.

Fig. 5 is an end view of the piston, with the follower removed.

Fig. 6 is a diametrical section through the central ring.

Fig. 7 is a central section through central ring, perpendicular to the axis of the cylinder.

Fig. 8 is a diametrical section through packing rings.

A and *B* constitute the body of the piston, *A* being the "spider," and *B* the follower, which parts are constructed and bolted together in the usual manner, as shown. When these two parts are bolted together, they form an annular space between their flanges for the reception of the central ring *C*, which is supported by the ends of radial arms of the portion *A*, so as to lap over the joint of the two parts *AB*, as shown in Fig. 4, and thus prevent the entrance of steam in the body of the piston. This ring, *C*, is fitted so as to be steam tight and immovable, when screwed between the spider *A*, and follower *B*. The ring, *C*, has a central rib which leaves on each side of it an annular space for receiving the expansible packing rings. This centre ring, *C*, forms one side for each expansible ring (the inner side), the other side being formed by the circular flanges of the piston as shown in Fig. 4. The inner corners of the circular flanges of the piston are bevelled as shown at *aa* for receiving the corresponding bevelled surface of the packing rings *bb*, portions of which are closely fitted within the annular chambers above described, so that their circumferences project a short distance beyond the circumference of the piston. The packing rings *bb* are made up of segments or sections, and are held out, so that their outer surfaces press gently against the inner surface of the steam cylinder, by means of springs *cc*, the ends of which are bent outward, as shown in Fig. 5, so as to act upon the *ends* of the segments *bb*, and thus to keep the ends of all the segments snugly together, except those segments between which the springs are bent outward. These springs, *cc*, not only act to expand the sectional packing rings, but they also operate to keep the ends of the sections together so as to form tight joints. To prevent the entrance of steam within the chambers between the sections, *b*, and the flanges of central ring *C*, there are recesses in the ends of those segments which receive the ends of the springs, *cc*, and, inserted into said recesses, short pieces, *G*, which break joints with the joints of said segments. The segments *G* may be screwed, or riveted to the segments *b* on one side of the joint. For horizontal working pistons, where the weight of the piston is supported upon the lower inner surface of the cylinder, there is a segment, *g*, which is inserted into a recess formed in the ridge of ring *C*, and acted upon by a spring or set screws, *m*, or both combined, which supports, or nearly supports, the weight of the piston upon said piece *g*. This piece, *g*, may be made of hard brass or any other suitable metal, and as its outer surface wears away, the spring will force it outward, or, from time to time, it can be set out in proper place by set screws, *m*, so that the axis of the piston, and axis of the cylinder, within which the piston works, will always coincide. This will prevent the piston, stuffing box, and its rod, from wearing untrue. Where the weight of the piston is supported by the piston rod, as in upright cylinders, the piece *g*, with springs and set screws *m*, can, and should be dispensed with. It will be seen by reference to Fig. 4, that the circumference of the packing rings projects sufficiently beyond the circumference of the piston flanges, to allow steam to pass those flanges, and act upon the projecting bevelled surfaces of the packing rings, and expand them against the surface of the cylinder with a pressure commensurate with the force of steam. The springs, *cc*, are designed merely to keep the packing rings expanded and in position to be acted upon by steam, and forcibly expanded thereby. This invention is not confined to steam engine pistons, as it is applicable to pistons for air and water engines. The segments *b* may be made rectangular, or of other suitable shape, in cross section.

Construction.—16 ins. may be assumed as the diameter of this piston, and it may then be drawn on a scale of one-*sixth*, showing a little more than half of Figs. 5 and 7.

Example XXI.

Thirty-six and Fifty-four-inch Pistons.

Description.—In Pl. XI., Fig. 1, R represents a vertical section of a piston, with partly spherical upper and under surfaces. All of its horizontal sections being circles, a plan view of it could easily be added, to make a separate example.

Pl. IV., Fig. 3, represents a fifty-four-inch propeller engine piston; *h*H*b* is half of the part called the "*spider*," and H′A′K′ a vertical section of it on KH, so that all below A′B′, except E′*d*D′ and F′G′, should be filled with shading lines. The correspondence of the letters will show the relative heights of the different points above A′*e*″. Thus, the tops of the arms and rim, *b*, *ae*, *h*, are in the plane, D′I′. LM is the *cover* which closes on to the spider, its under face resting on the plane, D′I′. The bore for the piston rod is slightly conical.

In the annular space, H′G′I′, between the spider bottom and the cover, is inserted the packing. Formerly this consisted of two or more rings cut once or more, so as to be adjustable to the inner surface of the cylinder, and breaking joints, so that the cuts not lying together, steam could not pass through them from one side of the piston to the other. These rings were then set out by springs, bearing on the inside of the rings, and pressed against the rings by screws bearing at their inner ends against some of the unyielding parts of the piston.

At present *steam-packing* is very generally used. In this example, OO′ is a part of the *skeleton* ring, resting against the outer ends, IJ, of the spider arms, *n′n′* ; *n*, N shows the packing itself, consisting of two rings, of which the inner one, $\frac{3}{8}''$ square in section, fits within the other, as at *n′*, *n″*. Shallow cuts filed away, as at *f*, admit the steam over the edges of the piston body, which does not quite fit the cylinder, into the space between the skeleton ring, O′, and the packing. Steam can also be admitted through small holes through the piston bottom and cover, near their edges.

Construction.—A larger section of the packing, on a scale of $\frac{1}{4}$ or $\frac{1}{2}$, might be made, including the adjacent parts of the piston body.

b—Developable Receivers.

Example XXII.

A Fourneyron Wheel Plan.

Description.—This example, Pl. IV., Fig. 5, does not include the finished wheel, but only what is most essential, viz.: the laying out of the bucket and guide curves, as seen in plan, where their true curvature is shown. O, near the bottom of the plate, is the centre of the wheel, whose extreme radius is $49\frac{1}{2}$ ins.; radius to outer ends of buckets, 49 ins., and to their inner ends, 40 ins. There are 44 buckets, $8\frac{1}{4}$ ins. apart at the outer edges, and $\frac{9}{64}$ of an inch thick, made of polished Russia iron.

The water enters the wheel from above, through a trunk of nearly the same diameter as the inner radius of the wheel, filling the guide channels, and issuing thence against the buckets, and producing rotation in the direction of the arrow. As the wheel gives way by its revolution from before the water, the latter does not bend round and run out as if in a fixed channel, AGBE, but goes directly on, as indicated, in a general way, by the lines KM and *h*N, carrying the bucket with it, and hence the wheel. Still, as a particle of water, relatively to the bucket, follows its curve, AFB, the form of the bucket is not a matter of indifference.

The guides are fixed, and their number somewhat arbitrary, but usually taken at from half to three-fourths the number of buckets. To avoid the injurious pulsation which might follow if many guide edges should coincide at once with bucket edges, it is doubtless best to have the bucket and guide numbers prime to each other, that is, with no common divisor except 1. Hence we have proposed 31 guides to 44 buckets in the present example.

The regulating gate is a vertical thin cylinder, which shuts vertically downward in the annular space, HG, all around the wheel, between it and the guides. Water, therefore, enters the guide passages throughout the entire circumference of the guide case, deducting only the thickness of the guides themselves; while in the Jonval wheel the guide openings, as they would be, without a gate, are half closed by the gate. This, however, is not an essential point, for in both cases water issues against the bucket from the entire outer circumference of the guide case, with the above deduction, and the whole structure can be designed with such dimensions as to give any desired area of guide opening.

Construction.—With a scale of *one-tenth*, describe the principal circumferences with the dimensions given. The circumference containing the bucket ends, divided by the number of buckets, will give the distance BE. The shortest distance, EF, between the buckets being fixed by the designer, the following form and construction has been proposed: Put EF$=a$, and the thickness of the bucket$=b$. Make BC$=5a$, and draw the radius, CO, to determine A, the inner end of the bucket. Draw AD, tangent to the circle OA, and the arc E-F, with a radius equal to $a+b$. Then the direction of B*a* must be found by trial, so that AD being marked on the edge of a slip of paper, shall be applied with D always on B*a*, and the segments *te*, or *r*F, etc., constant and equal to BD. *a*BO may vary from 10° to 12°. Otherwise, alter the length of BC. The curve thus passing through A*eb*, etc., should be just tangent to EF, and will be the inner face of the bucket. If it be not so, assume a new position for B*a*.

For the guide curves, describe the arc, K—*f*, with the least opening of the guides for a radius, and H*g* with double this radius. Then describe *hg*, through *h*, and tangent to the two arcs just noted. This can easily be done by trial; determining the centre P (in Fig. 3). This gives the essential portion of the guide curve. The remainder, *gk*, is drawn from any convenient centre, as Q, such that *gk*, produced, would pass through the centre, O, of the wheel.

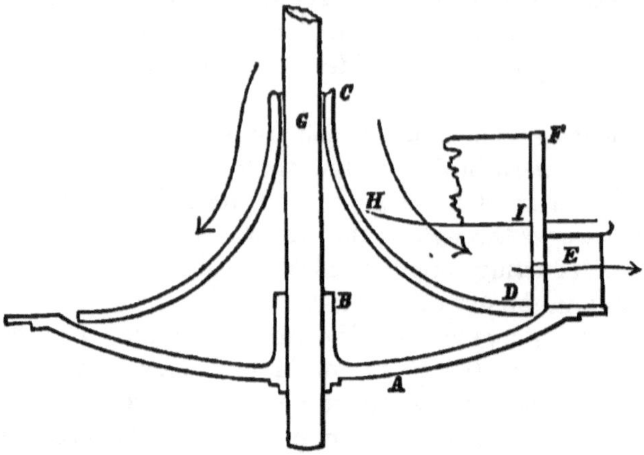

FIG. 21.

The centres for all the other guide curves corresponding with *hg* will be on the circle with radius OP, and those for the curves of which *gk* is one, will be on the circle with radius OQ.

The above figure, 21, will give a sufficient idea of the general arrangement of parts. A is the wheel plate, keyed to the shaft G, by a collar B, and bearing the vertical buckets, E, on its outer rim. CD is the fixed guide bottom, carrying the guide curves, of which HID represents one. F is a fragment of the thin cylindrical gate, shown partly open, which lifts out of the annular space between the guides and buckets, and admits the water from above through the trunk, to the guides, and thence through the buckets as shown by the arrows.

c—Warped Receivers.

Example XXIII.

A Jonval Turbine Wheel and Bucket.

58. *Introductory Explanations.*—Water wheels may be classified as *vertical* and *horizontal*. The former have *horizontal* axes; the latter, as in the last example, have vertical ones.

Vertical wheels are *overshot*, *undershot*, or *breast* wheels. These are so generally figured in the elementary books, or exemplified along many mill streams, as not to need illustration here. Overshot wheels are those to which the water is delivered, at or near their highest part, into trough-shaped buckets, deep and narrow, so as to retain the water longer. It escapes at or near their lowest line. Thus all the buckets on one-half of the wheel are loaded with water, the unbalanced weight of which causes it to descend, and the wheel to revolve.

Undershot wheels are like the paddle wheels of a steamboat; armed with floats at the circumference, parallel to the axis, so as to be acted upon by water rapidly running against them. They of course utilize but a very small part of the power of the water, but can be improved by curving or inclining the floats up stream.

Breast wheels are those to which the water is delivered at or near the height of their axes. They may have buckets like an overshot wheel, or floats like an undershot, but moving with their outer edges very near to a curved *apron* of wood or stone closely conformed to the curvature of the wheel on the lower descending quadrant, so that the floats will partly act as buckets, to retain the water.

With a low fall of water, a long breast wheel may be better than an overshot of diameter equal to the height of the fall, since its longer diameter may enable it to retain the water longer.

The principal *horizontal* wheels are *turbines*. They work under water upon a vertical axis.

The power given out by a wheel, other things being the same, depends on the quantity of water which it can dispose of. Accordingly, turbines, being small, usually revolve with a very high velocity, while the ponderous overshot wheels revolve slowly.

Again, for practical purposes, turbines may be distinguished as cheap and empirical ones, arranged according to partly fanciful notions, perhaps, of their designers; and costly highly finished ones, carefully modelled in every part by the results of the soundest theory and most decisive experiments. Of the latter class, in this country, are the Jonval Turbine, in the form known as the "Collins wheel," made at Norwich, Conn., the Fourneyron Turbine, as arranged in the "Boyden wheel," and made at Lowell and Chicopee, Mass., and elsewhere, and the Swain or centre discharge wheel; also made near Lowell. All of these wheels are so good as compared with many empirical wheels that we should not here wish to discriminate between them. There is only room to partially illustrate them; but the main difference between them will be well understood by observing that, in the "Jonval," the stationary *guides* are *above* the buckets, the bucket face is a warped surface whose rectilinear elements are *horizontal* and *radial*, and the water is discharged on the under side of the wheel in a stream of the radial width of the bucket, whereas, in the "Fourneyron," the stationary guides are within the revolving part; the buckets are vertical and cylindrical, with their rectilinear elements parallel to the axis of the wheel, and the water is discharged horizontally at the periphery of the wheel: and, finally, in the "Swain" wheel the guides are radially exterior to the buckets, and the water is discharged underneath, and around the axis.

There are, then, briefly, *bottom* discharge, *lateral* discharge, and *centre* discharge wheels; and these are the three radically different kinds.

Description.—Pl. XIV., Fig. 1, is a vertical section of the wheel, and Fig. 2 a side elevation, showing the general arrangement of parts very nearly as in the proportions of a two and a half foot wheel. AA is a section of the wheel, with the position of the top and bottom edges of buckets. It is keyed to the vertical shaft ST. A′A′ are developments of sections of buckets. BB are the guides, and B′B′ developed sections of them, showing them to be hollow to prevent the injurious eddying of the water which might occur if *only eo* were the wall of the guide. CCC is the gate, which consists of radial bars, CC, as long as the buckets are wide alternating with the openings C′C′. The gate thus is essentially a circular plate perforated next to the circumference and turning around the shaft. Such a gate might be thought to interfere with the proper flow of the water, but the perpendicular width of guide opening at R is so much less than at *ee* that half of *ee* can be taken for a gate bar without interfering with the best form for the vein of water flowing through the guides of the wheel. DD is the lighter plate made fast to the shaft. By receiving a part of the upward pressure of the water confined in the penstock, F, it relieves the footstep, E, and bridge, M, from a part of the weight of the wheel. The footstep, E, is of lignum vitæ. *aa* is a loose collar around the shaft to shut out water from the interior space N′N′. *bb* is the interior loose gland ring for the same purpose, and thus the two prevent the water pressure from acting on the wheel centre, or plate, PQ. This plate has openings, through which any water which may leak into the space N′N′ may escape. *dd* is the loose gland ring for the lighter plate, to prevent water from escaping there. The exterior loose gland ring, *cc*, is the one in which the wheel runs. It prevents the escape of the water from the guides and wheel, into the case without.

G is the cast-iron mouth-piece, and H the water trunk, through which, and the penstock or chamber F, the guides B, and wheel A, the water flows, as indicated by the arrows. But observe that as the buckets A′ are driven by the water rushing through the fixed guides B′, the water current is not really bent as indicated by the arrows on the bucket sections, but passes straight on, carrying the wheel circumference with it, as indicated approximately by the lines RR′. J is the upper bearing for the wheel shaft. K the gate shaft for turning the gate, and L its packing box. NN are the wheel and guide cases, and the whole is supported by the standards, O, in the bottom of the wheel pit.

Construction.—As nearly all the principal parts have circular horizontal sections, a plan could be constructed with sufficient completeness from the elevations.

Where so many thin sectional surfaces occur, it would be well to tint them instead of putting in so many short section lines.

A detailed view of the buckets is given on Pl. IV., Fig. 4. This figure was made from an actual bucket, but of an abandoned form. The great rapidity of motion of the wheel, and of the water flowing through it, causes the form of the bucket to be a very nice matter, so that we would not, if we could, give any improperly exact information concerning it, which had been obtained by long and costly experiments. The *drafting* operations of laying out a bucket are, however, fully shown.

The face of the bucket of a Jonval turbine consists of two warped surfaces, meeting in a common element, as *pq*. The lower surface, *r's'p'q'*, or *cdeg — c'd'e'g'*, is a portion of a *right helicoid*—the same as the screw surface of a square threaded screw —and has for its directrices, the central helix, CE—C'E', and the vertical axis, at O, of the wheel. The upper surface, *pq*XY, or *egha—e'g'h'a'*, is a portion of a *conoid*, whose directrices are the central curve, EA—E'A', and the axis of the wheel, as before. The elements, as *rs* ; *eg—e'g'* ; XY, etc., of both of these surfaces are horizontal, when the wheel is in position, having the horizontal plane for their plane director. Whence we have the following construction: With centre O, and the assigned radius, $= r$, describe an arc, CB, of the plan of the mid-line of the buckets, and wheel. We suppose the angles DC″B″ $= m$, and DAF $= n$ (where FA″ is horizontal), at which the water best enters and leaves the wheel, to be determined by investigation. Also, let the height, A″B″, of the wheel be given. The developed length, C″B″, of the buckets, may then be found thus: C″G, the horizontal distance from one bucket to the next $= \frac{2\pi r}{n}$, where $n =$ the number of buckets. Draw GE″ perpendicular to C″E″, and produce it. Draw E″A″, making QE″A″ $= \frac{m+n}{2}$. Then the perpendicular, DQ, through the middle point of E″A″, will meet GE″ at Q, the centre of the arc E″A″.

Let $m = 25°$, though it may be less than 20°, let $n = 100°$, and B″A″ = 8″. Having assumed C″E″ to be straight, it becomes the development of a helix. Therefore, divide it equally at pleasure, project its divisions on C″B″, and transfer them, in order, to CA in plan. Through the points, as *n*, thus located on CE, draw radii from O, which will be horizontal projections of elements, limited by the inner and outer circumferences, O*h* and O*a*, of the wheel. Then project up *m* and *i* to meet a horizontal line from *m*″, and *m'i'* will be the vertical projection of the bucket element *mi*. In like manner, we find other points of the front and back curves $d'e'a'$ and $d'g'h'$ of the bucket.

A″E″ is simply a circular arc, tangent to A″D″ at A″ and to C″E″ at E″, and hence DE″ = DA″. Having drawn the arc A″E″, proceed as before to find ea—$e'a'$, etc.

The flange R′S′M′N′ is for the purpose of fastening the bucket to the rim *pq*, Pl. XIV., Fig. 1, of the wheel, while the ear $a'k'$ stiffens the outer corner of the bucket.

d—Double Curved Receivers.

Example XXIV.

The Swain Central Discharge Water-wheel.

Description.—The following abridged description and figures are, by permission, from a paper by Hiram F. Mills, C. E., in the *Journal of the Franklin Institute* for March, 1870 :—

The central discharge iron wheel, known as the Swain wheel, has been in use in various parts of New England for several years, and has been justly gaining a reputation for good construction and efficiency.

Plate XV. represents a vertical section through the centre of the wheel and curb, and a plan of one-quarter of the buckets and guides, and a development of a portion of the outer surface of the wheel.

A flume, 6 feet wide and 6 feet deep, leads the water from a canal to the wheel. It enters the forebay, 6 feet square and 19 feet deep, from the side of which water is conveyed to the wheel.

The wheel-pit, upon the floor of which rest the cast-iron supports of the wheel and its case, is 14 feet wide, and 20 feet 6 inches long, with sides 5 feet high.

The cast-iron supply-pipe and the quarter-turn, D, Fig. 1, are 4·98 feet in diameter inside, and the radius of the central line of the latter is 3·75 feet. The former is 1·45 feet in length, and is firmly bolted to the side of the forebay, at AB.

The quarter-turn is bolted to the supply-pipe, and is supported by the cast-iron case, C, which enlarges to have an internal diameter of 6·98 feet. This case extends to a short distance below the bottom of the wheel, and is supported by six hollow cast-iron columns, Y, 1·5 feet long, 0·8 feet wide radially, and 0·3 wide tangentially, and presenting acute angles to the escaping current.

The columns also support the cylindrical disk E, which surrounds the lower half of the wheel, and upon the outside of which the regulating-gate, O, slides; and by means of three supports, one of which is represented at F, they sustain the annular chamber G, also the disk H, which cover the wheel, the lower shaft-coupling, and the step; and the shaft-pipe, I, which extends from the top of the disk through the upper part of the quarter-turn.

The disk K, cast with the six supports of the case, sustains the step of the wheel L, and the main shaft M.

The step L, of white oak, is a cylinder, 7 inches in diameter, terminated at top and bottom by cones, and having an extreme length of 8 inches. It is free to move with the shaft-coupling, or the latter may move upon its upper surface.

The step is always submerged, and to maintain its surfaces in as good a condition of lubrication as possible, several vertical holes are bored through it, and filled with a fine quality of plumbago.

By means of the set-screws N, the wheel may be adjusted to the proper height. These screws are reached through hand-holes in the curved disk H, which connects the wheel with the shaft.

The regulating-gate, OO, is represented as fully open. It consists of a cylinder of cast-iron, 7½ inches long, 3·51 feet in diameter inside, having at the bottom a narrow flange, against which a ring of leather packing is held by means of bolts and a cast-iron ring, by which leakage under its lower edge is prevented, and at the top, cast with it, and making an angle of about 80°, is a broad flange extending outward 7½ inches, with an edge turning downwards.

This flange serves as the lower disk, limiting the stream of water entering the wheel, and into it, projecting vertically from its upper surface, the guides PP are cast.

There are 24 guides, 21 of which are straight plates of wrought-iron $\frac{3}{16}$ inch thick, and 10⅝ inches long, having each end sharpened to a thickness of $\frac{1}{32}$ of an inch, with a bevel $\frac{1}{8}$ inch long. The mean distance of their inner edges from the centre of the shaft is 1·835 feet, and they are distant radially one inch from the outer edge of the buckets. Their direction makes an angle of 22°, with the tangents through their inner edges.

The remaining three guides are of cast-iron, of the form represented in plan at O in Fig. 2.

Through these guides pass the three wrought-iron rods, adjacent to the supports, F, and by which the gate is raised and lowered. Two of these are represented at R. They extend from the under side of the gate flange through stuffing-boxes in the upper part of the quarter-turn, and terminate in racks SS, into which work the pinions TT, which are driven, through the action of the worm U, by a crank.

When the gate is raised, by which movement it is also closed, all of the guides rise through slots in the upper disk, which limits the stream approaching the wheel, and which is similar in form to the lower disk already described; and are enclosed in the guide-chamber G. By this arrangement the stream passing between the guides is decreased in height in proportion to the closing of the gate.

The wheel, W, has an extreme diameter of 42 inches, and an extreme height of 15$\frac{7}{8}$ inches. It has 25 buckets of wrought-iron, $\frac{1}{8}$ of an inch in thickness, which are formed in a die, and are cast into the upper disk *aa*, which forms the crown of the wheel, and into the band *bb*, which surrounds the lower part of the buckets for a height of 8·42 inches, leaving, between the top of this ring and the under side of the crown, a space 5·35 inches high, through which the water enters the wheel.

The horizontal projection of the buckets, for a depth of 5 inches from the top, is shown by the heavy lines in Fig. 2, and Fig. 3 represents the development of a portion of the cylindrical surface containing the outer edges of the buckets.

The diameter of the cylinder, which would contain the inner edges of the buckets for a depth of 5 inches, would be 2·092 feet. Below this portion, the surface which would contain the inner edges of the buckets is generated by the revolution, about the vertical axis of the wheel, of a quadrant whose radius is 8·4 inches, and which is convex towards the axis.

The under side of the crown of the wheel is horizontal through a radial distance of 4 inches from the outside. For the remaining distance, it forms part of a surface of revolution, C, whose sections through the axis have a radius of 11 inches, which surface continues downward and towards the shaft, and forms part of the cylindrical socket which surrounds the step and its supports.

The following dimensions were measured after the wheel was removed from the pit. :—

Vertical distance from the under side of the crown to the lower edge of the buckets............	1·148 feet.
Vertical distance from under side of crown to top of band....................................	0·446 "
Mean shortest distance of the inner edge of one bucket from the adjacent bucket at 1 inch below the crown	0·065 "
Ditto, ditto at 5 inches below the crown...........	0·080 "
Ditto, ditto at 1 inch from the outside at the bottom.	0·135 "
Mean shortest distance from the inner edge of one guide to the adjacent guide	0·214 "
Mean shortest distance from the outer edge of one guide to the adjacent guide	0·332 "
Mean area of outlets of wheel	18·97 sq. ins.
Total area of outlets of wheel	8·29 sq. ft.

From a series of *ninety* careful experiments, the interesting details of which will be found in the journal referred to, it was found that the mean maximum efficiency of the wheel is as follows, viz. :—

With full gate, $81\frac{3}{10}$ per cent. of the power of the water. With three-quarters gate, $77\frac{3}{10}$ per cent. of the power of the water; and with one half-gate, $69\frac{3}{10}$ per cent. of the power of the water.

Construction.—Fig. 1 may properly be made on a scale of three-fourths of an inch or a whole inch to the foot; and with the sectional surfaces tinted, instead of being filled with shade lines.

CLASS III.—COMMUNICATORS.

59. These are pieces whose office it is to take up a certain motion at one point, and impart it unchanged, or modified in velocity, form, direction, or uniformity, at another point.

Most of the pieces interposed between the receiver and the operator in any machine are but a succession of communicators, as, for example, the wheels of a watch, which communicate the motion received from the main spring by the barrel to the hands. In this case, the original *velocity* is greatly changed. In the use of the crank, the *reciprocating* and *variable* velocity of the piston of a steam engine is changed into the *rotary* and *uniform* motion of the crank-pin; and, again, toothed wheels and *rocker arms*, that is, arms which vibrate about a fixed point, change the direction of the motion received by them.

A—Point Communicators.

Point communicators are not numerous. Only one example is given here, separately. The *crank-pin*, and the *cross-head*, will be found elsewhere, as parts of other examples.

Example XXV.

Collins' Shaft Coupling.

Description.—Pl. XXX., Fig. 1, is a sketch merely of this coupling. Shafts are necessarily of limited length; though a line of shafting may extend the whole length of a very long room. Some of the requisites of a good coupling are, ease of application, permanence of adjustment, lightness of separate parts, and neat finish, with absence of projecting parts. These conditions are well met in the coupling here shown. SS is one side of a sleeve, whose two halves, together, embrace the shaft. CR—C'R' is one of two cone rings which together cover the slightly tapered part, *ab*, of the sleeve; *rn*—*r'n'* is one of the ring nuts, to be screwed on to the screwed portions, S*a*, S*b*, of the sleeve, by inserting a pin in the holes *r'*, *n'*, etc. It thus drives the cone rings, whose inner bore is tapered the same as *ab*, upon the sleeve, and thus clamps the two parts of the latter tightly to the shaft. Sometimes, small pins, *p*, inside of the sleeve, enter corresponding holes, one in each piece of shafting, in order to give the coupling a firmer hold; but this is not necessary. The whole, when put together, is a smooth cylinder capable of being used as a band pulley if desired.

Construction.—The measurements attached to the sketch allow all parts to be accurately drawn; with an exterior view of the other half sleeve; or an end view of either half.

B—Line Communicators.

60. *Linear* communicators are distinguished as *flexible* and *rigid.* The former will, for convenience, be described first; on account of the connection of the latter with subsequent forms of communicators.

Band, Cord, and Chain Wheels.

61. The motion of one wheel, or of a curved sector, may be communicated to another by means of a band or belt, a cord, or a chain passing around both wheels; or fastened suitably to a point on the circumference of each sector. In either case, the wheel and the band are both of so simple a form as hardly to present any fit examples for graphical construction; except as the curved arms often seen in band pulleys may afford further practice in the nice construction of tangent arcs; and as a fully shaded projection of two or more band wheels, in different planes, might make an effective plate of shaded drawing. Such a plate the student could readily design and execute for himself, after having learned the following principles.

The present subject will, therefore, be treated only in relation to the leading practical principles of band wheels or pulleys, belts and chains.

62. *Angular velocity* is that at which the angular space around a centre is swept over by a radius turning about that centre.

For purposes of comparison it is estimated at a *unit's distance*, as a foot, from the centre. Thus if a point on the periphery of a wheel of 4 ft. radius be moving at the rate of 20 ft. per second, the velocity of a point at one foot from the centre will be 5 ft. per second, and this will be the angular velocity of the wheel.

63. The *direction* of the motion of a revolving body at any point is estimated on the *tangent at that point* to the circle described by that point.

The *direction of an angular velocity*, as to its *sense*, i. e., as forward or backward, right or left, is estimated by that of a watch having the same relative position to the observer as the given motion. The hands of a watch are said to move from left to right, since they really do so in passing through the upper or XII point. Any opposite rotary motion is then said to be from right to left.

Theorem I.

A rotary motion of two parallel axes may be maintained indefinitely, and in one and the same direction, for both, by a band, passing directly around cylindrical pulleys in the same plane, on those axes; but, if the band be crossed, the rotations will be in opposite directions; but in both cases the ratio of the velocities will be constant.

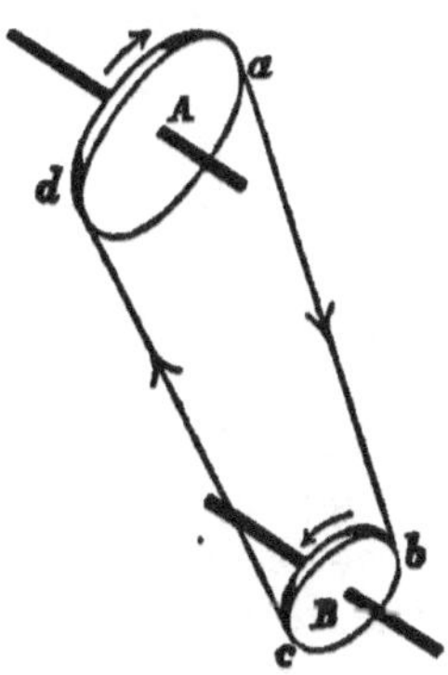

Fig. 22.

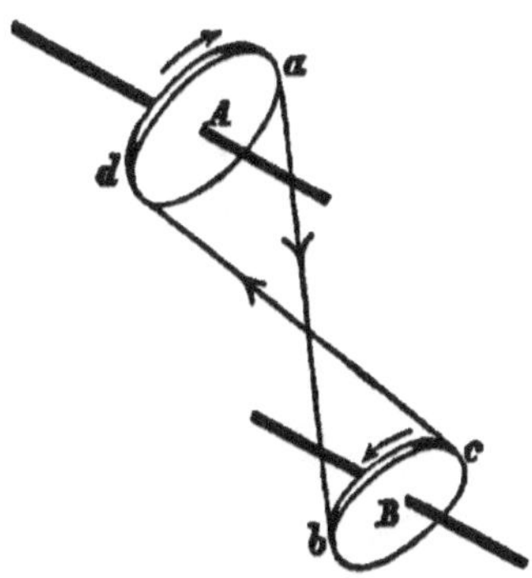

Fig. 23.

Let A and B, Fig. 22, be two thin cylinders in the same plane, secured to the axes A and B, also in the same plane. A leather or other flexible band, *abcd*, whose ends are united to make one piece, may then be passed directly and tightly around the two pulleys, as shown, and will then be in the same plane as the wheels. Its adhesion to the wheels will then be sufficient to communicate the motion of either to the other. And as no part of the band is hindered in passing any point touched by it on the wheel, the motion in either elevation, *ab* or *ba*, may be indefinitely continued. As the points *a* and *b* move in the same direction and on the same side of the axes, the angular motions of the wheels will be in the same direction. But if the band be crossed, as in Fig. 23, the points *a* and *b* being on opposite sides of the plane of the axes, the band will produce rotations in opposite directions, as shown by the arrows.

Finally, the ratio of the velocities will be constant, for all points of the band must move with equal velocity, else it would break; hence the peripheries of the wheels have equal velocities, which are equal to that of the band. Hence as *each* wheel has a constant radius, its angular velocity also will have a constant ratio to that of its periphery, that is, to that of the *band.* Hence the ratio of the angular velocities to each other will be constant.

Remark.—While this last result is theoretically true, it is not so practically, if there be any slipping of the band.

Theorem II.

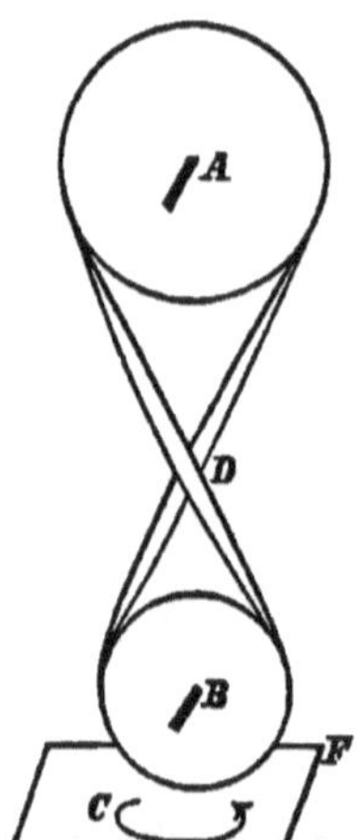

Fig. 24.

A band should be crossed by giving it a half twist, in a plane perpendicular to that of the wheels which hold it; it should be shifted, laterally, by operating on its advancing side, and if applied to a cone wheel, will work itself towards the larger end of the cone.

First, the band should be crossed as described; first, to bring the same side, and the inner side of the leather against the pulley; and, second, to make it cross itself flatwise, as at D, or face to face, instead of edge to edge. This is accomplished by taking hold of the band, as shown in Fig. 24, and carrying it half round, that is, with a half twist, as indicated by the arrow C, in the plane CF, parallel to the plane of the axes.

Second, the rigidity of the band in the direction of its width, together with its adhesion to the surface of its pulley, makes any point of it follow the line into which it is drawn at any point, in going on to the pulley from that point.

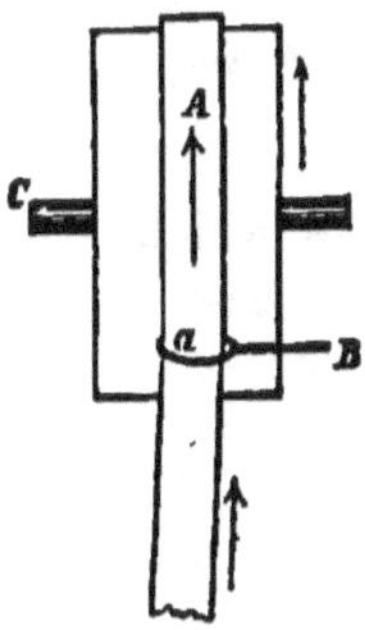

FIG. 25.

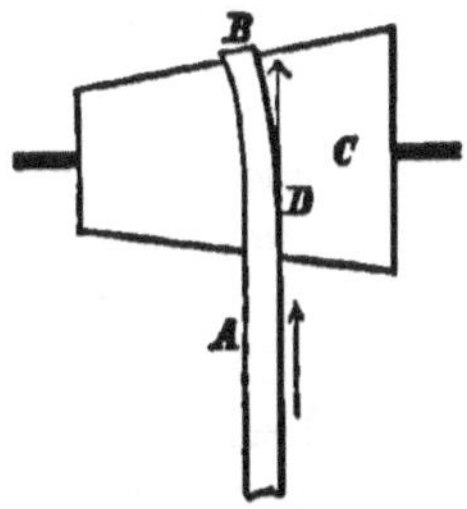

FIG. 26.

Thus a band A, Fig. 25, being drawn aside by a looped rod B, just before striking the wheel, will, by its own stiffness, follow the direction *a*A till it touches the wheel, and will then continue in that direction.

Hence, in all shifting of bands from one pulley to another, they must be acted upon on their advancing sides; and any ordinary lateral action applied to them on their retreating side, will not displace them on or from their pulleys.

Third, the resistance of the band to lateral bending likewise makes it work towards the larger end of a conical pulley. Thus the band AB, Fig. 26, in striving to fit the conical pulley, C, will bend; and the point D will, by the resistance of the band to lateral bending, tend to move in the line D*d*, while, by the continuance of these two actions, the band will work itself to the larger base of the cone.

FIG. 27.

Hence the *simplest method* of keeping bands upon their pulleys is, to make the latter consist of two conical frusta, with a *common larger* base.

Thus band pulleys, even when running with the highest velocities, are made as in Fig. 27, slightly coned from *a* and *c* to the larger central diameter at *b*. The band will thus keep itself upon the pulley, without either external guides or loops, or flanges, all of which would wear its edges and hinder its motion.

The principles now explained enable us to solve the following problem.

Problem I.

To connect wheels lying in different planes, by a band, when the intersection of their planes is also a common tangent to the two wheels.

This general problem includes several cases which will be taken up successively, as variations from a given case.

I.—*To connect the wheels so that they shall revolve in a given manner.* The band should be *led on* to each wheel in the plane of that wheel, and should *leave* each wheel at the point of contact of the common tangent to the two wheels. Let PN and

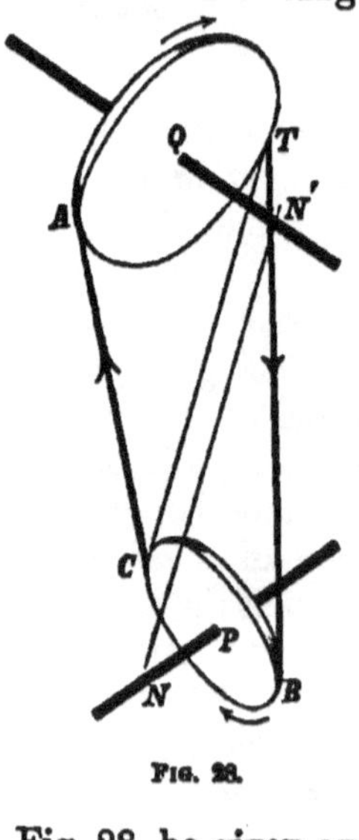

Fig. 28.

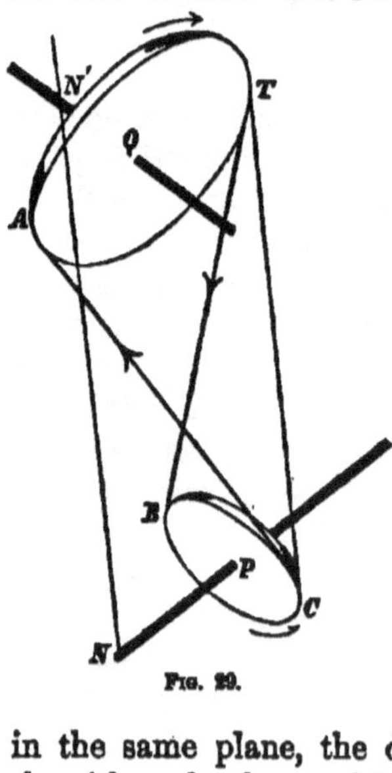

Fig. 29.

QN′, Fig. 28, be given axes not in the same plane, the directions of whose axes make any angle with each other, and let the motion be in the direction of the arrows. Let CT be the common tangent, which is the intersection of the planes of rotation, BTC and ATC. Then the band, moving as shown, must enter the lower wheel in the plane BTC, and the upper one in the plane ATC. Also, by the second principle, it must leave the lower wheel at C, and the upper one at T.

The same thing is shown in projection, in Pl. XVI., Figs. 6 and 7, where the same letters being repeated at like points, the figures explain themselves, by the help of the arrows. In these and the following figures, the horizontal plane of projection is taken parallel to the given axes, for convenience in constructing the projections of the wheels.

From this case as a starting point, two principal variations may be made: *first*, to reverse the motion of *either one* of the wheels; *second*, to reverse the motion of both of them.

II.—*To reverse the motion of either one of the wheels.* Let the motion of the upper wheel be reversed as in Pl. XVI., Fig. 8. This case is here illustrated with the directions of the axes perpendicular to each other. Observing the points of analogy and of difference between Figs. 28 and 29, it appears by mere inspection that, to reverse the motion of either wheel alone, the common tangent, CT—C′T′, to the two circumferences, which is the intersection of the planes of rotation, shifts to the opposite side of the wheel whose motion is reversed; while the other wheel, BC—B′C′, shifts to the other side of the common normal, N—N′N′, from where it was before.

In the corresponding change from Pl. XVI., Fig. 7, the wheel BC will move parallel to TA till its point C shall be under A. The student can construct the figure. If BC—B′C″ in this case were moved on its own axis, parallel to itself, it would have to increase in diameter in order to preserve a common tangent to the two circumferences, as before. But in this case, the velocity ratio between the axes would be changed, which might be undesirable. Any two wheels, whose central sections, AT and BC, were, as seen in plan, generated by any one point of NC, as *a*, would give the same velocity ratio.

Fig. 29 illustrates this case pictorially, with the motion of the *lower* wheel reversed; and we see, as before, that the intersection, CT, of the planes of rotation is tangent on the opposite side of *that* wheel, and that the *other* wheel, Q, is on the opposite side of the common normal from what it was before, in Fig. 28.

III.—*To reverse the motion of both wheels.* This is illustrated in Pl. XVI., Fig. 9, for the general case of the directions of the axes making *any* angle with each other. Like letters for like points still being retained, we see, by comparing with Pl. XVI., Figs. 6 and 7, that *each wheel goes to the other side of the common normal*, N—N′N″, and that the common tangent, CT—C′T, finds its contacts on the opposite side of each wheel from where these points were before.

Let x=the angle between the directions of the axes,

A and B the radii of the wheels denoted by the same letters, and

P and Q their respective distances, measured on their axes, from the common normal N—N′N″. Then it is easily found that for wheel A,

$P = B \operatorname{cosec} x + A \cot x.$

and for wheel B,

$Q = A \operatorname{cosec} x + B \cot x.$

which when $x = 90°$ reduces to

$P = B$; and $Q = A$; or, in this case, the distance of each wheel (the central plane of its width) from the common normal equals the radius of the other wheel, as in Pl. XVI., Figs. 6 and 8.

The twisting of the band at its departure from each pulley is undesirable, and the more so the nearer the wheels are to each other. We therefore next show how to avoid this evil to a great degree.

Problem II.

To connect band wheels in different planes, when the intersection of those planes is not a common tangent to the wheels.

Let I—I′I″, Pl. XVI., Fig. 10, be the intersection of the vertical planes, QI and CI, in which are the wheels TA—T′A′, and BC—B′C′. Take any two points, *mm′* and *nn′*, on I—I′I″, and from *mm′*, draw *m*A—*m′*A′, and *m*B—*m′*B′; also from *nn′*, draw *n*T—*n′*T′, and *n*C—*n′*C′. Then place guide pulley, *pp′*, in the plane A′*m′*B′, and another, *qq′*, in the plane T′*n′*C′, and a band led around the four wheels, as shown, will act in either direction, as indicated by the opposing arrows, since the band *enters* upon, and *leaves* each of the four wheels in the plane of that wheel.

By drawing *m′*T′ and *n′*A′, the lines to B′C′ remaining unchanged, and placing *p′* and *q′* in planes B′*m′*T′ and C*n′*A′, the *relative* direction of rotation of the wheels P and Q would have been changed.

The student can now design an arrangement with *one* guide pulley for this case, giving rotation in *one* direction only; also an arrangement either by one or two pulleys for giving rotation in either direction in the previous cases.

Notes on Band Wheels.

64. The foregoing are all the important geometrical principles connected with band wheels. Some notes are added, mostly from an extensive recent collection on the subject.*

Area.—100 square feet of belt per minute, per horse-power, at a belt speed of about 1,800 feet per minute, seems to be a fair average allowance of belting to power transmitted, according to several examples from various authorities and from actual practice.

One example gives only 24 sq. feet of belt per horse-power per minute, at a speed of about 750 ft. per minute; and another, 68 sq. ft. per horse-power per minute, at a belt speed of 1,000 ft. per minute; but the horse-power transmitted may have been variously or loosely estimated.

A more satisfactory average result is from a mean of twenty seven select examples, in which all evidently extreme cases, like that of 24 sq. feet just named, had been rejected, and the belt speed disregarded, as the various particulars, of *material, arc of pulley embraced, condition* and *tightness* of the belt, might neutralize or outweigh the effect of difference of speed. The mean of the twenty-seven examples, then, gives about 76 sq. ft. of belt per minute per horse-power, to pass a given point on the pulley.

65. *Material.*—Numerous authorities agree that oak-tanned leather is preferable in point of durability to all other substances for belting purposes. These other substances are woven rubber fabrics, gutta-percha, paper, and any of these, or like materials with metallic strands incorporated into them. As an example of a very large belt, there may be mentioned an india-rubber belt, 4 ft. wide, 320 ft. long, and of 3,600 lbs. weight, designed for a large grain elevator.†

* See "Belting Facts and Figures," in the Jour. Fr. Inst. 1868–70.

† J. F. Inst. Sept., 1869.

66. *Pulley surface.*—This is of cast-iron or wood. In either case the power transmitted may be greatly increased by covering the pulley with leather, between which and the belt the friction will be very much greater than between the uncovered pulley and the belt.

67. *Belt fastenings.*—The two ends of a belt are very often fastened by *lacing* with leather strings through two sets of holes, thus (Fig. 30):—

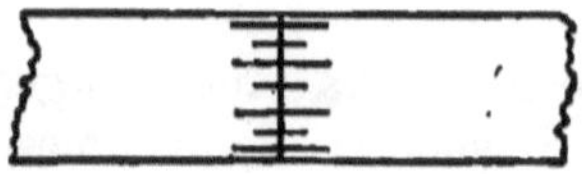

FIG. 30.

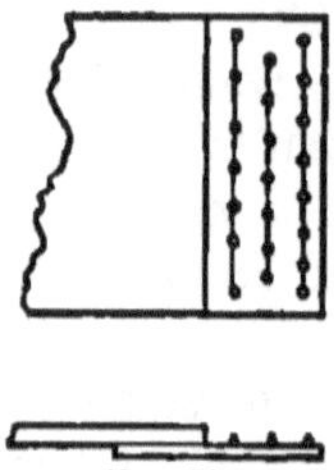

FIG. 31.

Also by *rivets*, and by "*belt hooks*," where the ends of the belts are hammered on to the thin plate, armed with numerous short, sharp points, which go through the belt, and can be clinched (Fig. 31).

Thin flexible *studs* are also used, thus:—

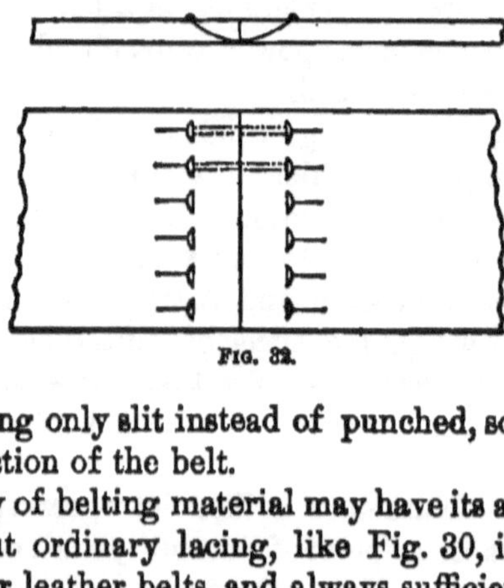

FIG. 32.

the leather being only slit instead of punched, so as to save loss of resisting section of the belt.

Each variety of belting material may have its appropriate belt fastening. But ordinary lacing, like Fig. 30, is by many considered best for leather belts, and always sufficient if the belt is not stretched too tightly. The belt hook, Fig. 31, is recommended for rubber and paper belts.

68. *Side against the pulley.*—It seems to be agreed on all sides, that the *hair* or *grain* side of the leather should be against the pulley, and for two reasons. First: that side is smoother and firmer, and therefore adheres more closely to, or has a better hold upon, the face of the pulley. Second: because the strongest

part of a leather belt is about one-third of the way through from the flesh side, and this bears the tensile strain of the outer half of the thickness of the pulley, while it is also freed from abrasion by the pulley.

69. *Paper belts.*—These are a new invention known as Graves' patent. They are made only for straight and unshifting belts, not less than five inches wide, nor to embrace pulleys less than six inches diameter, and are particularly recommended for heavy belts. Examples are recorded 12 to 14 inches wide, and from 50 to 120 ft. long.

70. *Proper tension.*—Belts should by no means be strained to anything near their breaking weight, as they would thus be so extended as to be too loose on their pulleys. Leather belts should not bear a strain of much above 300 lbs. to a square inch of cross section. Gutta-percha will bear about 400 lbs. to a square inch, its breaking weight being about 1,680 lbs. per square inch.

71. *Effective radius.*—Where nice calculations of speed are to be made, and where the belt can be relied upon not to slip, the effective radius of a pulley is estimated to the middle of the thickness of the belt, the particles within that point being compressed, and those without extended, as the belt bends around its pulley.

72. *Dressing.*—It is very generally recommended that the dressing for belts should not consist of a liquid penetrating oil, but of a stiffer composition, such as one of tallow and oil, further stiffened by a small portion of beeswax or resin.

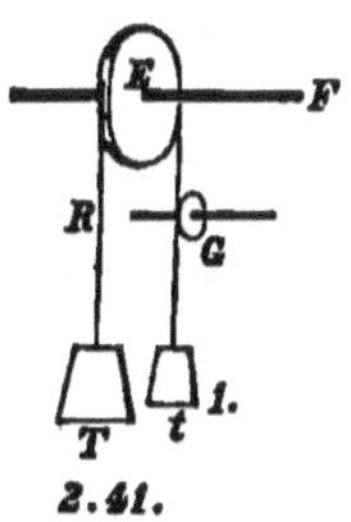

FIG. 33.

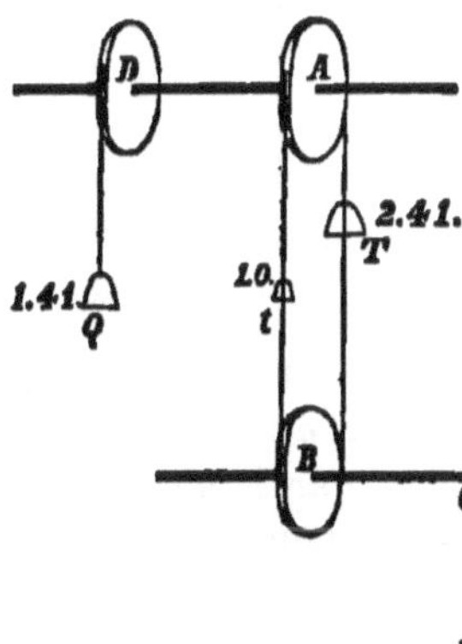

FIG. 34.

73. *Driving power.*—Although this topic belongs more appropriately to a treatise on the dynamics of machinery, yet an outline of the elementary considerations to be regarded, seems too interesting to be omitted. Let E, Fig. 33, be an experimental fixed pulley on a fixed axis, and

thus incapable of motion; and let T and t be weights attached to opposite ends of the same belt, R, passing over the pulley. Let G be a small guide pulley, freely turning on a movable axis, so that, as it is made to press the belt in or out, more or less than half of the circumference of E may be in contact with R. Then let T be adjusted to each different length of arc of E in contact with the band, until it be sufficient to descend and draw up t, by the slipping of the band.

The *difference* between T and t will vary, *first*, with the coefficient of friction for the material of the belt and the pulley; *second*, with the ratio of the arc of contact to the entire circumference of E, and with the width of the band, and consequent absolute values of T and t.

The *ratio* of T and t will, however, depend only on the two first particulars.

From a table of values,* found experimentally, as just described, the following examples are taken:—

BELTS IN ORDINARY CONDITION.

$$Q = T - t.$$

On Cast-Iron Pulleys.		Unit of Comparison $= t$.	Ratio of arc embraced, to the circumference.
T	Q		
1.69	.69	1.	0.3
2.41	1.41	1.	0.5
3.43	2.43	1.	0.7
2575.3	2574.8	1.	2.5
Ropes on rough wooden drums.			

Now, for the case of pulleys on freely revolving axes, see Fig. 34, where, to produce equilibrium, a weight, $Q = T - t$, must be hung from a second pulley, D, on the shaft AD, and of the same diameter as pulley A. If $A = B$, as shown, 0.5 of the circumference of each will be embraced by the belt, and we shall have from the table the *relative* weights of t, T, and Q marked in the figure.

* Practical Treatise on Mill-Gearing. London: Spon. 1869.

Next replace T and t by any sufficient motive power, applied at C, and it would raise the weight Q at a uniform speed without slipping the belt.

$n \times Q$ may in like manner be raised by treating it thus :—

$nQ = nT - nt$, and the belt must then be made n times as strong as before, when it withstood only the strain T. This can be done by making it n times as *wide* as before. If, however, it be made n times as *thick*, its stiffness may interfere with the friction of an equal area of contact being increased n times, by the n times increased strain nT. That is, the ratio of T and t may be changed so as to make $nQ = rT - nt$, and then the strength of the band must be rT times as great as before. This could be determined experimentally.

74. *Relative merits of belting and gearing.* Belting is much more quiet, and more readily allows a change of relative velocity. It only needs to be properly proportioned and applied, to give as good or better results, according to eminent practical authorities.

A belt too tightly stretched on its pulleys, not only injures itself, but produces an injurious friction of its pulleys thus drawn towards each other, on one side of the shaft of each. It is desirable, therefore, to have a long overhead line of shafting run through the centre of the width of the room, so that belts can pass from it down to machines on opposite sides ; and thus, partly, balance the opposite pressures upon the shaft.

Rubber or paper belts may be used for uncrossed and unshifted belts, and when running in the open air, and the former also in damp or wet places. Leather belts may be used in other cases, and with leather-covered pulleys, and, if of abundant width, will run quite slack without slipping, or loss of power, from excessive friction by undue pressure upon the axles.

Transmission by Ropes, Cords, etc.

75. Round bands of leather or other material are occasionally seen on a small scale, driving some of the subordinate parts of a machine, or as a substitute for belts in models, etc. They run in circumferences variously grooved in different cases to hold them in place.

Concave sided grooves, forming an angle at the bottom, are recommended in place of V-shaped ones.

Since 1850, however, the *transmission of power by wire ropes*,* on a grand scale, has often been adopted in Europe as a permanent arrangement, and might be so employed, advantageously, wherever the source of power and the place of its application are necessarily, or most conveniently, far apart.

One of the most prominent cases of this kind is that of a large water power in a spot so precipitous as to afford no good building site near it. In such a case, the great expense of building in a bad position, or of constructing a canal from the water power to the distant mill driven by it, may be avoided by the use of wire ropes from a water wheel, placed directly at the site of the water power.

The main propositions and facts concerning this topic can be briefly expressed in a few sentences.

The use of a round endless wire rope, running at a great velocity in a grooved sheave, constitutes the transmission of power by wire rope.

Thus, the power of a 100-horse power turbine has been transmitted 3,200 feet by a $\frac{5}{8}$-inch wire rope running on $13\frac{1}{2}$ feet wheels, making 114 revolutions per minute, and 400 feet apart.

The requisite tension is obtained by the weight of the rope in long reaches, where the deflection of the rope, at rest, will be about $\frac{1}{25}$ of the distance between the wheels.

Special care should be taken to set each wheel squarely on its shaft, and the latter truly perpendicular to the line of transmission.

When the wheels are at different heights, a smaller arc of the lower one will be effectually embraced by the rope, as the weight of the rope then makes it tend to drop away from that wheel. In such a case, the rope may be stretched tighter; or, if the inclination of the line from one wheel to the other be great, as 35° or more, guide pulleys *level with the upper* pulleys should be provided, so that both parts of the rope shall be first *vertical* from the lower to the guide pulleys, and thence horizontal to the upper pulley.

Chains.

76. Chains of various forms are sometimes used in place of bands, where great power is to be transmitted. They are usu-

* Trans. of Power by Wire Ropes. Van Nostrand, N. Y., 1869.

ally made either with toothed links to engage in notches on the periphery of the wheel, or in carefully riveted square links to embrace teeth upon the wheel. But in all cases the arrangement is expensive and troublesome, being liable to get out of adjustment by the wear of the rivets. It is therefore inferior to toothed gearing, or wire rope, for long transmissions.

For hoisting purposes, etc., where the motion is limited, and the chain passes around only one barrel or drum; if the latter be suitably grooved, a chain of common oval or of rectangular links will coil itself very evenly on the barrel.

Inflexible Linear Communicators.

77. The common crank may be seen, in its double form, in the cranked axle of any locomotive with inside cylinders. A single crank may be found in any stationary steam engine of the usual form, having a cylinder and reciprocating piston.

Two armed cranks, whose arms are perpendicular, or oblique to each other, go by the general name of bell-cranks, being most familiar in door-bell connections. They also enter largely into the mechanism of organ-stops.

78. "Rockers," Pl. XVII., Fig.10, *s* and *h*, are familiar in nearly all American locomotives, where they have a vertical position on each side of the engine, and communicate the motion of the outer end of the excentric rods, *e*, to the valve stems, *v*. A rocker or "rocker arm," *s*, vibrating in a vertical plane, about its lowest point, R, as a centre, may be called a *standing rocker*; one, *h*, which similarly vibrates about its highest point as a centre, may be called a *hanging rocker*. The valve rod, or stem, *v*, of a locomotive is attached to the upper point of a standing rocker, and the excentric rod, *e*, which moves it, to the lower end of a hanging rocker on the same rocker shaft, R.

EXAMPLE XXVI.

A Locomotive Parallel and Main Connections.

Description.—The parallel connection, Pl. XVII., Figs. 1–2, is the heavy bar that connects the driving wheels of locomotives having more than one pair of drivers. It always lies parallel to the line joining the centres of those wheels. Hence its name. The bar itself is a very simple affair, being straight and of round, or nearly rectangular, or sometimes, I section. But, with the adjunct parts, forming its attachment to either crank pin, it presents some useful features for study and practice.

R is the lighter and moulded part of the connecting rod. The square-cut part, at each end, as E *ss*, is the "stub-end," and S is the strap. BB are the boxes, of brass, lined with "Babbit" or anti-friction metal, within which the crank pin, P, works. In this example, the front face of each box is formed with a moulded cover, best understood from the plan, Fig. 2, at BB, also shown by its principal circular lines in Fig. 3, which shows a plain box without a cover. This cover shields the head, *p*, of the crank pin. As each half of the box wears out on the inside, it is set up against the crank pin, by wedge keys, K and *k*, which fit into vertical square cut grooves in the backs of the boxes as shown, where *ss* and *tt* are the backs of the boxes, into which, as at *gg* in the plan, the vertical edges of the keys set.

The rear box, being slipped into the strap from its open end, and drawn over the crank pin, the front box can then be likewise slipped in, and after it the stub-end. The keys can then be adjusted till the box is close to the pin,—the edges of the vertical joint, *hh*, being filed away, if necessary, to secure a proper fit—while also the bolt holes for the bolts, H and T, must agree in the strap and stub-end. These bolts, then, hold the whole fast. Their nuts, N, and jam-nuts, *n*, clamp each other to the bolt threads, and bear against a washer iron, W. The key, K, passes through W, and is clamped by the screw bolt *e*.

The similar washer, Y, is bolted to the strap by F, and receives the key, *k*, which is clamped by *e*. C is the oil-cup.

The action of the keys, in detail, is as follows: The boxes having become loose by wear, loosen the clamp-screws, *e*, and hammer down each key till its box is properly tightened. The tapering edge of each key bears against the unyielding strap; hence, when driven in, the vertical edges move towards the crank pin, and carry the boxes along with them. And the keys readily descend, since the slots, as *ab* and *cd*, by which they go through the strap, are wider than the keys.

Construction.—This should include the plan, Fig. 2, and a detailed view of a box, Fig. 3, all made in *proper relative position*, and to a uniform scale. An end view, or a section through *hh*, may be substituted for the plan. The scalo may range from one-*fourth* to one-*half* the full size.

Similar instructions will apply to the following case:—

The Main Connection-joint:—This, Fig. 4, which is merely a sketch, differs from the parallel rod-joint, just explained, mainly in being heavier. But the opportunity is here improved of showing a variation in the construction, such as is frequently seen. As before, R is a part of the main connecting-rod, from the forward driving-wheel to the cross-head; E, the stub-end; B and B, the boxes around the crank-pin O*p*. SS is the strap and K the adjusting key, really setting into the front box, as before. With *one* key, the bolt-hole for the bolt, H, is *slotted*, that is, made oblong or wider than the diameter of the bolt, as shown by separate dotted lines for each. Then, when this bolt is loosened and the key set in, it first sets up the front box, and afterwards acts to draw on the strap S, and thus close up the rear box. Moreover, the key, in this case, is a screw-key, passing through the stirrup T*t*, and actuated and clamped by the nut and jam, *n* and *n*.

79. Having now illustrated the principal members of the *main train* of a locomotive—the cylinder, piston, frame, axle-box, guides, cross-head, and crank-pin joints, it may be best to sum up with a few elementary theorems and observations about locomotive action.

80. We begin with the following principles. *First: Action and reaction are ever equal and contrary. Second:* The *moment of a force is its product into the lever-arm with which it acts. Third:* The point of contact, R, of the wheel with the rail, R*r*, is the fulcrum to which the power acting at the crank-pin, and the resistance acting at the axle, are referred. *Fourth:* Inertia is that property by which bodies resist either an acceleration or retardation of velocity.

THEOREM III.

The effective propelling power of a locomotive, taken at the axle, is the same, whether the crank-pin is above or below the axle.

Let the crank-pin be at *p*, Fig. 35; let the engine be moving forward; and let the radius, OR, of the driving-wheel be 3 feet, the crank-arm $Op = Op' = 1$ foot, and the pressure on the piston 9,000 pounds. Then a force of 9,000 pounds, acting as at p, 4 feet from R, gives a moment of 36,000, which, divided by 3, the lever-arm, OR, of the centre, gives 12,000 pounds as the equivalent force acting, as at Oa. To oppose this, there is the reaction of 9,000 pounds, against the back cylinder-head, communicated through the engine-frame and body to the *front axle-box*, and represented by Ob. This leaves a surplus of 3,000 pounds effective pressure forward, at O, acting to propel the engine.

On the other hand, let the crank-pin be at p', pressed backward at $p'q'$ with a force of 9,000 pounds, acting at 2 feet from R. The moment of this force is therefore $2 \times 9{,}000 = 18{,}000$, and dividing by 3 the lever-arm at O, we find 6,000 pounds as the equivalent force acting at O in the direction of Ob. To oppose this, there is the reaction of the steam pressure of 9,000 pounds upon the *front* cylinder-head, and thence through the frame, etc., to draw the latter *forward* against the axle. This leaves a free surplus force, as before, of 3,000 pounds acting at O to drive the engine forward.

It follows from this that it is just as easy to start a load with both crank-pins below the centre O, as with them both above.

Theorem IV.

The pressure between the axle and the front axle-box of an engine going forward, is double that between the axle and the back axle-box.

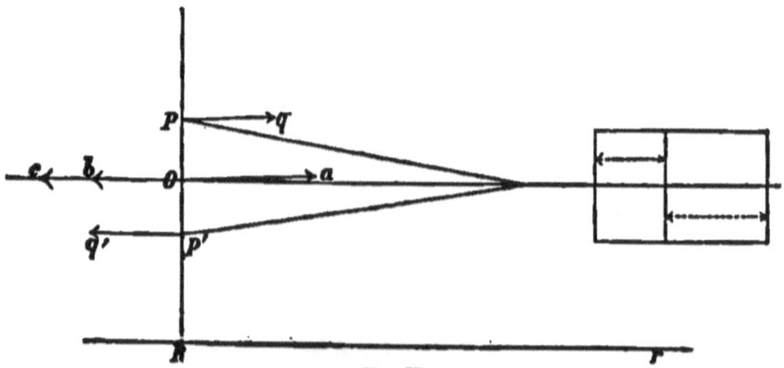

Fig. 35.

We have seen, Fig. 35, that there is a free force of 3,000

pounds out of 12,000 pounds, acting forward at O, when the crank-pin is at *p*. To this effective force is opposed the sum of all the train resistances, represented by the additional arrow *bc*. This resistance is a "hanging-back," felt through the frame and drawing the front axle-box against the axle. Thus the total pressure between the front axle-box and the axle is 12,000 pounds. Moreover, this result is constant under the given conditions; for if the train resistance in the form of friction, resistance of the air, and the jerky resistance of *lifting* the cars over the vertical unevenness of the track are less than 3,000 pounds, the balance will be made up by opposing *forces of inertia*, developed in *increasing the speed*, until the former resistances, *alone*, equal 3,000 pounds; when the equilibrium of uniform motion will ensue.

Again, the 3,000 pounds free effective pressure forward at O, when the crank-pin is at *p′*, is opposed by the train resistances as before, which act to draw the back axle-box *from* the axle. Hence the pressure between the axle and the back box is 6,000 pounds, or one-half as much as there is between the axle and its front box, which satisfies the enunciation.

As this diminished pressure, and, consequently, diminished axle-friction, occurs in the lower half of the crank-pin circle, it follows that if there is any sensible advantage in starting a train at any time, it is when both cranks are below the axle, though this happens to be just opposite to the notion said to be held by some engine-drivers.

Theorem V.

The piston, etc., move in space faster than the engine, going forward, in the forward stroke, and slower in the backward stroke.

The piston, piston-rod, and cross-head all move together, and in the same direction. Let the motion of the piston, therefore, represent them all.

Now the piston of a locomotive has a compound motion. 1*st*, its general motion forward, in common with all points of the engine. 2*d*, its own *proper motion* relative to the cylinder, which is the same as it would have in a stationary engine having an equal cylinder, and working with the same rapidity. The joint effect is, that in its forward stroke the piston is moving forward *in space*, or relative to any fixed object, *faster* than the engine; but, in its backward stroke, slower than the engine. Thus a piston of two feet stroke, making three double strokes per second, moves in its cylinder at the rate of twelve feet per second. Acting on a seven-foot driving wheel, the three corresponding revolutions of the latter would carry the engine forward very nearly sixty-six feet per second, which, by the way, is forty-five miles per hour. Then, during the forward stroke, supposing, for convenience, that the piston's proper motion is uniform at all points of the stroke, though it really is not so, it will move forward in space at the rate of seventy-eight feet per second; but in its backward stroke it will move forward relative to the ground with a velocity of fifty-four feet per second.

And the above example fairly represents all actual cases, and thus the theorem is proved.

Theorem VI.

The crank-pin has an accelerated motion in space, from its lowest to its highest point, and a retarded one from its highest to its lowest point. Also, it moves faster than the engine during the forward stroke of the piston, and slower during the backward stroke, the engine having a forward motion.

To demonstrate this theorem, refer to Pl. IX., Fig. 2, where AP is the rail, AQ the initial, A′G the middle, and PQ′ the final position of the driving-wheel for one revolution. AGP is then the cycloid described in space by the point A of the circumference of the wheel; and *agp* is the prolate cycloid described by the crank-pin in making a complete revolution from its lowest point. The cycloid may be constructed as in Prob. III. The prolate cycloid, *agp*, is also found in a very similar manner. Thus, when the diameter, M′F′, becomes vertical at *m*, *g* will be at *f*, as far to the left of the vertical line at *m* as *g* now is to the left of a vertical at f''. Or, what amounts to the same thing, $fo = f''n$, and the like is true for *e*, *d*, etc. Thus the distance of *d* from A′G equals that from d'' to the vertical at *k*, which is the vertical position of the diameter K′D′.

This being understood, observe that the horizontal spaces, $a'b'$, $b'c'$, etc., corresponding to the *equal* angular motions $g''m''$, $m''l''$, etc., of the crank-pin, increase from *a* to *g*, that is, from the lowest to the highest point of its path. But these equal angular spaces indicate the uniform advance of the engine. Therefore the crank-pin motion is accelerated, as described. And as the remaining half, *gp*, of its path is equal to *ag*, its motion is retarded as described.

Again, *dk'* is the horizontal measure of the space described by the crank-pin during the forward stroke of the piston, corresponding to the semicircle, *rq*D, of the crank circle; while *kk''* is the corresponding forward motion of the engine, since *k* and *k''* mark the two vertical positions of the horizontal diameter K'D'. But, making $a'd''' = a'd'$, we have *d'''ad'* for the path of the crank-pin during a backward stroke of the piston, and *d'''d'* for its corresponding horizontal motion, which is less than *kk''*. Hence the relative velocities of the crank-pin and the engine in the two piston-strokes are as enunciated.

THEOREM VII.

The wear of the two crank-pin boxes is equal.

This result follows from the equal pressures on the crank-pin in the upper and lower half of its circular path, as found in the last two theorems. In the upper semicircle, the pin being *drawn* forward by the piston, etc., the back box is drawn against the crank-pin; while, in the lower semicircle, the pin is *pushed* by the connecting-rod, the front box is pressed against it, but with the same pressure—of 9,000 pounds, in the last example supposed—as before.

But as the difference of taper of the keys, Pl. XVII., Fig. 1, K and *k*, seems to indicate an opinion that the front box wears faster, since its more rapidly tapering key would set it up faster, the question may be more fully examined.

When the crank-pin has a *retarded* motion, the inertia of the connecting-rod, etc., resists this diminution of speed by striving to go on uniformly, and it thus increases the pressure between the *back* crank-pin box and the pin. But the crank-pin also resists an acceleration of its motion, and thus presses back the *front* box against the pin.

Let us now examine the results for both strokes.

From *d''* to *g''*, corresponding to *d'''a* in space, Pl. IX., Fig. 2, the *steam* pressure of the *front* box against the crank-pin is *diminished* by that of the *back* box, caused by the resistance of the connecting-rod, etc., to *retardation;* and from g'' to s'', corresponding to *ad* in space, the same steam pressure is increased by that of the *front* box, caused by the resistance of the same parts as before, to *acceleration.*

On the other hand, from s to g, corresponding to *dg* in space, the *steam pull* of the *back* box against the crank-pin is *diminished* by the continual resistance through the *front* box to *acceleration* of the connecting-rod, etc.; while from g to d'', corresponding to *gk'* in space, the same pull is *increased* by the resistance, through the *back* box, to retardation of the connecting-rod, etc.

Thus, as the curve *agp* is symmetrical with respect to AG, the pressure of the crank-pin boxes against the crank-pin are equal during the *forward* stroke, corresponding to sgd'' or dgk', for the *back* box; and, during the backward stroke, corresponding to $d''g''s$ or $d'''ad'$ for the *front* box.

Hence we conclude that the difference of taper of the keys K and *k*, Pl. XVII., Fig. 2, is arbitrary, or only because there was not room for so large a key as K behind the crank-pin.

80. A few fundamental principles concerning the mutual relations of boiler, cylinder, driving-wheel, speed, and load, being doubtless of interest to students, the following is here added:—

It must be premised that locomotive power does not reside primarily in a large driving-wheel, or steam-cylinder, but in the boiler capacity to deliver steam of given pressure, at a certain rate. This, again, depends on the proportion and position of the fire-box and flues relative to the water spaces of the boiler; and, last, without going back of causes acting in the engine, to the primeval sunshine as the power which produced the vegetable growths from which coal came; the power in question resides ultimately in the coal consumed by the engine.

Let the duly related fire and water spaces, taken together, be regarded as the given boiler capacity for steam delivery at a given pressure and rate, and we shall have the following theorem. In it the further mechanical principle is employed, that the *work* of a force is the product of its intensity, or the *pressure* it can produce, by the *space* through which it acts.

THEOREM VIII.

With a given boiler capacity and size of cylinder, the larger the driving-wheel, the greater the adaptation to a light load at a high speed; and, conversely, the wheel being given, the larger the cylinder the greater the adaptation to the moving of a large load at a low speed.

This theorem can be better established by a numerical example than by abstract reasoning. Then let the following suffice.

Suppose a boiler capable of supplying 20 cubic feet per second of steam, at a *mean pressure* of 100 pounds per square inch, *in the cylinder*, and that each cylinder has a capacity of 2 cubic feet. Then 5 double strokes, or 10 single strokes, per second for each cylinder, with the steam cut off at half stroke, will consume 10 cubic feet of steam for each cylinder, or 20 feet for both. Next, suppose the driving-wheels to be 7 feet in diameter; then the circumference of each will be about 22 feet, and the five double strokes will advance the engine 110 feet. Let the piston have an area of 150 square inches, then the stroke will be the volume of the cylinder divided by 150 =

$$\frac{3456 \text{ cub. ins.}}{150 \text{ sq. ins.}} = 23 \text{ inches, very nearly.}$$

Then the 10 single strokes of each piston = 230 inches = 19.2 feet very nearly, which is the space passed over by the piston under the steam pressure of 150 × 100 = 15,000 pounds. Now, if the train have a uniform velocity, the work of the steam will be in equilibrium with that of the resistances, and we have, considering the two cylinders,

$$2 \times 15{,}000 \times 19.2 = x \times 110 \text{ ft.},$$

$$\text{whence } x = \frac{2 \times 15{,}000 \times 19.2}{110} = 5236.+, \text{pounds} \ldots . (1)$$

as the sum of all the resistances. A velocity of 110 feet per second = that of a mile in 48 sec., or $\frac{4}{5}$ of a minute, = 75 miles per hour. Now, suppose the resistance of the atmosphere at 30 miles an hour to be equal to all the other resistances on a level, and to increase as the square of the velocity; and let 10 pounds to the ton be the force required to overcome these other resistances on a level. Now $75 = \frac{5}{2}$ of 30 and $\left(\frac{5}{2}\right)^2 = \frac{25}{4}$.

Then let y = total train resistance, other than from the atmosphere, here assumed to be uniform.

or, y = atmospheric resistance at 30 miles per hour, and

$\frac{25\,y}{4}$ = atmospheric resistance at 75 miles per hour.

Hence, $\frac{25\,y}{4} + y = \frac{29}{4}\,y =$ total train resistance of all kinds

$= 5326$ pounds,

whence $y = \frac{5236 \times 4}{29} = 722.2$ pounds.

Now, at 10 pounds per ton of power, this force would move $\frac{722.2}{10} = 72.22$ tons on a level. Allowing 30 tons for the weight of the engine, the balance of 42.22 tons would be equal to about *two* heavy and well-filled passenger cars; a very light load evidently for such an engine, at ordinary speed.

These results are, of course, not given as perfectly accordant with facts. Train resistances, including those peculiar to the engine, other than atmospheric, are not really uniform at all speeds; the total atmospheric resistance depends partly on the number of cars, and on their distance apart; and, finally, there are little or no definite and understood results of experiment with very high speeds. Still, the above example sufficiently verifies the theorem.

Without going through the other cases in a similarly detailed manner, it is sufficiently evident that if the driving-wheel be reduced to four feet, the only result would be to increase the load and decrease the speed. For the number of piston strokes must be unchanged, else steam would either waste at the safety-valve, or fail to supply the cylinder at the required pressure. A four-foot wheel will advance the engine about 12.6 feet at each revolution, or 63 feet for the 5 double strokes of the piston; and as the work of the steam in the cylinder is unchanged, that of the resistance will be so also, and for the diminished *space* through which the resistance is felt during these five double strokes we shall have in place of (Eq. 1):

$$x = \frac{2 \times 15{,}000 \times 19.2}{63} = 9142.8 \text{ pounds,}$$

to be spent on the train, and atmospheric resistances, from which, by a similar calculation to the previous one, and remembering that 63 feet per second = about 43 miles per hour, we find, after deducting the engine as before, a load approximately of 16 cars at 18 tons each, loaded.

Finally, with the driving-wheel of fixed size, and the cylinder variable, let us again take a four-foot driving-wheel, and let the cylinders be of 4 cubic feet capacity each, or of about 19 inches diameter and 24 inches stroke. The capacity being thus doubled, the number of strokes necessary to consume the given volume of steam at the given pressure will be halved, giving 2½ double strokes per second, = 10 feet per second piston speed, which will advance the engine 31.5 feet.

The piston area of each cylinder = 288 square inches, whence in place of (Eq. 1) we should have $x = \frac{2 \times 28,800 \times 10}{31.5} = 18,286$ pounds.

Also 31.5 feet per second = about 20 miles per hour. Hence the given suppositions about atmospheric resistance would now give

$y \times \frac{4}{9}y = 18,286$

or $y = 12,660$ very nearly, which at 10 pounds per ton, to overcome other than atmospheric resistances, gives a load of

1,266 tons, or, deducting 36 tons for engine,

leaves 1,230 tons of load, or

70 cars at about 17 tons each, including their load.

None of these results may correspond very nearly with practice, though they may usefully indicate the path of calculation and experiment to be followed in actual cases.

These principles concerning the adaptation of engines, of certain design, to a certain load and speed, do not imply that an engine of different design, but with the same boiler capacity, cannot handle the same load at the same speed. For if, as we reduce the driving wheel, the cylinder be also reduced, so as to consume the same quantity of steam at the same pressure as before, by means of the more numerous strokes which a smaller wheel would require to maintain the *given speed* of the train, the work of the steam in the cylinder will be the same in both cases, and this work may thus be expended in producing the required high speed of a small load. Accordingly, within the past ten years there has been a very general reduction of the driving wheels, from 6½, and 6 ft., diameter, to 5½, and 5 ft., *the cylinder remaining the same*, instead of an enlargement of the cylinder, in order to move the heavier trains of later years at an undiminished speed. In either case, however, either an enlargement, or an improved proportioning, of the boiler, would be required to provide the increased quantity of steam demanded.

Some of the advantages of small cylinders and drivers are, reduced height, and consequent increased steadiness upon the track, greater lightness of running parts, and increased facility of handling them in the shop. And, by an elegant analogy, as the human frame, the most inimitable of all machines, performs better with full, muscular body, and finely moulded limbs, so we may expect that improved locomotive action is to be next sought in *steel boilers*, safely carrying 200 lbs., or more, per square inch, of steam pressure.

EXAMPLE XXVII.

A Working-Beam.

Description.—Engines may be divided, in regard to the relative position of the cylinder and shaft, into those in which the axes of the cylinder and of the main shaft are in the same plane, and those in which they are not in the same plane. The ordinary horizontal engines and vertical engines are of the *former* kind. In these, Pl. XVI., Fig. 4, the crank, OC, connecting rod, CH, and piston rod, PH, are all in one and the same horizontal or vertical line together at two positions of the crank. In the *other* class, the shaft, O, Pl. XVI., Fig. 5, is at right angles to the direction of the piston rod, LP, and the connecting rod and piston rod are parallel in the two vertical positions of the crank, OR. In this case the piston rod communicates its motion to the connecting rod through a *working-beam*, WB, oscillating on a fixed centre, C.

Pl. XVI., Figs. 1, 2, represents, with a trifling alteration of some of the mouldings, which it would be quite tedious to draw on a small scale, the beam of the Brooklyn Water-Works Pumping Engine No. 2. Its ponderous character may be seen from its main measurements, about 31 ft. 8 in. extreme length, 7 ft. 4 in. extreme width, and a thickness varying from 6 in. in the web to 2 ft. 4 in. at the hub.

The hub is chambered out, at *gh*, to avoid so great an extent of turned bearing as would otherwise have to be made. CC—C′

is the main centre; DD′ is the point of attachment of the connecting rod, or of the link, as BL, Fig. 5, to the piston rod; *bb′* is the point of attachment of a pump rod.

This beam is not of particularly fine outline, being bounded by circular arcs. A more elegant form would be given by a parabolic outline as sketched in Fig. 3, where, if too clumsy when *d* is made the end of the beam, and *abcd* is bounding curve, the latter can be prolonged, as in the curve *ce*, making *d*, the vertex of the parabola, the centre of the pin.

Indeed, by merely proportioning the thickness of the beam properly for strength, any other similarly curved outline may be taken for its elevation, as a single circular arc, or an ellipse, for each long side AG or A′F.

The great weight of each half of the beam itself, acting at its centre of gravity and supported at C, acts to separate the particles of iron at A′ and to compress them at A. To resist this tendency, the lower flange, A,A″, is made vertically thicker, as shown at those letters.

Construction.—Let the exercise be varied by adopting some of the outlines just mentioned. The plan will be a sufficient guide for showing the mouldings, as *mn*, and the end of the beam, in the section.

EXAMPLE XXVIII.

A Stephenson Link.

Description.—There is some difficulty in either clearly explaining the separate members of the train of parts which compose a locomotive valve motion before explaining the whole, or in explaining the whole concisely, before describing its several parts.

In adopting first the former course, the relation of each part to the whole will be touched as briefly as possible, and only to make each part as intelligible as may be.

Locomotives are required to go either forward or backward at will. Hence it is plain that if, when the engine is at rest, the relative position of the slide valve T,T′,T″, Pl. IV., Fig. 2, and piston, P, is such that steam will be admitted to a certain side of the piston, through one of the steam passages, and will make the engine go *one* way; then, to make the engine go the *other* way the valve must be shifted to such a new position as to open the opposite steam port and so admit steam to the *other* side of the piston, instead.

It is thus also plain, without tracing here all parts of the train of pieces from the driving shaft to the valve, that the valve must be actuated, in the two cases, either by two different positions of one piece made to take hold of the valve spindle or its rocker-arm; or else by two different suitably disposed pieces. In the latter case, which is the one employed in locomotive, and many other engines, these pieces may be either separately or simultaneously acted upon, to put one of them *in* and the other *out* of gear with the valve.

In Stephenson's and other similar link motions the latter course is pursued. The two Stephenson, or "shifting" links, one on each side of the engine, are raised and lowered simultaneously by one lever in the engineer's cab, and thus the valves are actuated so as to produce forward or backward motion at pleasure.

The link and its attachments are thus formed, Pl. XVII., Fig. 5. LL is the link, which is an open or slotted curved bar. The link is embraced by a saddle block, SS, which is sometimes hung, or, as in the figure, borne up by the supporting link A. In either case the link A is attached to one arm of a "bell crank," which turns on a fixed shaft, called the "tumbling shaft" (see T, Pl. IX., Fig. 3, or any locomotive), the other arm of which is operated on from the "foot-plate," where the driver stands, to raise and lower the link.

F is the link block, attached to the lower end of the hanging rocker, *h*, Fig. 10, by the pin *g*, and on which the link freely slides as it is raised or lowered. The link is thus rigidly connected with the three moving points P and P′, the eccentric rod pins, and Q, the saddle pin, and has also a movable connection with the link block pin, *g*, whose motion depends on that of the link.

Construction.—Leaving further details, which here require too much to be imagined, except to persons already familiar with the locomotive, to the next example, and mainly until valve motions in general shall be described, we only add that the three projections in Fig. 5 are only sketches not drawn to scale A scale of from one-*third* to one-*sixth* would be appropriate.

C—SURFACE COMMUNICATORS.

a—Plane Communicators.

EXAMPLE XXIX.

A Circular Eccentric, Strap, and Rod.

Description.—A circular eccentric, Pl. XVII., Fig. 6, gives rise to a variable rectilinear motion.

The figure shows two elevations of the eccentric, which is made in two pieces, strongly bolted together, at *bb*, on one side, through the projecting halves of the sleeve IIIIAF. E—E is the eccentric, bearing the rib R—R, which prevents the strap, Fig. 7, from slipping off laterally. In this rib is the oil groove *g*. Quite small eccentrics are solid plates, except where the shaft to which they are fastened goes through them. This one, which belongs to a locomotive, is open, and hence stiffened by an arm, *a*, at its widest opening. In putting together a locomotive valve motion, the eccentric has to be set, as will be more fully explained further on, so as to give the desired motion to the slide valve. This is done partly by a series of trials. It is therefore secured to the driving shaft, when properly adjusted, by set screws through the holes *p* and *q*.

We will now show that the eccentric is a substitute for a short crank. A heavy cranked axle, as at Fig. 9, for giving a very short motion as a valve motion $= 2ac$, or only about equal to the diameter of the shaft itself, would be a very clumsy and costly device, and indeed quite impracticable; first, for want of room if the axle were cranked again for the connecting rod, as in engines with "inside connections;" and second, because the direction of the valve crank arms could not be exactly enough determined in advance.

Now let O, Fig. 6, denote the centre of the driving shaft, and centre of *motion* of the eccentric; and let *o* be the centre of *figure* of the eccentric. Then *o* will describe a circle about O, with the radius O*o*; and any *fixed* point on the eccentric will describe a circle about O, with a radius equal to its distance from O. Hence, if the strap, SS, and eccentric rod, *a—a*, Fig. 8, were rigidly fastened to the eccentric, the pin *a* would describe a circle of six feet radius, more or less. But if, as is done in practice, the strap is secured on the inner circumference, so as to slide freely on the rib, R, of the eccentric, then the pin *a* will be actuated back and forth a distance equal to the difference of O*n* and O*m*, Fig. 6, which, since the eccentric is circular, = 2O*o* This result is the same as that due to a crank of the length O*o*.

An eccentric is most readily conceived of as essentially a crank, by considering it simply as a crank-pin so large as to em brace the axle, and include the crank arm within it. The eccen tric strap is thus seen to be the counterpart of the strap S, Fig. 1, which couples the stub-end of a connecting rod to a common crank-pin.

The strap, Fig. 7, consists of two irregular semi-rings, SRP and SOK, bolted together at *bb*. To the upper belongs the shoulder P through which is the oil passage *mn*, while to the lower segment there belongs the oil well O, which is chambered out to form a reservoir inside; and the bearing, K, for the back end, *r*,*r*, of the eccentric rod, *r*—*a*. From the two measurements of the thickness, an edge view may be made by the student.

CC′ shows a portion of the front end of the eccentric rod where it takes hold of the link, Fig. 5, as at P or P′. Here, C is a plan or top view, and C′, an elevation or side view.

Construction.—After the full description just given, it is enough to add that the scale may range from one-*half* to one-*eighth*.

Summing up now the valve train of a locomotive in a skeleton sketch, we have, Fig. 36, A, the driving axle; the eccentrics

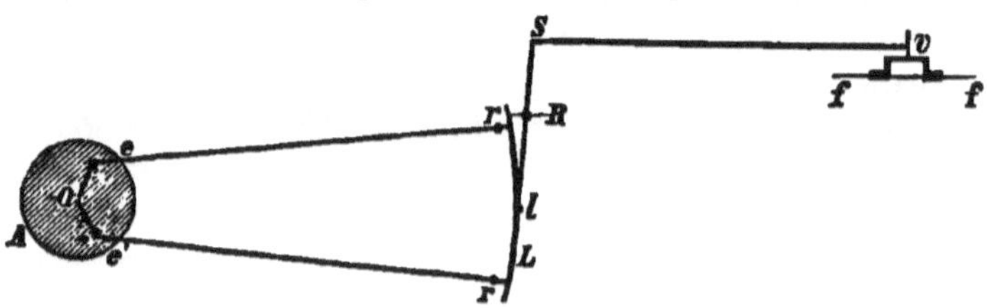

FIG. 36.

represented by their centres *e* and *e*′, and crank arms *eo* and *e*′*o*; the eccentric rods, *er* and *e*′*r*; the link, L; link block, *l*; fixed rock shaft, R; hanging and standing rockers, R*l* and R*s* (78); valve stem or spindle, *sv*; and valve, *v*, on its seat *ff*. Ex. XLII. Further on it will be shown how to lay these out in their proper relative positions; though the student can exercise himself on this, with measurements from practice, and with certain fixed data, in each case, learned from a mechanic, or engine driver.

Example XXX.

A Heart Cam or Eccentric.

Description.—This cam is one for producing a *uniform rectilinear motion from a like rotary one.*

The distinction between a cam and an eccentric is perhaps not very closely defined.

It may be said, in the absence of fixed usage, that a cam wipes or pushes againt the piece which it acts upon, without being actually joined to it by a material connection. Hence it is often also called a wiper. An *eccentric*, on the contrary, may take hold of that which it actuates.

The term heart eccentric, above, is quite indefinite, since various heart-shaped curves of different properties may be devised. A heart eccentric of the kind here required will be bounded, on its acting circumference, by parts of two opposite spirals of Archimedes, since the definition of that curve is, that for *equal angular* motions of its generating point, the same point has also a *uniform radial* motion.

Construction.—Hence, Pl. XVII., Fig. 11, let it be required to lift and let fall a bar vertically through a space of 9 inches, from M upwards. Then lay off from M, downwards, the least radius, MO, in this case 9 inches, and thence the greatest radius, OG, which must be 9 inches greater, as required, or 18 inches. Divide the 9 inches from G upward into any convenient number of equal parts, as eight (not shown), and through the points draw arcs from O as a centre, making any one of them a complete circle. Divide each half of this circle into eight equal parts, each way from OG, and draw radii through the points of division. Then M, or the 0 point of each branch, will be the intersection of the innermost or 0 circle, with OM, the 0 radius; the point 1 of the cam will be the intersection of the next, or circle 1, with the next radius, O 1, etc., on each side of OG. Through the points thus formed, the curve can be drawn by an "irregular curve," or by circular arcs, if desired, by a repeated application of the problem "to draw an arc through two given points," as for example through 3 and 4, and tangent to an arc through the points 0, 1, and 1.

Having thus found the outer and essential curve of the cam the parallel inner one ACD can conveniently be drawn tangent to numerous little arcs having their centres in M46G, and a common radius of 1 inch, the thickness of the rim.

The construction of the feathered arms is obvious enough on inspection.

The vertical section through MG is made by simply projecting over from the curved elevation, and laying off the widths of the different parts. The similar letters at like points will sufficiently explain the relations of the two figures.

For further practice let the student make a horizontal section through O.

b—Developable Communicators.

Gearing.

81. Before entering upon the construction of toothed wheels, either separately or as acting together, a few preliminary explanations of the principles of gearing will be given, which may make the subject more intelligible, yet without entering into the theory of the subject so fully as would be done in a work on analytical Cinematics.

Gearing is the term commonly applied to any combination of toothed wheels; that is, wheels fitted with teeth, so formed and disposed upon their circumferences as to engage each other in regular order, giving a desired motion to the wheels.

82. Gearing may be classified: *first*, according to the relative positions of the axes of the wheels; *second*, according to the disposition of the teeth relative to the elements of the convex surfaces of the wheels.

According to the former classification, the axes may—

I.—Intersect at an infinite distance, or be *parallel.*

II.—Intersect at a finite distance, or simply *intersect.*

III.—Intersect nowhere, when they will *not be in the same plane.*

In the first two cases the axes *will* be in the same plane.

83. In the first case, the wheels mounted on the parallel axes, form *spur gearing*, and are of cylindrical form, Pl. XX., Fig. 4.

In the second case, the wheels, having converging axes, are of conical form, and constitute *bevel gearing*, Pl. XX., Fig. 5.

In the third case, the convex surfaces of the wheels are frusta of hyperboloids of revolution having the given axes for their axes, and tangent to each other along an element; that is, they have a common generatrix. See Pl. XX., Fig. 6, where the two hyperboloids represented by AB and CD are tangent along the common element *mn*.

In Fig. 4, A and B are the parallel axes of two thin cylinders, in contact at T, and from which the finished spur wheels are made.

In Fig. 5, V is the common vertex, and VT the common generatrix, or element of contact, of two tangent cones, VCT and VDT, whose axes are VA and VB. From tangent frusta, as T*n*C and T*n*D, of these cones, a pair of bevel wheels may be formed.

In Fig. 6, AB and CD represent two hyperboloids, whose axes are AB and *pq*, and whose common generatrix and element of contact is *mn*. Then a pair of thin tangent frusta, as those having C and A for their bases, and fitted with teeth set in the direction of the elements of the respective hyperboloids, will act together, though less smoothly than in the two preceding cases, since the tangents *m*N and *m*K, to each, at their point of contact, as *m*, will not coincide; while the direction of the revolution of each is that of its own tangent.

84. By the second classification the teeth are disposed:—

I.—In the direction of the elements of the surface of the wheel. This is the usual case. See Pls. XIX., XX., and XXI.

II.—Spirally, as in Pl. XXXI., Fig. 1, 2, where each longitudinal edge of a tooth of the wheel forms part of a very long helix, that is, a helix of very great pitch.

85. Among special and older forms of gearing are the *lantern*, Fig. 37, where pins, generally included between two disks, take

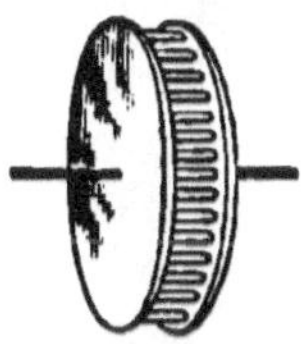

Fig. 37.

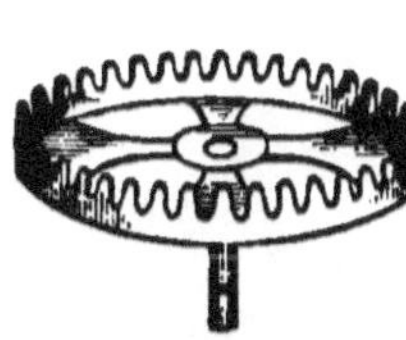

Fig. 38.

the place of teeth of the usual form. These may be seen in common brass clocks. Also the *crown wheel*, Fig. 38, where the height of the teeth from top to bottom, instead of their length, is parallel to the axis of the wheel. These may be seen in old watches, and cider mills.

Small toothed wheels engaged with much larger ones, are often called *pinions*, and their teeth are called *leaves*. In small work, especially, the axis of a tooth wheel is often called its *arbor*.

THEOREM IX.

The number of revolutions in a given time, and the angular velocities of toothed wheels, are inversely as their radii.

Let A and B, Fig. 39, be two cylinders tangent to each other, in close contact along the element whose projection is C, and mounted on the shafts also indicated by A and B.

So long as there is only a rolling motion, without slipping, between their circumferences, both of *their convex surfaces will move with equal velocities.*

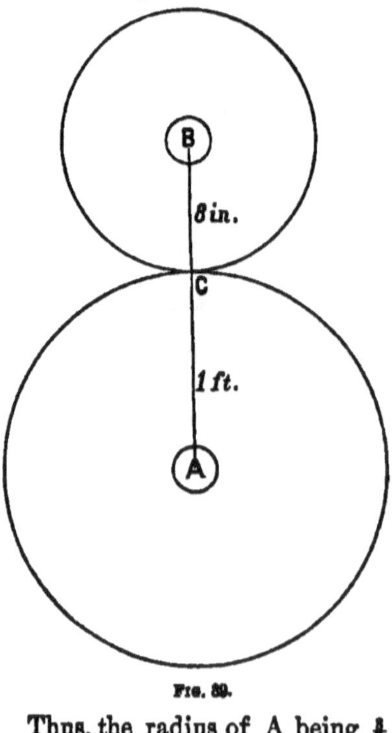

FIG. 39.

Now the length of the circumference varies directly as the radius. If, then, as in the figure, BC$=\frac{2}{3}$ AC, the circumference of wheel B will be two-thirds of that of wheel A. Therefore, when the two wheels are made to revolve, by virtue of the friction between them, until B has made one entire revolution, its entire convex surface will have been in contact with two-thirds of the circumference of A. That is, A will make *two-thirds* of a revolution, for *one* revolution of B; that is, *the number of revolutions of each wheel is inversely as its radius.*

Thus, the radius of A being $\frac{3}{2}$ of that of B, its number of revolutions will be $1 \div \frac{3}{2} = \frac{2}{3}$ as many as those of B.

The angular space swept over by a given radius of each wheel in a unit of time measures its *angular velocity;* and, from the explanation just made about the comparative revolutions of two wheels, it is evident that their angular velocities are also inversely as their radii.

86. It is quite evident that but little power could be transmitted from A to B, or the reverse. In other words, if B offered great resistance to being turned, A would either merely slip upon it without turning it; or, if the two wheels were very severely pressed together, their surfaces of contact would be speedily ground away.

In the light of these facts, the object of gearing is to enable either wheel to turn the other against a great resistance, and also to preserve the same relations of motion that have just been stated.

The gearing itself consists in suitably-formed, alternate ribs and grooves on the surfaces of contact of the two cylinders, so that they shall not merely be tangent to each other, but shall take hold of one another. Let us proceed to see how this can be done.

87. Since cylinders are each of equal cross-section throughout, let the revolving tangent cylinders be represented by their circular sections, which will be sufficient for all purposes of explanation. Then, with A as a centre, Fig. 40, which is double the size of Fig. 39, strike two circles, AD and AE, equidistant from the circle AC, within and without. Do the like with B, as a centre with respect to the circle BC, so that D and E shall be points of contact, as shown.

Now conceive teeth to be formed, as shown, on each wheel, and so shaped as to be in contact at C, *where their action is most effectual.* Also let them be so shaped as to begin and end their contact, as at *n* and *o*, at equal distances within the circles AC and BC. Thus the *average* distances from A and B, of all the points of contact of the teeth which are in contact at once, are AC and BC. Hence, wheels armed with such teeth will move with the same equal velocity at C, and with the same relative number of revolutions as did the original tangent cylinders, AC and BC.

Again, the teeth must be equal in width, and equally distributed on the two wheels, in order to preserve the same relative velocity between the two wheels. *Hence the number of teeth on each wheel* will be directly as its radius. Thus, if BC = ⅔ of AC, and if the wheel A has 24 teeth, the wheel B will have 16 teeth.

88. The several parts of the teeth and their governing circles may now be defined.

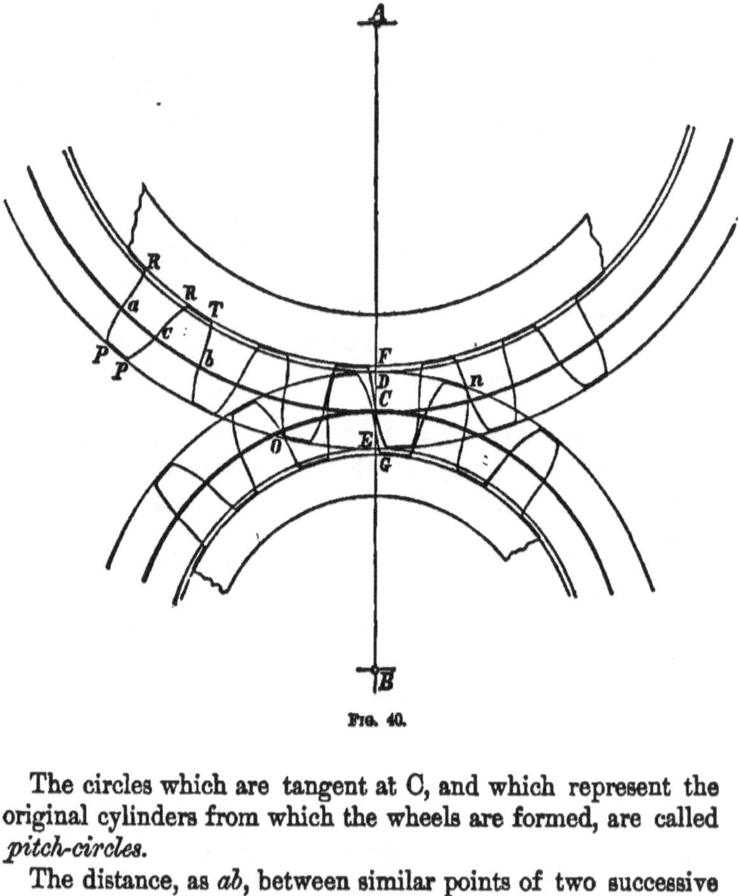

FIG. 40.

The circles which are tangent at C, and which represent the original cylinders from which the wheels are formed, are called *pitch-circles.*

The distance, as *ab*, between similar points of two successive teeth is the *pitch.* The pitch is necessarily the same on any two wheels that will work together.

The parts, as *pp*, are called the *points* of the teeth; while the lines, as at R and R, are their *roots.*

The circles AF and BG are the *root-circles.* They are a little within the circles AD and BE, so that the points of the teeth of one wheel shall not strike the rim of the other wheel on the spaces, RT, between the teeth.

The circles AE and BD are the *point-circles.*

The portions of the sides of the teeth, as at *pa*, are the *faces* of the teeth. The parts, as from *a* to R, are the *flanks* of the teeth.

The spaces, as *cb*, between the teeth, are a little greater than the widths, as *ac*, of the teeth; both distances being taken on the pitch-circle. This prevents the teeth from getting wedged together.

89. It is now necessary to describe the construction and some of the properties of a class of curves, which, as will soon be seen, are of use in giving the true forms to the teeth of wheels.

These curves are the cycloid, epicycloid, hypocycloid, and involute.

When any curve, A, Pl. XVIII., Fig. 1, rolls upon any other curve, B, any point, as T, of the rolling curve *generates* or describes a curve of the class called *Trochoids.* The above-named curves are merely the most familiar and practically useful trochoids. In their formation, either the rolling or the fixed lines, or both, are circles. Either one, but not both at once, may be a straight line.

90. The *common* cycloid, or simply the *cycloid*, AK, Pl. XVIII., Fig. 2, is the curve generated by any point, as A, of the circumference of a circle, as OA, which rolls on a fixed straight line, AB. If the rolling curve be any other than a circle, the curve generated will still be a cycloid, but the above is the *common* cycloid. A similar remark applies to each of the following cases.

91. The *epicycloid*, AD, Pl. XVIII., Fig. 3, is generated by a point, A, of a circle, AO, which rolls upon a fixed or *base* circle, AB. The two circles may have any relative size. The epicycloid has two varieties, the *external* epicycloid, Fig. 3, where the centres, Q and O, of the given circles are on *opposite* sides of their point of contact, A; and the *internal* epicycloid, Fig. 4, where those centres are on the same side of the point of contact. In the latter case, the rolling circle is the larger one, and its concavity rolls upon the convexity of the fixed circle.

92. The *hypocycloid*, AK, *ak*, AK′, Pl. XVIII., Figs. 5, 6, is the curve generated by a point, as A, or *a*, of a circle, OA, which rolls on the interior of a larger fixed circle, AB.

93. The involute, AC, Pl. XVIII., Fig. 7, is the curve gen-

erated by any point, A, of a straight line, AD, which rolls upon a fixed circle, AB. It is thus the curve described by any point of a thread when unwound from a circular plate. This is, strictly, the *common* involute. If the fixed curve be any other than a circle, the involute would be an involute of that curve.

The construction of these curves by points will now be explained.

Problem III.

To Construct a Cycloid.

First Construction.—Let AB, Pl. XVIII., Fig 2, be the base line, and OA the rolling circle. By the definition of the curve set off from A, on the circle, spaces 0—1; 1—2, etc., equal to 0—1; 1—2, etc., on AB. When the corresponding points, as 2, 2, coincide and become the point of contact of the circle and base line, the centre, O, will be at the point 2 on the line OD, parallel to AB. This line is therefore divided in the same manner as AB. Also the point A of the circle will then have risen to the same height above AB that the point *2d* now has, and hence will be found somewhere on the line *cd*, through 2, and parallel to AB. The like is true for other points. Hence take the successive points 1, 2, etc., on OD as centres, and the constant radius, OA, and describe arcs, which will be parts of the successive positions of the rolling circle, and where these arcs intersect the horizontal lines 1—1, 2—2, etc., will be points 1, 2, 3, etc., of the cycloid.

Second Construction.—To avoid acute and indistinct intersections, as at 1 and 7 on the curve, make $ab = cd$, $ef = gh$, etc., to find the points of the curve, *instead* of drawing the successive positions of the circle OA.

Fundamental properties, apparent on inspection. *First*, The distance from A to where the point 0 of the circle again coincides with AB is evidently equal to the length of the circumference of the rolling circle. *Second*, The extreme distance, 8—8, of the curve from the base line is equal to the diameter of the rolling circle. *Third*, When the centre O moves uniformly on OD, the generating point, 0, moves most rapidly, both on the curve and in the direction of AB, when at K, as is obvious on comparison of the spaces 0—1; 1—2 . . . .

7—8, etc., on the curve, which correspond to the equal spaces 0—1, 1—2, etc., on OD.

Problem IV.

To Construct an Exterior Epicycloid.

Let Q, Pl. XVIII., Fig. 3, be the centre, and QA the radius of the base circle, and let OA be the rolling circle. The construction, as may now be seen by inspection, is so exactly analogous to that of the cycloid that detailed description is unnecessary. 0—1, etc., on the circle OA = 0—1, etc., on AB. When 1, 2, etc., on AB come to be points of contact, 1, 2, etc., on OC are the corresponding positions of the centre of the rolling circle. OC is drawn with QO for a radius, and Q 1—1; Q2—2, etc., are its successive radii, through 1, 2, etc., on AB, to find 1, 2, etc., on OC. Then, as in the cycloid, the point 2, for example, on the epicycloid AD, is found at the intersection of the arc 2—2*b* with the arc having the centre Q, and radius Q*d*; or by making *ab* = *cd*.

The fundamental properties are the same as those mentioned in the last problem.

Problem V.

To Construct an Interior Epicycloid.

Here, again, Pl. XVIII., Fig. 4, the construction is so similar to that in the two preceding cases, that it is sufficient to point out the given parts.

The circle QB is the fixed base circle. The circle OA is the rolling circle. As it rolls, its centre O describes the circle QO. The points, 0, 1, 2, etc., on the circle QO, are the successive positions of the centre O, when 1, 2, etc., on the circle OA come to be the points of contact. AD is the epicycloid, and, as before, when 2, on circle OA and 2 (*n*) on QA unite as a point of contact, A will have removed as far radially from the circle QA, as 2 on OA now is from the circle QA. Hence A will be found on the arc 2—2, having Q for its centre, and at its intersection with 2—2 having (*k*) 2, on circle QO for its centre. Or, as before, make *ab* = *cd* and *ef*=*gh*, to find the points 3 and 4, for example, of the epicycloid.

The second property in Prob. IV. must here be modified to read, that the extreme distance of this epicycloid from the base circle will be equal to the difference, BK, of the diameters of the rolling and base circles.

Problem VI.

To Construct any Hypocycloid.

The construction, Pl. XVIII., Figs. 5, 6, is again so closely analogous to the preceding cases, that it only seems necessary to point out the different cases. AK, Fig. 5, is the hypocycloid generated by the point A of the circle OA rolling from A towards B on the inner side of the circumference QA. The circle OD is the path of the centre, O, of the rolling circle. Equal distances, 0—1, etc., = 0—1, etc., are laid off from A on both circles, and any point, as 4, of AK, can be found at the intersection of the arc as 4—4, with centre Q, and the arc, as 4—4, with centre 4 on OD; or by making $ab=cd$. Observe that when, as in this case, the radius, OA, of the rolling circle, is *more than half* of QA, that of the base circle, the hypocycloid, AK, and the motion of the circle OA, are on *opposite sides* of the initial radius OA.

In the same figure, *ak* is the hypocycloid, constructed just as before, generated by the point *a*, of the circle *oa*, which rolls towards *b*, on the inner side of the circumference Q*a*. Here, where the radius of *oa* is *less than half* of Q*a*, the motion of the circle O*a*, and the hypocycloid, *ak*, are both on the *same side* of the initial radius *oa*. This result prepares the mind to apprehend the intermediate case, shown in Fig. 6, where the radius, OA, of the rolling circle is *just half* of QA, that of the base circle, and the hypocycloid becomes a diameter AK.

The fundamental properties mentioned in Prob. III. hold good for the hypocycloids.

Problem VII.

To Construct the Involute of a Circle.

Let the circle OA, Pl. XVIII., Fig. 7, be the fixed circle, and AD the straight line which rolls upon it. As in the preceding problems, the rolling spaces of AD, and the corresponding circular arcs rolled upon, are necessarily equal. Hence set off equal distances, from A on AB; draw a tangent to the circle, at each of these points; and then, with 1 on AB as a centre, and 1A as a radius, describe the arc A*a* to intersect the tangent 1—1 at *a*. With 2, on AB, as a centre, and 2—*a* as a radius, describe the arc *ab*, giving *b* on the tangent 2—2. In like manner any number of points on the involute AC may be found. AC, as thus found, is, of course, not a perfectly true involute, since the radius of the latter should change at each instant; but neglecting the slight error of taking the chords 1—A; 2—1, etc., on AB as equal to their arcs, the points *a*, *b*, 3, etc., of the involute are exact.

THEOREM X.

In all the curves just described, the tangent, at any point, is perpendicular to the line from that point to the corresponding point of contact of the rolling and fixed lines.

This theorem becomes evident by simply considering that the rolling circles are of an invariable form and size, and hence, as in Pl. XVIII., Fig. 2, afford a geometrically inflexible connection between any point, as 4, of the cycloid, etc., and the corresponding point of contact, 4, on AB. That is, the chord *mt* (*ft*, Fig. 3) is a fixed line for the instant that *t* is the point of contact of the rolling circle and base line. In other words, *m* (*f*) is at the same instant about to describe a circular arc about *t* as a centre. Hence a perpendicular to *mt*, at *m*, will express the direction of this instantaneous circular effort, and will, therefore, be the tangent at *m*.

Like reasoning applies to Figs. 3, 4, and 5. In the case of the involute, the point *b*, for example, evidently tends for the instant to describe a circular arc about 2, on AB, as a centre. That is, the tangent to the involute at *b* is perpendicular to the line *bc*.

THEOREM XI.

The relative position of two circles is the same, whether one rolls over a certain arc of the other, which is fixed, or both re-

volve on fixed centres till they have had the same amount of contact as before.

Let the circle A, Pl. XVIII., Fig. 8, roll over the arc G′C′, on the circle B, which shall be fixed. C′D′, etc., $= c'd'$, etc.; C′G′ $= c'g'$, and the radius AG′ will be found at $A'g'$. Now turn the entire system about B as a centre till the line of centres BA′ has returned to its original position, BA. The point g' will then appear at g; A′ at A; $c'g'$ at Cg; C′G′ at CG; gAB will be equal to g'A′B, and G′g = CG. But the latter result is just that which follows from the revolution of A and B on their centres, while in contact, without slipping, at C.

Thus the *relative* position of the two circles in the two cases is the same, as stated in the theorem.

Theorem XII.

The relative position of three circles, which maintain a common point of contact, is the same, whether one of them is fixed, or all revolve on their centres.

Let B, C, and A, Pl. XVIII., Fig. 9, be the centres of the three circles. If A roll on B, from D to D′, we shall have $d''d'$ = DD′, and every point of $d''d'$ will have been in contact with DD′. If the circles C and A maintain the same point of contact, d, that point will be found at d', and C at C′. If, then, the circle C roll from its position C′d' to C″d'', we shall have $d''d'''' = d''d'$ = D′D. Hence, if circle C roll on B from D to D′, we shall have $d''d''''$ = D′D $= d''d'$. Thus, when the centres B, C, and A have in this manner reached B, C″, and A′, the circle C has virtually rolled on the exterior of B, and the interior of A, and over an equal length of arc on each.

Now revolve the entire system about B as a centre, from the position BC″A′ to BCA, and D′D will appear at DD″; $d''d'$ at dd''', and $d''d''''$ at dd^{v}, and we shall have DD″ $= dd''' = dd^{v}$. But the result DD″ $= dd^{v}$ is just what follows from the revolution of B and C on their centres with purely rolling contact at d. Also DD″ $= dd'''$ expresses a like motion of B and A, and $dd^{v} = dd'''$ a like motion of C on the interior of the circumference of A; *which satisfies the enunciation.*

THEOREM XIII.

In the rolling of three circles, with a common point of contact, any point of the inner circle will describe an epicycloid upon the circle on which it rolls, and a hypocycloid within the remaining circle. These curves will be the proper curves for teeth, acting tangent to each other, to give a rolling motion to the circles to which they belong.

The separate motions of the pairs of circles, B and A, Pl. XVIII., Fig. 10, being seen, when analyzed, to be as explained as in the last theorem, it is clear that the point of contact, *m* (*d*, Fig. 9), of the inner circle, A, will describe an epicycloid, as Q′*h*′ on the circle B, which, after revolution of the system from BR′ to BR (BA′ to BA, Fig. 9), will appear at Q*h*. Also that the same point of contact, *m*, when circle A rolls from A″ to A′ (circle C, Fig. 9, from C′ to C″, or, what is the same, from C‴ to C), will describe the hypocycloid *d′h′*, which appears at *dh*. But note that for the three circles to have constantly a common point of contact, is for A to roll simultaneously upon B, and within R. Hence the point *m* of the circle A *simultaneously* generates the epicycloid Q′*h*′ (Q*h*) and the hypocycloid *d′h′* (*dh*). These curves, thus having at each instant one common point, will therefore be constantly tangent to each other at the position for the moment of that point. And as their common normal constantly passes through the common point of contact of the three circles by Theorem X., as *mh* passes through *m*, it follows that the constant contact of Q*h* with *dh* will maintain a constant contact of the circles B and A. For *h* and *q* are always at the same distance from *m*, on the same line *m* (*qh*).

We have now the following fundamental theorem of gearing.

THEOREM XIV.

When any circle, less than either of two given pitch-circles, rolls on the exterior of both, and on the interior of both, it will, in the former case, generate the faces of the teeth of both wheels, and in the latter their flanks.

This theorem is little more than a slight expansion of the preceding, expressed in technical terms. The epicycloid Q*h*,

Pl. XVIII., Fig. 10, being external to the pitch-circle, B, any small portion of it, limited by a *point-circle*, concentric with B, and a little larger than it, would be a *face* of a tooth of B. Likewise the hypocycloid, *dh*, being internal to the pitch-circle R, a small portion of it between that circle and a concentric *root-circle* (88) within it would be the *flank* of a tooth of R.

Now it is obvious that, if the same circle A were to roll upon R and within B, similar results to the foregoing would follow; which proves the theorem; since, by the last theorem, these curves would always have a point of contact.

The circle, as A, Fig. 10, which thus carries the generating point of the tooth-curves, is called the *describing circle*.

Theorem XV.

Involutes are proper curves for the teeth of wheels.

A general proof of this would be that the rolling straight line EN, Pl. XVIII., Fig. 11, any point of which generates an involute, is but the extreme case of the rolling circle, viz., that in which the radius is infinite. But the generating line does not roll directly upon the pitch-circles, since the teeth would then be wholly exterior to the pitch-circles. Hence the generating line is taken, as at EN, tangent to *base-circles* within the pitch circles of the wheels, and containing the point of contact. Any point, as *d*, on the line EN, will then generate the involute, *de*, of the base-circle, BN, of the wheel B; and the involute *df* of the base-circle, AE, of the wheel A.

Any other point, as *n*, will generate the involutes F and G. Now, as all involutes of the same circle are equal curves, and as the common generatrix of both involutes, whether *d* or *n*, is on the line EN, the point of contact of any given pair of involutes will be on EN, and thus F and G may simply be regarded as new positions of the involutes *f* and *e*, caused by a rotation of the circles having A and B for their centres.

Now this result secures a rolling contact, or an equal velocity of circumference, at C, and hence an angular velocity of each wheel inversely as its radius, for we have by the definition of the involute

$$ek = eN - kN = dN - nN = En - Ed = EF - Er = rF;$$

or, $ek = rF$; that is, the circumference velocities of E and N are equal. But, by the similar triangles, AEC and BNC, the radii of the pitch-circles are to each other as the radii of the base-circles. Hence the circumference velocities of the pitch-circles also are equal, which secures their rolling contact at C, as stated and required.

Theorem XVI.

The teeth act by sliding contact, and their point of contact is on the generating line.

First, That the teeth of wheels act by sliding contact is evident from Pl. XVIII., Fig. 10, where, when the wheels A and B revolve through the arc MQ=*md*, the portion, Q*q*, of the face curve of the teeth of B has been in contact with every point of *dh*, since both are, as before shown, generated simultaneously by the same point *m*. But Q*q* being obviously much longer than *dh*, there must have been a sliding contact between them, equal to the difference of their lengths. The same property is evident in the action of involute teeth, Pl. XVIII., Fig. 11, since, for the instant in which all parts have the position there shown, the velocities of *d* as a point of *d*N, generating *de*, and as a point of *d*E, generating *fr*, are as the momentary radii *d*N and *d*E.

Second, The point of contact of the tooth curves is always on the generating circle, simply because one and the same point of that circle is the common generatrix of both.

Theorem XVII.

Teeth, formed by either of the preceding methods, give a constant angular velocity ratio to the wheels which carry them.

This has, in effect, been already shown for wheels with involute teeth, Pl. XVIII., Fig. 11, by means of the similar triangles AEC and BNC, in which EN constantly passes through C, and therefore BC and AC, the radii of the pitch-circles, are constant segments of the line of centres, and always proportional to BN and AE, which are constant.

For the case of epicycloidal teeth, see Pl. XVIII., Fig. 10.

Now, first, by Theor. X., and since the tooth curves are in contact at the position, for the moment, of the common generatrix *h*, of *dh*, and *q* of Q*q*, their common normal constantly passes through M, the point of contact of the pitch circles.

Second, to maintain contact as at *qh*, all points of the indefinite common normal, *he*, must at each instant be moving with the same velocity in the direction of that line. Hence the angular velocities of *r* and *e* are as the radii, R*r* and B*e*, drawn perpendicular to that line. But the ratio B*e* : R*r* is constant, since those lines are proportional to BM and RM, which are constant. That is, the *ratio* of the *angular velocities* of points *r* and *e*, whose velocities are determined by the action of the teeth, is constant; and therefore that of the pitch circles is so also, because the common normal divides the line of centres into constant segments, by passing always through the point of contact of the pitch circles.

94. The wheel which actuates the other is called the *driver*. The other is called the *follower*.

When the wheels revolve so that B, Pl. XVIII., Fig. 10, turns to the right, the pitch circle arc, as QM, through which action takes place between the tooth curves, beginning at *q*, is called the *arc of approach*. If the wheels should revolve so that B should turn to the left, the same arc would be termed the *arc of recession*, since it is the arc through which action takes place while the points M and *m* are receding from the line of centres to the positions of Q and *d*. The sum of the two is the *arc of action*.

Theorem XVIII.

Within certain limits, the face of a driver acts best upon the flank of a follower, and during the arc of recession; but for either of a pair of wheels to be a driver, the teeth of each must have both faces and flanks.

Since the *face* of the *driver* acts upon the *flank* of the *follower*, during the arc of *recession*, it follows that, if the motion be reversed, so that in Pl. XVIII., Fig. 10, for example, the circumferences of the pitch circles, R and B, shall roll *to the right*, through M, the former *arc of recession*, when motion was *to the left* through M, will become the *arc of approach*, and the flank, *dh*, of the wheel, R, will push the face, *Qq*, of the wheel B.

Now: *First.* It is found experimentally that the friction between the teeth is, at least within certain limits, more abrasive and vibratory during the arc of approach than during the arc of recession. Hence, if one of a pair of wheels were to be only and always a driver, it would be desirable that its teeth should only have faces, and the other wheel's teeth only flanks. But it is also important that any one pair of teeth should not quit contact with each other before another pair should come into contact. Indeed, to distribute the pressure better, two or more pairs of teeth should be in contact at once. Now, if this last condition be fulfilled in the arc of recession alone, it might require teeth so long as to make the friction between them as severe as during the arc of approach. This is evident by comparing the spaces 5—6, on the curves, in Pl. XVIII., Figs. 3 and 5, with the point *q*, Fig. 10. For the generatrix of the epicycloid, Fig. 3, moves as there seen with rapidity, but that of the hypocycloid, Fig. 5, with slowly increasing velocity. Hence, Fig. 10, there is more sliding contact between *h* and *q*, than when Q and *d* are in contact.

Hence, as a balancing of advantages of action, as well as to allow either wheel to drive, both wheels have teeth having both faces and flanks.

Second. That the action of a *face* of the driver's tooth upon the *flank* of a follower's tooth takes place in the arc of recession, is evident from Fig. 10. Let the wheels, R and B, revolve, so that their circumferences shall move to the right through Q′, until the initial points, Q and *d*, of the indefinite tooth curves shall have passed beyond Q′. Their curves will then no longer be in contact, but let the wheels revolve to the left, and these curves will begin contact at *m*; Q′*q*′ acting against *dh*, then at *mk*, since *m* is the initial position of their common generatrix, and will continue in contact in departing from the line of centres AB, to the left, till one or the other curve is limited.

95. The proportions practically adopted by millwrights are grounded, whether intentionally or not, on a proper balancing of the foregoing principles.

As differently given, these proportions are as follows :—

$$\text{Pitch} = 1.$$

$$\text{Width of tooth on pitch circle} = \frac{5}{11} \text{ or } \frac{7}{15}$$

$$\text{" space " "} = \frac{6}{11} \text{ or } \frac{8}{15}$$

$$\text{Radial length of tooth face} = \frac{3}{10} \text{ or } \frac{5\frac{1}{2}}{15}$$

$$\text{" " " flank} = \frac{4}{10} \text{ or } \frac{6\frac{1}{2}}{15}$$

96. For wheels of very few teeth, the teeth should be longer than these proportions give, in order to afford a sufficient arc of action. The common generatrix of the tooth curves being on the describing circle as A, Pl. XVIII., Fig. 10, or straight line, as EN, Fig. 11, the teeth may be limited by simply assuming any point, as *n*, Fig. 10, or *d*, Fig. 11, according to the direction of rotation considered, where contact shall begin or end, and drawing the point circle of the wheel, as B, through that point.

97. When the describing circle of the teeth is equal to either given pitch circle, the hypocycloid generated by a given point of it reduces to that point itself, since evidently no rolling motion of the describing circle can, in that case, take place within the equal pitch circle.

98. We have now all the data necessary for determining the forms of the teeth of wheels, on any given pitch *circles*, or *lines*, as in the case of a rack. Moreover, if we consider a bevel wheel as composed of indefinitely thin laminæ, of decreasing size, and perpendicular to the axes of the wheels, any two corresponding laminæ may be regarded as a pair of spur-wheels, whose teeth, at the principal point of contact, *i.e.*, on the element of contact of the pitch cones, will be in the same plane. And if we further regard a spiral gear (84) as composed of such laminæ, each set an indefinitely small distance, angularly, in advance of the preceding one, these laminæ, also, will be merely thin spur wheels. And, finally, plane sections of hyperboloidal wheels, perpendicular to their axes, and through a common point on the element of contact of the pitch hyperboloids, will be simply spur wheels, only not lying in the same plane.

Hence the foregoing principles are sufficient for every case with which we have to do.

99. These cases are as follows, for all forms of spur gearing;

and the solution in each case follows directly from Theorems XIV. and XV., or from the simple modifications which result from making the describing circle equal to *half* of a pitch circle, or *equal* to a pitch circle, or infinite, *i.e.*, a straight line, as in involute teeth.

I.—*General Solution.*

1. *Common Spur-wheels*—Pl. XXI., Fig. 4.—In the general case, use the same describing circle, D, for both wheels, making its diameter less than the radius, BT, of the least pitch circle, Theor. XIV., in order that convex faces may act against *concave* flanks. Then the *faces* of their teeth will be the *epicycloids* generated by a point, as T, of D, when rolling on the *exterior* of each pitch circle; and their *flanks* will be the *hypocycloids* generated by a point, as T, of the same circle, when rolling on the *interior* of the pitch circles.

2. *A Spur-wheel with an Annular Wheel.*—The teeth of the spur-wheel would be formed as in the preceding case. The pitch and point circles of the annular wheel, Pl. XXI., Fig. 7, would be within its root circle, and the faces of its teeth will be hypocycloids, and their flanks, epicycloids.

Thus, the face, T*e*, of the spur-wheel is generated by the point T of the circle D, rolling on the exterior of the pitch circle T*p*′. The *flank*, T*c*, of B, is generated by T as D rolls within T*p*′. The *face*, T*b*, of A, is a hypocycloid generated by T as D rolls *within* T*p*, the pitch circle of A; and, finally, the flank, T*a*, is generated by T as D rolls on the *convex* side of T*p*.

3. *Spur-wheel and Rack.*—The spur-wheel teeth being as before, both faces and flanks of the rack teeth will be cycloids, generated by the rolling of the same describing circle on both sides of its pitch line. The student can, therefore, readily construct a diagram of this case, which should be made to scale from assumed measurements. This can be done after reading the example of the spur-wheel.

II.—*The Describing Circle equal to half the Pitch Circle.*

1. *Spur-wheels.*—In this case, the flanks of the teeth of both

wheels will be *radial* and *straight*, as in Pl. XIX. The faces will be epicycloids, and, by Theor. XIV., the *faces* of either wheel, and the flanks of the other, will have the same describing circle.

Such wheels may be more easily made, but they have the disadvantage over those formed by the general solution, which is shown in the following—

Theorem XIX.

Any two wheels of the same pitch, formed by the general solution, with a constant describing circle, will work together; but one made by the second solution will work perfectly only with those of one other number of teeth, and the same pitch.

To prove this, suppose a wheel, A, of 40 teeth, to be adapted to work with one, B, of 50 teeth. The describing circle, *a*, for the faces of B to act upon the flanks of A, Theor. XIV., will have for its diameter the radius of A. Now let C be a wheel of 70 teeth, and of the same pitch as A and B have.

Then the faces of B, to act upon the radial flanks of C, should be epicycloids generated by a point of a different describing circle, *c*, whose diameter will be equal to the radius of C. Thus the faces of B will be different, according as it is to drive A or C. But by the general solution both faces and flanks of all the wheels are formed by a point of the same describing circle, hence any two of them, of the same pitch, will work together.

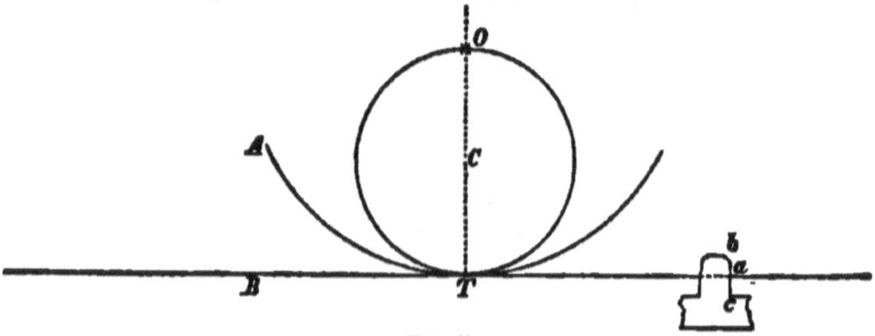

Fig. 41.

2. *Spur-wheel and Rack.*—Let A be the pitch line of the wheel, and B that of the rack. Let CT be the describing circle for

the *faces* of the *rack* and *flanks* of the *wheel*. Then the former, *ab*, will be *cycloids*, and the latter, radii of the circle OA.

Now, under the same solution, the radius of the rack pitch line being infinite, the radius of its describing circle will be infinite also, and this circle will coincide with BT. Hence the *radial flanks*, *ac*, of the *rack* will become perpendicular to BT, and the *faces* of the wheel, being generated by a point of BT, will be *involutes* of the pitch circle OA. By Theor. XVI. the contact of the teeth will be in the line BT, during the *arc of approach*, when the *rack drives*; and during the *arc of recession* when the *wheel drives*. Thus, during a part of the action, the teeth come in contact only at a single point of each, which is disadvantageous, by occasioning unequal wear.

The application to annular wheels is too obvious to need detailed rehearsal. The student can construct the case.

III.—*The Describing Circle, equal to one of the Pitch Circles.*

1. *Pin-wheel and Spur-wheel.*—Let A and B be the centres of two pitch circles, in contact at T, and let the circle A also be the describing circle, and *p*, a position of the generating point of a pair of tooth forms. Circle A, attempting to roll within itself, will have no motion, and the resulting hypocycloid *flank* of A will be only the point *p*. But let the same circle roll on the exterior of B, and the same point, *p*, will generate the epicycloidal *face*,

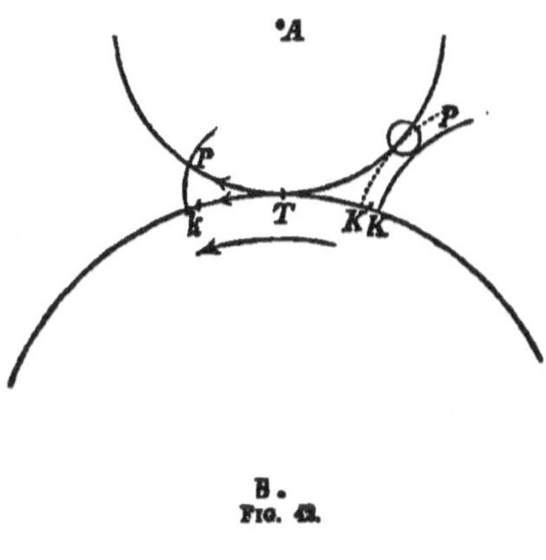

FIG. 42.

pk, of B, which will always be in contact with a linear pin, perpendicular to the paper at *p*, the contact beginning at T and continuing to the left, motion being in that direction, as shown.

Complete teeth being thus formed, B would drive A in either direction.

Were A to drive B, the pin p would push against the face pk during the *arc of approach* towards T.

If, as in the practical case, the pins of the pin wheel are of sensible diameter, as at P, the face curve, PK, of the teeth of B will be within the former one, K′, and normally equidistant from it by the radius of the pin P.

It is found in practice, that the friction during the arc of approach is more injurious than during the arc of recession. Hence the pin wheel is usually the follower.

2. *Pin-wheel and Rack.*—Either may drive, and, by the last principle, the pins will be given, in each case, to the follower.

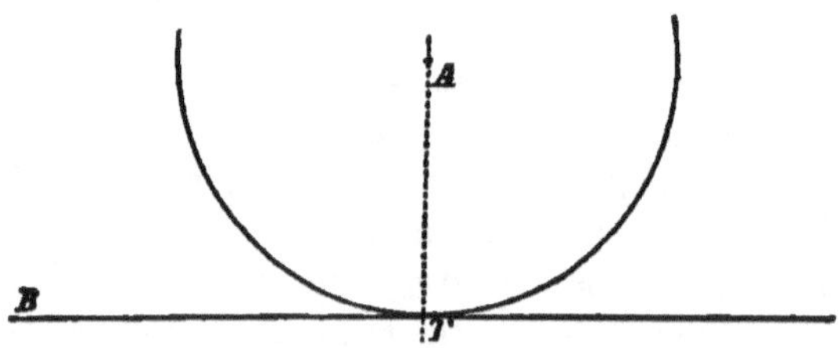

FIG. 43.

Then, first, let the rack be the driver. The pitch circle, AT, will then be also the describing circle of wheel A, whose teeth will be linear pins, as before, and the teeth curves of the rack B will be *cycloids*, whose base line is BT.

Second: let the wheel drive. The describing circle will then coincide with the pitch line of the rack; and hence the tooth curves of A, being generated by the rolling of BT upon the pitch circle, will be *involutes.*

In each case, the tooth curve corresponding to a pin of sensible diameter could be obtained as before; but in the last case the derived involute would be identical in form with the primitive one.

3. *Annular Pin-wheels.*—It is readily obvious, that if the annular wheel *drive*, its teeth curves will be hypocycloidal. If

the annular wheel be a *follower*, the tooth curve of the other wheel will be the *internal epicycloid* generated by rolling the *concave* side of the pitch circle of the annular wheel upon the *convex* side of that of the driver. The final curves in every case being at a normal distance from the primitive ones, equal to the radius of the pin.

In every case, of course, *the rim of the wheel, or rack, must be hollowed out between the teeth*, to allow of the passage of the pins.

IV.—*Solution by Involute Teeth.*

1. *Spur-wheels.*—These have been mostly explained already (Theor. XV.). It need only be recalled here that outlines of involute teeth consist of only a single curve, instead of a separate face and flank, uniting at the pitch circle. These teeth are also stronger than others, being wider at the base. They also possess the valuable property of allowing the same wheels to work together with a constant velocity ratio, though the distance between their centres be changed. This arises from the facts that all involutes of the same circle are equal curves, and the *base* circles may be constant while the *pitch* circles change; and that the common generatrix of both involutes which are in contact, is the same point on the same line, which rolls successively on the constant base circles. This may be easily made perfectly evident by a figure similar to Pl. XVIII., Fig. 11, only in which the distance AB shall be changed.

Since, however, the pressure between involute teeth is always in the direction of the generatrix, EN, that being their common normal, a component of this pressure will act *in the direction* of the line of centres; while in other forms of teeth their pressure, where most effective, that is, in passing the line of centres, is *perpendicular* to that line.

Involute teeth are therefore considered less advantageous for the transmission of great pressure than for the transmission of motion, for which they are especially adapted.

2. *Spur-wheel and Rack.*—Let AT be the pitch circle, Fig. 44, of the wheel, and AC its base circle. Let BT be the pitch line of the rack, and HT*n*, tangent to AC, the generating line. Then the involute, HK, will be one of the tooth curves of A.

The base circle of the rack is at once concentric with BT, and tangent to HT. That is, it is infinite, and its contact with HT

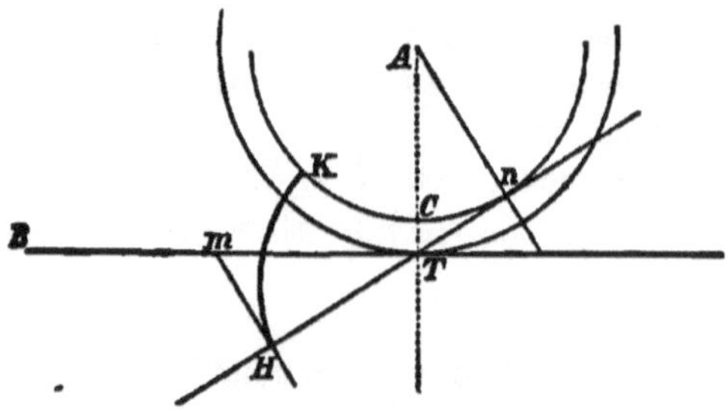

FIG. 44.

is at an infinite distance from T. Hence, when HTn rolls upon this base circle, any point, as H, will merely describe a rectilinear involute, as Hm, perpendicular to HT, which will therefore be the form of the side of the rack teeth.

Since the rack moves in the direction BT, while the contact of the teeth is on HT, the contact of the teeth will *not* be at a single point as in the second solution (II.).

EXAMPLE XXXI.

To construct the Projections of a Spur-Wheel, seen first perpendicularly and then obliquely.

After the preceding explanation of general principles, Pl. XIX. may serve as a full example of the operations of drawing a spur-wheel. It may be asked: why make an oblique elevation? Fig. 2. It would not be necessary in case of a single wheel, but some of the different wheels of a machine may sometimes be in planes which are oblique to each other. Hence the draftsman should by all means be competent to make any oblique view of a spur-wheel, or of any other part of a machine.

Pl. XIX. represents a spur-wheel of 28 teeth and a pitch of $2\frac{1}{2}$ inches.

Then the radius of the pitch-circle will be—

$$\frac{2\frac{1}{2} \times 28}{2\pi} = 11.14 \text{ inches.}$$

Let the scale be from 2 to $2\frac{1}{2}$ inches to the foot, according to the convenience of the student; or, from one-fifth to one-sixth of an inch to an inch.

On some accounts, as in taking off unusual fractions of the inch more exactly, it may be better to construct a diagonal scale of feet, inches, and 8ths of inches.

The scale of the plate is $2\frac{1}{2}$ inches to the foot.

Description and Construction of the Circles of the Wheel.— $C'a'$, Fig. 1, is the radius, $2\frac{1}{4}$ ins., of the inner circle of the hub. $C'b'$ is the radius, $4\frac{1}{4}$ ins., of the outer circle of the hub at its end, and is seen in plan, Fig. 2, at $C''b$. $C'c'$ is the radius, $4\frac{5}{8}$ ins., of the outer circumference of the hub, $= B''c$ in Fig. 2.

The circle of radius $C'c'$, or, simply, the circle $C'c'$, is also the projection of the circle of junction, $d''H''$, Fig. 2, of the hub with the feather, or rib, which connects the hub, arms, and rim together. $C'e'$, $5\frac{3}{8}$ ins., is the outer radius of this rib, where it surrounds the hub, that is, of the two circles containing the points I'' and K'', Fig. 2. Likewise $C'g'$ is the radius, $8\frac{5}{8}$ ins., of the edges of that part of the rib which is attached to the under or inner side of the rim. These edges contain the points T'' and P', Fig. 2, and lie in the planes gg and hh, Fig. 2, plan. $C'i'$ is the inner radius, 9 ins., of the rim, also of the circle jj, in Fig. 2, and through U'', Fig. 2, elevation, where the rib joins the inner surface of the rim. Finally, $C'l'$, 11.14 ins., $=$ $11\frac{1}{7}$ very nearly, is the radius of the pitch-circle. The other radii are found by construction.

In making the drawing, a tooth may have any position relative to the vertical centre line RS; but, if bisected by that line, as in the figure, the equidistant teeth on each side will be on the same level; which will much lessen the labor of making Fig. 2.

It is only strictly necessary to make one quadrant, as XS, of Fig. 1, and half of either plan, since $M''Y$ is an axis of symmetry of Fig. 2, El., the portion above and below being just alike, while the vertical diameters of the several circles of the wheel, above described, and projected up from C'',B'',A'', D'',D''',A'''', etc., divide the projections of those circles, in Fig. 2, symmetrically; so that, for example, the point on the left side of the wheel, corresponding to Q'' on the right, would be as far to the left of the vertical diameter through D'',D''' as

Q'' is to the right of that diameter. But it will be more elementary and easy for the learner to employ the whole plan; though either upper quarter of Fig. 1, El., is sufficient.

With the scale and size of plate here shown, take C', not over 13½ ins. by the scale from the top and left-hand borders, and then draw, very finely and accurately, the circles above described.

Divide the quarter pitch-circle, *l'l''*, into seven equal parts; or, to avoid puncturing this quadrant by repeated trials, make the division on any exterior and concentric quadrant; and draw segments of radii, very fine, through the points of division.

Take one of these divisions of *l'l''* as a scale of pitch, and divide it into 15 equal parts; first into five equal parts, and then each fifth into three equal parts, all very exactly.

Let the width of a tooth on the pitch-circle, as at *l''*, be $\frac{7}{15}$ of the pitch, and the space between the teeth the remaining $\frac{8}{15}$. Then, on each side of each point of division of the pitch-circle, as *l'*,*l''*,*l'''*, lay off 3½-fifteenths of the pitch; and test the widths of the teeth to see that each space is just $\frac{1}{15}$ of the pitch scale wider than a tooth. Or the whole pitch may be set off successively, on the pitch-circle, from each side of the tooth *l'* towards the one at *l''*, or from *l''*, likewise, towards the tooth *l'*. Thus all the pitch-circle points will be located.

For mere purposes of graphical construction, the faces, as *v* and *v'*, of the teeth may be circular arcs, whose radius is the pitch, and whose centres are on the pitch-circle. These arcs extend from the pitch-circle points, just found, to the *point-circle*, which is thus found. The extent of the teeth, beyond the pitch-circle, is 5½-fifteenths of the pitch. Then make *l'm'* equal to 5½-fifteenths of the pitch, and draw a new circle through *m'*, with C' as a centre, for the outer limit of the faces, as *w'w''*.

The flanks, as *w'y'''*, are here radial lines, whose inner limit is the *root-circle*, C'*k'*; found by making *l'k'* equal to 6½-fifteenths of the pitch.

At this stage of the construction, the entire rim, with the circles of the hub, making the key-seat of any suitable size, may be inked, observing the obvious position of the heavy lines.

This done, proceed with the construction of the arms and connecting-rib as follows: Let the taper of the arms be such that, if produced to the centre lines, XY and RS, they would *there* be 5 ins. wide, and 4 ins. wide at a distance, as C′*o*, of 5¾ inches from C′.

The ribs, as R′, upon the arms, are uniformly 1¼ in. thick throughout their straight part, as seen in Fig. 1.

The constructions, as at *s*, *q*, *p*, and *r*, for finding the centres of the little arcs, by which the several intersecting edges of the wheel are rounded into each other, mostly explain themselves. It is only necessary to notice that the short portions, as *ut*′, of an arc which is rounded into a straight line, as *un*′, are treated as straight; *t*′*s* is then a radial line, and *ts* is perpendicular to *un*′, and *ut*′ and *ut* are equal. The other constructions are similar.

Construction of the Plans.—Only one finished plan, Fig. 2, is necessary. To make it, project down all the necessary points of each circle in the elevation of Fig. 1, separately, upon separate lines, parallel to XY. Then construct the outlines of Fig. 2, plan, by the following measurements: The face of the wheel, as at M*m*, is 6⅜ ins. wide. Then, first, locate the rectangle M*mm*″, far enough to the right to separate Figs. 1 and 2, suitably, and make its width 6⅜ ins., and its length equal to twice C′*m*′. Draw its transverse centre line, C″A″″, and make the hub plan 8¾ ins. for the extreme length, and 7¾ ins. for the length of the thicker portion *cc*‴.

Construction of the Oblique Elevation.—Every point of this is found by the elementary operation shown in Fig. 11; on the one principle that each point, as *m*″, is on a perpendicular, *mm*″, to the ground line from the plan, *m*, of that point, and at its intersection with the parallel, *m*′*m*″, from the primitive elevation, *m*′, of the same point.

Where many separate groups of points are to be constructed, and each group to be shown in separate projections, it is of prime importance, in order to secure ease and accuracy, and to avoid confusion, to construct each group separately, and in Fig. 2, Elev., where each point is found from Figs. 1 and 2, plan, ***the whole construction for each point should be complete before beginning that of a new point.***

To illustrate, let us take the outermost ring of points *m*″S″Y.

Draw any line parallel to XY, below Fig. 1, Elev., and upon it project down C′, and all the points of the point circle *m*′S, only. Then transfer these points carefully to *mm'''*, Fig. 2, by the dividers, or *by a slip of paper having these points exactly marked on its edge.* Then draw the point lines of the teeth, as 1—1; 2—2, etc. Then project up all the points as 1, on *mm'''*, *one at a time*, and project over the same points, as *w''*, from Fig. 1. The intersection of these projecting lines will give 1', etc., on the oblique elevation, Fig. 2. In the same way find all the points of the point circle, *m'*S—*mm'''*—*m''*S''. Then take a new line below Fig. 1, Elev., and parallel to XY, and go through a similar round of operations for finding the pitch circle points. In like manner, also, every point of Fig. 2, Elev., is found. It is only necessary to transfer the horizontal projections of the different circles of Fig. 1, each to its proper plane in Fig. 2, plan. Thus, *by* is the plane of the circles C'*a'* and C'*b'*, of Fig. 1. B''*c* is that of C'*c'*, Fig. 1. The latter circle is also laid off from D'' as centre, in *gg*, in Fig. 2, plan, to give the intersection, through II'', Fig. 2, Elev., of the hub with the rib. This rib is shown in plan, Fig. 2, by the dotted lines *gg* and *hh*, $1\frac{1}{4}$ in. apart. Then the circles C'*e'* and C'*g'* are both laid off, both on *gg* and *hh*, to give the front and back edges, through I'' and K'', and through T'' and Q'', of the visible parts of the rib. C'*i'* is laid off on *mm'''* for the front circle, *i''i''*, of the rim; on *jj*, for the intersection of the web and rim, and on MP''', for the back circle through P', of the rim.

The horizontal arms which are only seen edgewise in elevation, Fig. 1, are seen flatwise in plan, and are thus drawn: At their junction, as *s''s'''*, with the hub, they are $6\frac{5}{8}$ ins. wide. They taper from a width of $6\frac{5}{8}$ ins., if produced to the axis, as at A''A''', to a width of $4\frac{5}{8}$ ins., at the distance of $8\frac{5}{8}$ ins. from the axis. Their extremities are rounded as shown in Fig. 2, plan. To find their oblique elevations, use their principal points, on Figs. 1, Elev., and 2, plan, that is, such points as *t* and L, where their straight and curved edges join.

Further directions might only confuse the subject. By fully understanding from the foregoing how any one point of Fig. 2, Elev., is found, all can be found, since all are constructed in the same way.

We only add some ways of checking the work. *First.* If lines pass through a point in space, their projections, on any plane, will contain the similar projection of that point. Hence all the flanks on Fig. 2 should radiate from A‴, the projection of C′,A″, the centre of the face of the wheel. *Second.* Each circle of the wheel will be vertically projected on Fig. 2, in an ellipse, which can be readily found, by construction from its axes, as in (26, 27), and then points of these circles, as OO″, or NN″, can be projected up from Fig. 2, plan, or over from Fig. 1, Elev., upon these ellipses. Portions of the curved edges of the web were thus found in Fig. 2, as that through *e′,e,e″*. Compare also corresponding radii, as D‴H″, which is the same line as the radius C′G′, Fig. 1. *Third.* The thickness of the rib being everywhere the same, the lines *gh*, *ef*, IK, etc., Fig. 2, plan, are parallel, hence their vertical projections, as I″K″, or Q″T″, are of constant length. That is, having any front circle, the points of the corresponding back circle, in Fig. 2, Elev., are found by laying off a constant horizontal distance from its front points. Thus the back points, M″,O″, etc., of the teeth, so far as visible, are found by laying off the constant distance, M″*m*″, from their front points. In like manner P′ is found from P″. Also, G″H″, the vertical projection of GH, is constant; as so is P″U″, the vertical projection of PU.

100. Introductorily to the next example, we will here explain more in detail than hitherto the formation of a bevel wheel.

Let V, Pl. XX., Fig. 7, be the common vertex of a pair of bevel wheels, and V*b* the generatrix of the primitive cone from a frustum of which the wheel is formed, and let VV″ be the axis of this cone and of the wheel. This cone will then be the *pitch cone* of the wheel. Let V*d* be the generatrix of the *point cone;* and V*f*, that of the *root cone.* Let the ends of the teeth, whose length is *bo*, be conical surfaces, normal to the pitch cone. Then *d*V″ and *n*V′ are the generatrices, and V″ and V′ the vertices of these cones. Now let the five generatrices, thus far described, revolve about the common axis VV″. The points *d*, *b* and *f* will respectively generate the *outer* point, pitch, and root circles; and *n*, *o* and *p* will generate the *inner* point, pitch, and root circles; and all of these circles will be perpendicular to the common axis VV″.

EXAMPLE XXXII.

To construct the Projections of a Bevel Wheel whose axis is perpendicular to the vertical plane.

Let VV″—O′ be the axis of the wheel; let the following be its measurements, laid off to a scale of $2\frac{1}{4}$ inches to 1 foot, = $\frac{3}{16}$ of an inch to 1 inch, or of $\frac{1}{8}$ or $\frac{1}{6}$ of an inch to 1 inch, as may be most convenient. Make $Va = 10\frac{3}{8}$ inches, and $ab = am = 11\frac{5}{8}$ inches, and draw bV for the extreme element of the pitch cone. mb is then the horizontal projection of the outer or greater pitch circle. Draw bV″ perpendicular to bV, and bV″ will be the extreme or horizontal element of the cone containing the outer or larger ends of the teeth.

With centre V″ and radius V″b draw an arc bC, which will be the development of a portion of the pitch circle, mb, considered as the base of the cone V″mb.

With centre, O′, and radius, O′b′ = ab, draw the vertical projection, $m'u'b'$, of the outer pitch circle, and divide each quadrant of it into five equal parts, to obtain the pitch; since the wheel is supposed to have 20 teeth. Lay off this pitch, which by measurement is $3\frac{5}{8}$ inches, once, on bC, from b, which will give the point B.

Adopting the finer proportions given in (95), divide bB into fifteen equal parts, and lay off seven of these parts from b and B, to give the widths of the teeth bA and BC. Next, lay off $bd = 5\frac{1}{2}$ of these fifteenths, to give d, a point on the point circle dF. Also $6\frac{1}{2}$-fifteenths from b to f, giving f, a point of the root circle fD. Make $fh = 1\frac{1}{4}$ inches, for the thickness of the conical rim which bears the teeth, and draw the arc, hE, of the development.

We now have this set of four parallels to the ground line, viz.: dd'', the outer point circle; bm, the outer pitch circle; ff'', the outer root circle; and hh'', the outer rim circle. Project up the three foremost of these, only, since the rim circle is hidden in vertical projection by the rim itself, giving the point circle, O′d′; the pitch circle, O′b′, already projected, and O′f′, the root circle.

Returning to the plan, draw dV, fV, and hV, and on bV make $bo = 4\frac{7}{8}$ inches, for the length of a tooth. Then draw V′on, parallel to V″b, for the extreme element of the cone containing the inner ends of the teeth. The intersections of this line, with those already drawn to V, will be the extremities of the inner point circle, *n*G; the inner pitch circle, *o*H; the inner root circle, *p*I; and the inner rim circle, *q*K. The vertical plane face, *q*K, of the wheel, is, however, rounded into the cone of the inner ends of the teeth by a concave double curved surface, as indicated by the curve at *qp*. Hence project up only the three foremost circles, giving the inner circles, O'*p*', O'*o*', and O'*n*'.

Now, each point of division, as *y*, made in finding the pitch, is the middle point of a tooth, on the outer pitch circle, in order that, for convenience, the vertical line, O'V'', shall be an axis of symmetry of the vertical projection. Then make $yN' = yu' = \frac{1}{2}bA$, the width of the tooth in development, and do the same at each other similar point of division. Also, for each tooth make the point at *s*, where the outer point circle is cut by the radius, as O'*y*, and make $sM' = sv' = \frac{1}{2}FF''$, the width of a tooth at the point. Since the development shows the true form of the ends of the teeth, the curves, as M'N' and *u'v*, may be drawn with sufficient accuracy, as circular arcs with their centres found by trial on the pitch circle *m'u'b*, taking care that all of them shall have the same radius.

The flanks of the teeth being radial plane surfaces, as indicated in the development, their planes will all contain the axis VV''—O', and hence their vertical projections will simply be the straight lines, as *u'*S', limited by the inner root circle O'*p*', and converging to O'. Add the point lines, as M'Q', also converging to O', and the inner face curves, as Q'*t*'', drawn as M'N' was, that is, with centres now on the inner pitch circle O'*o*', and the vertical projection of the teeth will be complete. To find the elevation, only draw the eye of the wheel with a radius O'*r*' of $3\frac{5}{8}''$. L'*r*' is the key seat, of any suitable size.

Before constructing the plan, it must be understood that it represents the wheel with the upper right-hand quadrant cut out; so that the part to the right of VV'' shows the lower right-hand quarter, not seen in the elevation, and a section in the horizontal plane O'*d*'.

This being understood, project down the outer points, as M' and *v*', of the *left-hand* half of the elevation, upon the outer point circle *dd*'', as at M and *v*, and through these points draw the point lines, as MQ, towards V. Project down the pitch points, as N′ and *u*′, upon the outer pitch circle, *bm*, and draw the outer flank lines, as NP, towards V″, giving the root point, as P and R. Thence draw the visible root lines, as RS, towards V, and limited by the inner root circle *p*L. Having gone thus far, the inner pitch points, as *t*, may be found in three ways: first, by projecting down from their elevations, as *t*; second, by drawing the inner flank lines, as S*t*, radiating from V′; third, by making the intersections of the imaginary pitch lines, as *ut*, drawn towards V, with the inner pitch circle, *o*H.

As the teeth are seen under various angles, no two on the same quarter of the wheel will appear alike, and the face curves, as NM, must be sketched by hand.

As the teeth on the right-hand *lower* quarter are vertically under those of the right-hand *upper* quarter, the latter will serve equally well for purposes of projection. Thus, project down the points, as T′, of the inner ends of the teeth to find T, etc., on the inner point circle. Likewise project down the inner points, as U′, to find U, etc., on the inner pitch circle, *o*H. Then draw the face curves by hand; and the flank lines, as UU″, radiating towards V′; and the visible portions, as TT″, of point lines, radiating from V.

To complete the plan, make the length, K*i*, of the hub=6 inches; its greater radius, *kl*, $6\frac{5}{8}$ inches, $ik = \frac{1}{2}$ an inch, and the radius at $ij = 6\frac{1}{8}$ inches.

The portion *rr*″*pfhj*, is the generatrix of the united hub and rim, by revolving about the axis VV″. The rim is further bound to the hub by four arms, which bridge the annular open space generated by *l*″*f*″*h*. These arms are 2 inches wide, therefore make *kk*″ = 1 inch; *lh*″ and *l*″*h* are edges of arms; and *l*″*w*, a minute distance, strictly, to the right of *l*, is the intersection of the side of the arm with the cylindrical surface of the hub. It can be constructed by projecting down from a fragment of the hub and horizontal arm, easily made in vertical projection.

EXAMPLE XXXIII.

To construct the Projections of a Bevel Wheel, seen obliquely relative to the vertical plane.

See Pl. XX., Fig. 2, where like points have the same letter in both projections, and on Fig. 1.

Begin with the plan, which is simply a copy of that in Fig. 1, except in being in a different position relative to the ground line, and in showing the parts on both sides of VV″ alike. VV″ is thus an axis of symmetry from which the various points on the several circles of the wheel are laid off, each way, on O*n*, *ot*, *p*S, *ab*, etc., which are at the same distance apart, and from V, as are the same lines in Fig. 1.

The plan being thus simply made, the elevation may be made wholly, or in part, in any one of three ways, but best by a combination of any two of them; either serving as a check upon the other, and some points being found best by one method, and some by another, according to their positions; having regard to the principle that the lines of construction, which determine any point, should form as large an angle as possible with each other.

First Method.—Every point of the elevation may be found by the elementary operation of projecting up the points of the plan, as P, into horizontal projecting lines, as *p′p′*, from the same point on the elevation in Fig. 1. In this case, the lines of construction will always meet at right angles; but two things must be noticed: first, to avoid confusion, only one point at a time, as MM′, for example, should be projected up, and projected over from the elevation, Fig. 1. It is a bad practice to draw a great number of projecting lines from the plan, and then a great many from the elevation, Fig. 1, *before* noting any of their intersections. Second, by this method quite a number of invisible points of the inner ends of the teeth must be marked on the plan; by finding them first on the plan, Fig. 1, by projecting down from the elevation.

Second Method.—By any of the familiar methods, construct (26–27) the six ellipses, as *m′*N′*b′* and G′T′*n′*, in the elevation, which will be the vertical projections of the six visible tooth-circles of the wheel. Thus the semi-ellipse *m′*N′*b′* has for its semi-transverse axis K′Y″ = *ab*, and for its conjugate axis *m′b′*, the vertical projection of *mb*. After constructing these ellipses, the points of the teeth found upon them, as M′, N′, P′, T′, U′, S′, may be conveniently projected up from the plan, for the teeth near the highest one; and over from the elevation, Fig. 1, for those near the ground line. And *the points* thus constructed *should be joined as fast as found.*

Third Method.—After constructing any two rings of points, but preferably those of the larger ends of the teeth, by the first or second method, all the remaining points can be found by projecting them from the plan, or from the elevation in Fig. 1, upon the lines meeting at O′, O″, and O‴. For these points are the vertical projections of the three vertices, V, V′, and V″, to one or another of which every straight line of the wheel tends.

Thus, having found M′ and N′, for example, by the first method, Q′ will be the intersection of M′O′, with a projecting line from Q, or from Q′, in Fig. 1.

By this method, certain invisible points of the elevation must be temporarily found. Thus, find the invisible back root point, P″, and then S′ will be the intersection of P″O′, with a projecting line from S, or from S′ in Fig. 1.

But some of the points may be found by the intersections of these converging lines with each other. Thus, having found S‴ as before, t'' will be the intersection of O″S‴, produced, with N′O′. Or, better, finding t'' first, S‴ will be the intersection of P′O′ and t''O″.

Fourth Method.—This merely consists in applying the principle, in connection with the other methods, that *for the same side of the same tooth*, lines as O″N′ and O″t'' are parallel. Thus, having N′O‴ and N′O′, the point t'' may be found as the intersection of O″t'', parallel to O″N′, with N′O′.

101. *To construct the oblique projection of the foregoing, or any other object, by using only three projections,* proceed as illustrated for a semicircle only, in Pl. XX., Fig. 3. Then let AB be the horizontal projection of a vertical semicircle. Let $a'b'$ be the ground line of an auxiliary vertical plane, parallel to the plane of the semicircle, and on which the latter will therefore be shown in its true size, as at $a'e'b'$. Let A′B′ be the ground line of a vertical plane oblique to the semicircle, and on which the required projection is to be found. To do this, it is only necessary to project up the several points, A, C, E, etc., of the plan, and make their heights, M′C′, O′E′, etc., equal to $m'c'$, $o'e'$, etc., on the principle that the different vertical projections of the same point are at the same height above their respective ground lines.

102. Minor modifications of the construction just given are

obvious. Thus, AB might have been parallel to the plane A'B', and then the oblique elevation would have been at $a'e'b'$. Again, the plane A'B' might have been made parallel to AB, and the plane $a'b'$ parallel to the present direction of A'B', in which case, again, the highest of the three figures would have been the oblique elevation. Finally, let $a'b'$ be brought down parallel to itself, near to AB; let A'B' be carried up parallel to itself; and then let AB be placed parallel to the present position of A'B'; and then the middle figure will be the oblique projection.

Practical Forms of the Teeth of Wheels.

103. After all the foregoing statement of only the main points in the theory of the teeth of wheels, it must be acknowledged that in practice they are bounded by circular arcs. In *empirical* practice these arcs are taken arbitrarily, and even absurdly. In *scientific* practice they are taken so as to conform as closely as possible to the theoretical outlines.

104. The theory as above given is thus abundantly useful, as leading to the determination of proper approximate arcs. And, on the other hand, the length of an epicycloidal or involute arc forming the limits of the side of a tooth in a real wheel is so small, that, except for very large wheels, the circular arc, however finely traced, would sensibly coincide, except, perhaps, under a magnifier, with the theoretical curves, within these small limits.

A not uncommon empirical method for constructing the *faces*, the *flanks* being radial, is to describe them with the pitch for a radius, and the centre in the pitch circle, as in Ex. XXXI.

105. Among scientific practical methods, the following are the principal, if not the only ones:—

First: Construct *templets*, as T, Fig. 45, that is, thin pieces of hard wood, carefully shaped to the exact curvature of the intended pitch circles and describing circle. The latter of these templets, C, carries in its circumference a firm, sharp tracing-point, p. Then, by rolling it successively upon the pitch templets, the correct *face* curves of teeth are traced mechanically, and by rolling it on other templets of the same curvature, but *concave* on their curved edges, the *flanks* of the same teeth will be traced; in both cases upon a board on which the pitch templets are held, so as to coincide with the pitch circles traced upon it.

A templet can then be cut to the form of the tooth, and used in tracing the outlines of the ends of the teeth on the rough pattern of a wheel.

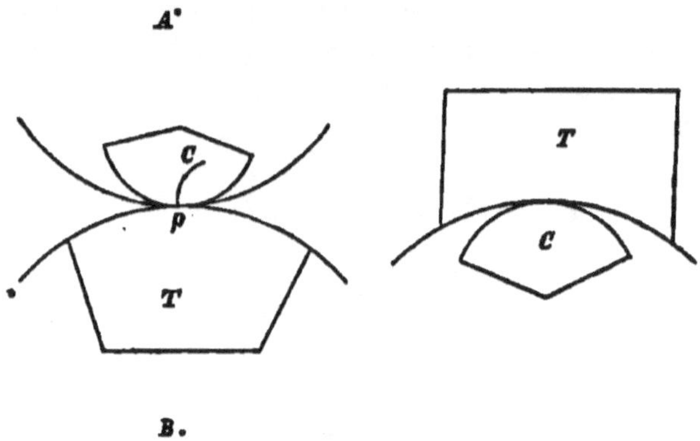

FIG. 45.

Second: Having traced a face and a flank, as above, take three points on each, and by the familiar method construct the circular arc passing through those three points, and it will approximate very closely to the theoretical curve.

106. Both of these methods would evidently be nearly or quite impracticable, except for quite large teeth; while the second is rather vague, owing to the somewhat arbitrary assumption of the three points. Besides, by the insensible slippings of the templets, and minute instrumental errors, the supposed true curve might be more erroneous than a circular approximation made from a single centre by some simple rule. Therefore—

Third: Let the approximate circular arc be described from a mean centre and radius of curvature of the theoretical teeth. These data can be determined analytically, and thus each tooth-curve may be struck at once with a single centre and radius. This method has been proposed by Euler and Redtenbacher, and one adapted to practice was founded upon it by Prof. Willis (Principles of Mechanism), and conveniently embodied in an instrument called the *odontograph*, the theory and use of which will next be briefly described.

THEOREM XX.

Circular tooth-curves, with centres on a line through the point of contact of the pitch circles, will give a sensibly constant velocity ratio to those circles.

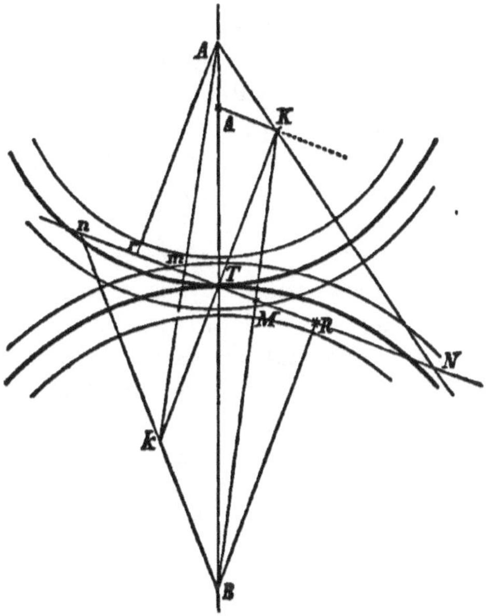

FIG. 46.

Let A and B, Fig. 46, be the centres of the pitch circles, whose contact is T. Through T draw Nn, and make TK = Tk, perpendicular to Nn, and less than either pitch circle radius. Draw AK to N; Ak, giving m; BK, giving M; and Bk to n. Also draw BR perpendicular to Nn. Now conceive the system Bn; nmR; Am to be jointed at n and m, and on the point of turning about the fixed centres, A and B. For the instant of occupying the position shown in the figure, we have

$$nk : k\mathrm{T} :: n\mathrm{B} : \mathrm{BR}.$$

But as kT is a fixed line, this proportion simply shows that T, alone, is the fixed point of Nn for the instant, or, in other words, is the intersection of two consecutive positions of Nn. And, as T is on the line of centres AB, while nN is a common normal to a pair of tooth curves, when their centres, as M and N, or *m* and *n*, are on that line, it follows, from Theorem XVII., that such tooth curves will give a constant angular velocity ratio for the instant in which T is fixed as above described. These curves being quite short, this velocity ratio will be sensibly constant during the short period in which they are in contact.

TK = T*k* is less than either AT or BT, merely to throw both of a pair of centres, as *m* and *n*, on the same side of the tooth curves B, so that one of the latter shall be convex towards the concavity of the other. Then, this being understood, the remoter centre is that of the *concave* tooth curve, which, of course, is a *flank*. But *each* wheel must have both faces and flanks. Hence for one face and flank pair, acting together, the wheels are momentarily represented by the linked arms A*m* and B*n*; giving *m*, the centre of a *face* of A, and *n*, the centre for the corresponding flank of B. Conversely, for the other pair, the wheels are represented momentarily by the arms AN and BM; giving M the centre of a *face* of B, and N the centre for the corresponding *flank* of A.

106. If we take the contact of the tooth arcs a little way from T, on the opposite side from *m* and *n*, there will be a corresponding pair, at the same distance to the left of T, so that the action will be exact at two points in the total arc of action; which will give abundant accuracy of action at all points.

107. Assuming the total arc of action to be about twice the pitch, in all cases, the contact of the first pair of tooth curves may be at a distance from T equal to half the pitch.

Finally, the angle *n*TA is found experimentally to be best fixed at 75°.

By calculating and tabulating the distances, as *n*T and MT, for wheels of various sizes, and by graduating them on the odontograph, Pl. XXI., Fig. 2, the final adaptation of this system to practice will be made.

These distances are easily found, as follows.*

* Willis' Principles of Mechanism, p. 125.

PROBLEM VIII.

To find the Radii of the Tooth Curves.

Let TK = a ; BT = R
TM = c and BTR = α
Then from the similar triangles, BRM and KTM.
TM : MR :: TK : BR.
Whence TM × BR = MR × TK
But MR = TR − TM; BR = R sin α and
TR = R cos α.
Hence, by substitution, TM × R sin α = R cos α × TK − TM × TK.
Or TM (R sin α + TK) = TK × R cos α.

$$\text{From which TM} = \frac{\text{TK} \times \text{R} \cos \alpha}{\text{TK} + \text{R} \sin \alpha} \ldots\ldots\ldots\ldots (1.)$$

It only remains to show how the length of TK should be governed. Now, in every systematic manufacture, a certain series of *pitches*, and *numbers of teeth*, will be adopted, as sufficient for all ordinary cases. The greatest pitch and number of teeth will determine the greatest wheel, and the least of both elements will determine the smallest wheel, and by various combinations of the two elements, wheels of almost any intermediate size can be made, while the radius can be immediately found from the formula

$$\text{R} = \frac{\text{P} \times \text{N}}{2\pi}$$

where P = pitch, N = number of teeth, and R = radius. TK, then, is so taken that for the least wheel of a set, AK shall be parallel to TN, thus giving a flank centre at an infinite distance from T, and hence a straight flank for the least wheel.

Denote the least radius, A′T, by r, then

$$\text{TK} = r \sin \alpha,$$

which, substituted in (1), gives

$$\text{TM} = \frac{\text{R}r \cos \alpha}{\text{R} + r} \quad (2)$$

In like manner, beginning with the similar triangles nkT and nBR, we shall find

$$nT = \frac{TK \times R \cos \alpha}{R \sin \alpha - TK}, \text{ and finally,}$$

$$nT = \frac{Rr \cos \alpha}{R - r} \ldots\ldots\ldots (3)$$

where $\alpha = 75°$.

108. By assuming a series of values of *pitch* and *number of teeth*, the value of R, for *each* number of teeth, with *each given pitch*, can be found, and substituted in (2) and (3) where r and α are constant; and thus a series of values of MT and nT may be obtained and tabulated, expressed in twentieths of an inch as on the edge of the odontograph, Pl. XXI., Fig. 2.

Such tables accompany the instrument, which may be obtained from mathematical instrument dealers.

109. Teeth thus formed are analogous to those of the *first* or *general solution*, in having both *faces* and *flanks;* but more closely in that *any two wheels of the same pitch, having teeth thus formed, will work together.* This appears from Fig. 46, where, if AT, the pitch radius of one wheel, be changed, it will only change m and N, the centres of its own tooth arcs.

M and n, being the centres for a *face* and a *flank* of B, all the other faces and flanks will have the same radii MT + $\frac{1}{2}$ *pitch* and nT + $\frac{1}{2}$ *pitch*, and their centres will be in circles, through M and n, with B for their centre.

Problem IX.

To find Centres for approximate Involute Teeth.

By making KT = kT infinite, Fig. 46, BM and Bn will coincide with BR; and the two centres, thus united at R, will become that of a single arc. A tooth profile of two arcs can have a point of exact action for each arc (106), but now, with one arc, there can be but one such point. Let T be that point, then RT the tooth radius = R × cos α, where R = the radius of the wheel. Now as the angle nTA = α, is somewhat arbitrary, assume it at 75° 30′, which is otherwise convenient, and cos α = $\frac{1}{4}$ very nearly, which gives

$$RT = \frac{R}{4}, \text{ a very simple value.}$$

That is, in approximate involute teeth, let the base circle be tangent to a line, BC, Fig. 47, making an angle of 75° 30′ with

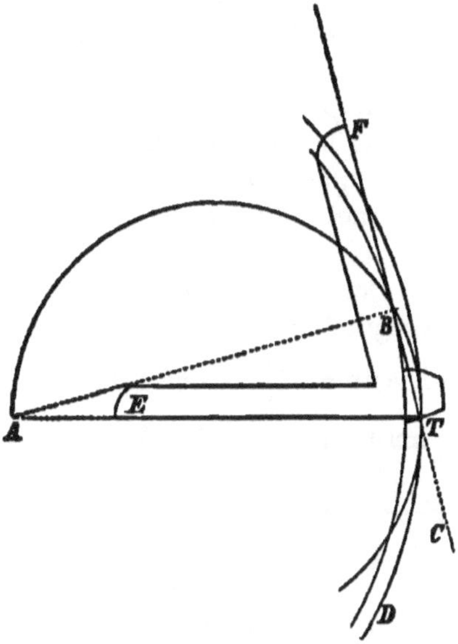

FIG. 47.

the line of centres, AT; and let the tooth radius, BT, = ¼ AT, the radius of the pitch circle DT.

111. An odontograph, ETF, for this case, accompanies the larger one. Its angle ATB is made = 75° 30′, and its edge TB is graduated in *quarter* inches; so that the radius of the wheel being given in *whole* inches, the *same* number of quarter inches, read off from T, when the arm AT coincides with the radius of the wheel, as in the figure, gives the tooth centre B.

Having divided the pitch circle, for the teeth and spaces, all the other tooth curves will have the same radius BT, and their centres will be in the circle AB.

EXAMPLE XXXIV.

To construct Teeth, having separate Faces and Flanks, by the Odontograph.

Let the two wheels be denoted by their centres A and B, Pl. XXI., in each illustration. Let F*a* denote face, and F*l*, flank in the notation of the figures. The general proportions of the teeth given in the example of the spur-wheel are followed as closely as need be for illustration, remembering that 5½-15ths = a little more than one-third, and 6½-15ths, a little more yet (of the pitch), while a tooth occupies a trifle less than half the pitch on the pitch circle.

The odontograph angle, OTH, is not given here of its true value, 75°, so that the student by only following the *text* carefully, and using the figures of Pl. XXI. as guiding sketches, will be able to construct the figures accurately.

First: In Fig. 1, let the wheel A have a radius of 6 ins. and 20 teeth. Then the pitch is equal to the circumference divided by the number of teeth $= \frac{2\pi r}{20} = 1.88$ ins.

As the pitch must, of course, be the same for both wheels, and as pitch and radius cannot be given, lest they should afford a fractional number of teeth, which it would be impossible to have, we must have given the *pitch* and *number of teeth*, as 12, for the other wheel. The *product* of these gives the new circumference, which, divided by 2π, gives the radius,

$$\frac{P \times t}{2\pi} \qquad \frac{1.88 \times 12}{3.28} = 3.6 \text{ ins.}$$

Now draw any line, AB, for the line of centres, and on it place the point T, the point of contact of the pitch circles. Make TA, 6 ins., *scale* ½, and A the centre of wheel A; also TB, 3.6 ins., and B the centre of wheel B, and describe the two pitch circles with radii AT and BT. From T, lay off *each* way, on *each* pitch circle, 1.88 inches, for the pitch. Then let a tooth of A be just below T; and, accordingly, lay off, *downward*, as at T*s*, from the *upper* end, as T, of each pitch, as T*d*, of wheel A, a distance equal to a little less than half the pitch, say 0.9 ins. A tooth of B will thus extend from T *upward*, hence lay off 0.9 ins. *upward*, as at T*g*, from the *lower* end, as T, of each pitch, as T*e*, of wheel B. The two sides of each tooth on both of the pitch circles will thus be marked. Next draw the point and root circles of each wheel as usual (95).

The figure is now ready for the application of the odontograph; directly, if drawn in full size, or, otherwise, indirectly by laying off the tooth-centres to scale. Consulting the table of tooth-centres, which accompanies the instrument, and remembering that for *any* given series of wheels having the same pitch, it contemplates no wheel of less than twelve teeth, and that then the radius of the flanks is infinite, Prob. VIII., we at once make the radial flanks, as *qp*, of B.

For the *faces* of B, we have from the table T—*Fa*,B, $\frac{9}{20}$ in., which gives the centre, *Fa*,B (on QO), of the first face, *qr*, on the other side of T. All the other faces of B have the same radius, *q*—*Fa*,B, just used, and their centres on the arc with B as a centre, and B—*Fa*,B as a radius.

For the *flanks* of A, we have from the table $\frac{45}{20}=2\frac{1}{4}$ ins., which is laid off from T to the point on OQ, marked *Fl*,A, to give the centre of the first flank, *mn*, of A, on the opposite side of T. Then, as before explained (109), all the flanks of A will have radii equal to *Fl*,A—*m*, and with centres on the arc through the first flank centre, *Fl*,A, and with A for its centre.

For the *faces* of A, we have from the table 0.6 in., to be laid off from T to *Fa*,A, to give the centre for the first face, *sk*, below T of wheel A. Then, again, all the other A faces have their centres on the arc of radius A—*Fa*,A, and radii equal to *Fa*,A—*s*.

After this minutely detailed description of one figure, the others may be understood almost wholly by mere inspection, the point, pitch, and root circles, and the divisions of the pitch being laid out as before.

Second: In this case, Fig. 3, the relative position of the teeth in contact at T is the reverse of that in Fig. 1; the tooth of the *right*-hand wheel being here just above T. This brings the 75° line, as it may briefly be called, which represents the graduated edge of the odontograph, into the position *Fl*,A—*Fl*,B. This would require the instrument to be graduated on both sides on the edge MN. But this is unnecessary, since it can be applied separately to the two wheels in the position similar to OQ, Fig. 1, by applying it to any radius, as B*b*, ending on the upper side of a tooth of B, and to any radius, as A*c*, ending on the under side of a tooth of A; applying the edge TA, first on B*b*, and then on A*c*.

But we will proceed with the figure as shown, it being drawn, as before, on a scale of $\frac{1}{2}$.

In both wheels, the teeth and pitch are given, from which, as before, we find the radii, 5.1 ins., and 7.64 ins. The centre of A is lost in Fig. 1, the point A on Fig. 3 merely indicates it without showing its true position.

We have then from the table, for the given pitch and number of teeth, T—Fl,A $= \frac{40}{20} = 2$ ins., for the centre of the *flank*, *ca;* and T—Fa,A $= \frac{13}{20}$, for the centre, Fa,A, of the *face*, *de*, of A. Likewise, T—Fl,B $= \frac{78}{20}$ for the centre, Fl,B, of the *flank*, *fg*, of B; and T—Fa,B $= \frac{11}{20}$ for the centre of the *face*, *bh*, of wheel B.

Having thus found four initial centres, one for a face and a flank of each wheel, the remaining faces and flanks are drawn in the same manner as in Fig. 1.

Third: Fig. 4 is varied from the two preceding by using a smaller pitch; and a much larger radius for one of the wheels. Also AB is placed differently.

Having the radius, *r*, and number of teeth, *t*, of the upper wheel, we find its pitch 1.6 in., which thus becomes that of the lower wheel, which, having 24 teeth, has a radius of 6 ins.

Lay out the pitch, point and root circles, and the pitch points as before, and draw OQ' to make QTB = 75°. Then, from the table we have for the given pitch and numbers of teeth, T—Fl,A $= \frac{18}{20}$ in., for the centre, Fl,A, of the flank *ab*, and T—Fa,A, for $= 12\frac{1}{2}$-20ths for the centre, Fa,A, of the face *cd*. Likewise, T—Fl,B $= \frac{32}{20}$ ins. for the centre, Fl,B, of the flank, *ef*, of wheel B, and T—Fa,B $= 4\frac{1}{2}$-20ths, for the centre, Fa,B, of the face, *gh*, of the wheel B. The remaining tooth curves can be completed as before.

Example XXXV.

To Construct approximate Involute Teeth by the Odontograph.

This case, Fig. 5, is drawn in full size, and represents approximate involute teeth, as given by the small odontograph, Fig. 47.

As in previous figures, the *given* data are included by a brace. The line OQ is drawn so as to make OTB (B, the centre of the wheel PR, is in Fig. 4) = 75°:30'. Then, having set out the various circles as before, with the pitch 0.75 in., each way from T, and divided as before, read off from T, on the graduated edge of the instrument, lying on OQ, the same number of quarter inches, Prob. IX., that there are of whole inches in the radius BT, to give *a*, the centre of the curve, *bc*, of a tooth having a single arc from point to root, the part of which, exterior to the circle B — *ank*, represents the involute of *ank* taken as a base circle. In like manner, T*d* = 2.6 quarter inches, gives *d*, the centre of the tooth arc, *ef*, of the other wheel.

110. No finished example is here given of hyperboloidal wheels (83). Approximations to them are known in practice as *skew-bevels*, and consist of a pair of thin conic frusta, each tangent on a circle of contact, to one of the given hyperboloids, Pl. XX., Fig. 6. Teeth are then set on these frusta, not in the direction of their elements, but in that of the common generatrix of the hyperboloids. Also the cones to which these frusta belong have not a common vertex, since their axes are the same as those of the hyperboloids for which they are a substitute.

111. Pl. XX., Fig 9, is a fragment of bevel gears, of very unequal diameters. Fig. 10 shows a form occasionally seen, where the teeth of B are on the *interior* of the pitch cone generated by *m*V; extending from its surface *towards* its axis, and gearing with teeth on the exterior of the wheel A.

112. Pl. XX., Fig. 8, shows a method of communicating motion between two given axes, AB and CD, which are not in the same plane, *indirectly*, by bevel gear, instead of directly by hyperboloidal wheels, as in Fig. 6. In Fig. 8, OV is an intermediate axis, intersecting both of the given ones; then O is the common vertex of the bevel wheels, *m* and *n*; and V is the common vertex of the bevel wheels, *p* and *q*. As the figure is drawn, AB and OV are in the plane of the paper, and CD is out of it.

The parts, *n* and *p*, of the intermediate double frustum need not be in one piece as shown, but may be separate, and any where on OV, to suit the positions of *m* and *q* on their axes.

EXAMPLE XXXVI.

Projections of Bevel Gearing.

Pl XXII. shows a very beautiful example of bevel gear.

By lettering like points with the same letters, capital or small, on this and on Pl. XX., but little is left to be explained.

As the flanks are radial, the generating or describing circles in the development are drawn with radii, AT and BT, equal to half of the elements, V″*b*, and *v*″*b* = *v*‴T, of the cones containing the outer ends of the teeth and having the pitch circles, *mb* and M*b*, for their bases.

The pitch cones (100) of the two wheels have a common vertex V—V‴, and element of contact V*b*—V‴*m*″.

Designating the wheels by their pitch circles, or centres, the wheel *mb*—*m*″*a*″ is drawn by first making its circular projection O′. Likewise, M*b* is made from the circular projection, V‴—M′*d*′, of the same wheel.

The student may assume a scale, and from that determine the measurements.

Wheels of given radii act together better, the more teeth they have; hence, as the arc of action of bevel teeth, at their outer extremities, for example, is in the plane V″*bv*″ (perpendicular to the paper) and virtually with radii V″*b* and *v*″*b*, the action of the bevel wheels, with nominal radii *ab* and *db*, is equivalent to that of spur wheels with radii V″*b* and *v*″*b*.

c—Warped Communicators.

Example XXXVII.

The complete Projections of a Screw and Nut.

Description.—If a square, A*a*—A′*a*′*r*′*s*′, Pl. XXIII., Fig. 1, revolve uniformly around an axis, O—O′O″, and at the same time move parallel to the axis, and uniformly, it will describe the winding or spiral rail, *thread*, or solid ADG *gda*—A′*a*′*r*′*s*, G′G″ *g*″*g*′, M′M″ *m*″*m*′. This surface is bounded as follows: Its *outer* surface, generated by A—A′*s*′, during the movement of the square, is cylindrical and vertical in this case. Its inner surface, generated by *a*—*a*′*r*′, is also cylindrical and vertical. Its upper and lower surfaces generated by A*a*—*s*′*r*′, and A*a*—A′*a*′, respectively, are called *right helicoids:* called *helicoids* from the form of the bounding curves, as ADG—A′D′G′, which is a *helix;* and called *right* helicoids, because A*a*—A′*a*′ and A*a*—*r*′*s*′ are perpendicular to the axis O—O′O″.

Now let equidistant threads, like that just described, be formed, solid with an interior cylinder or core, *adg*—*a*'TU'V', and the result will be a square threaded screw. The height, AM', to which the point A ascends, in one revolution around the axis, O—O'O'', is the *pitch* of the screw. As the spaces, as *t*'G, between the threads must be equal to the threads, in order to receive the corresponding threads of the internal screw, Fig. 2, it follows that A'M' must be some even number of times A'*s*', the height of a thread. Here A'M'=6 A'*s*', and the figure represents a three-threaded screw, the threads W' and X' being separate and distinct, intermediate between the thread, A'*a*'*s*' G', which reappears at M'M'' *m*'' O''.

Fig. 2, as indicated in the plan, shows the interior of the back half of the internal, or hollow or concave screw, within which the solid screw works. It, of course, has the same pitch as the screw.

From this description, and remembering that both motions of the generating points and lines of the threads are uniform, we have the following construction:—

Construction.—For the screw, begin with the concentric semicircles, OA and O*a*, of the plan, using a scale of not less than *one-half*, and divide these semicircles into six equal parts, as shown, to indicate the uniform angular motion of A*a*—A'*a*'. Lay off XX = 3'' from the ground line, A'N', on a vertical line, and divide it into *twelve* equal parts indicating the uniform ascent of AA', etc., during a whole revolution. Then A, being projected at A', on the ground line, and on every second line, as *s*'*c*', above it; B will be projected on the *first*, *third*, *fifth*, etc., line from the ground line; C on the ground line and every second line above; D, like B, and so on.

At first it may be better to construct only one thread in elevation as indicated by A'B'C', etc., which will guide the eye in constructing the other threads. In any case it will be better to complete the outer helices before beginning the inner ones, since only certain portions of the latter are visible. Also, after completing the outer helices, the threads are to be distinguished from the spaces, by marking the former in pencil in some way on both sides, as by the letters W' and X'.

The elements A*a*—A'*a*'; B*b*—B'*b*', etc., of the helicoidal surfaces being horizontal, project up *a* at *a*'; *b*, at *b*' on the *first* line, B'*b*'; *c*, at *c*', etc., and for *one* thread, construct, completely, all the four helices of that thread, beginning as at A′a′r′ and s′; which will show clearly the number of distinct threads in the screw.

For the remaining threads, construct only the visible portions of the inner helices, viz.: the *upper* portion, as d″g″, on the *right*-hand half of each thread; and on the *lower* side, as at a′d′, on the *left*-hand half.

Small portions, as t′F′ and M′L′, of the *back* half of the *outer* helices, or the same portions of the threads as just named, will be visible. And portions of the extreme elements, as from u′ upwards, and from m′ downwards, of the cylinder of the screw are seen.

The curves of the elevation are to be mostly drawn with an irregular curve, *using the same portion reversed* for D′G′ that was used for drawing A′D′, making the two parts unite smoothly at D′; and observing that A′D′, a′d′, and all like parts are *convex upward*, while their counterparts, D′G′ and d′g′, are convex *downward*. Also, as the helix is a *continuous curve*, and *not pointed*, it is of *prime* importance to have the helices truly *tangent*, as at G′, G″, u′, s′, m′, etc., to the vertical elements, as G′G″ and a′r′.

The plan is inked in strict agreement with the elevation, showing a horizontal section of a thread at CceE.

To vary the exercise, the student should make a *one*, or a *two*-threaded screw, or divide ADG into eight instead of six equal parts, or cut off the screw by a different horizontal plane, as M″m″, or L′l′.

The Nut, or Internal Screw, Fig. 2.—This construction requires little further explanation. Having the same dimensions of threads as the screw, and the same pitch, the projections of its visible helices in the inside of the *hinder* half are found as before, and must be inclined in the same manner as threads G′M″ on the *back* of the screw. Here the *inner* helices are visible all across the elevation, while the outer ones, as G′J′M′, are only visible on the *under left*-hand sides of the threads, and on their *upper right*-hand sides, as at G″l″.

Note that a *space*, as A″r″, of the nut corresponds with a *thread*, as A′r′, Fig. 1, of the screw, while a *thread*, as R′S′, of the nut is between the same horizontal lines as the space G′t′ of the screw. The figure otherwise explains itself.

EXAMPLE XXXVIII.

The Abridged Drawing of Screws.

When either the *scale* or the *pitch*, employed in representing a screw, is so small as to make the apparent curvature at points 00′ 6, 6′, etc., Pl. III., Fig. 9, so sharp as to be sensibly pointed, the helices may be thus pointed in vertical projection, or, in *general terms*, in the projection on a plane parallel to the axis of the screw.

The figure illustrates this modification in the drawing of a *one*-threaded screw of a screw, of 8 inches outside diameter, on a scale of *one-fourth*.

Similar points in the two projections are numbered with the same figure, a very convenient method in many cases.

Pl. III., Fig. 10, represents a *triangular* threaded screw of *two* threads, to the same scale and dimensions, except pitch, as the last one. Here the thread is generated by the *isosceles triangle*, *a*′6′6″, whose base, *ao*6″, is vertical. In this case, the space between two threads extends from the middle of one thread to the middle of the next, as from 0′ to *b*′. Hence, although the screw is plainly *two*-threaded, as seen by following an outer helix, 0′,6′,12′, yet the pitch, 0′,12′, is but *twice* the height, *a*′6″, of a thread; while in the square-threaded screw, Fig. 9, a pitch, 0′*b*′, of double the height, *a*′*b*′, of a thread gives only a *one*-threaded screw. To have made Fig. 10 *one*-threaded, we should have made 0′*b*′ the pitch, and made the horizontal lines half as far apart as now, and projected 6 upon the horizontal line through *a*′. The helix beginning at 0′ would then have reappeared at *b*′.

The student should construct a triangular-threaded screw on a large scale, as partly shown on Pl. XXIII., Fig. 3; also the internal screw for the same. Strictly, the visible contour of a helicoid is curved, being tangent to the successive elements, but is nearly straight for so short a distance as *ab*; hence it is a sufficient refinement of construction, unless the scale is very large, to make *ab* straight, but tangent, as at *a* and *b*, to the outer and inner helices, instead of running from A to *n*, as in Pl. III., Fig. 10.

All the helices shown, but by straight lines.—Pl. III., Figs. 11 and 12, illustrate this abridgment. The curvature of a helix, as seen in these figures, is so slight that the idea of a screw is well suggested by making the helices straight. In Fig. 11 AC is the pitch, and the screw is two-threaded. Points, as *c* and *e*, are in the same horizontal line, as in Pl. XXIII., Fig. 1; and generally all, parts are shown just as in that figure, except that they are made straight, and therefore only their extreme and middle points, as A,B; *e* and *r*, and *n*, need to be noted before drawing them. This is sufficiently done by drawing horizontal lines at a distance apart, B*b*, equal to the thickness of a thread, together with the five vertical lines through A*c*, *e*, *dn*, *ar*, and B; where $ce = a\text{B} = \text{B}b$, and $\text{B}d = df$.

In like manner, Fig. 12 is like Fig. 10. AC is the pitch $= 2\text{A}f$; $\text{A}c = de$ and $\text{AD} = \text{D}e$, and the horizontal lines need only be drawn at a distance apart equal to *ca*.

The student should repeat these constructions for a *one* or a *three*-threaded screw.

AC, Fig. 13, $= 4\text{C}c$ for a two-threaded, $2\text{C}c$ for a *one*-threaded screw, $6\text{C}c$ for a three-threaded one, etc.

AC, Fig. 12, $= 2\text{A}f = 4ac$ for a two-threaded screw; A*f*, for a *one*-threaded screw, etc.

Outer Helices only shown.—Fig. 14 shows a further abridgment, where so much of the *outer* helices, only, as are on the front half of the screw are shown.

Smaller Triangular Screws.—Fig. 15 illustrates a screw with triangular threads, in which the greater steepness of the inner helix is neglected, and the outer and inner helices are made parallel, the former being inked heavy.

Very Small Screws.—These are represented in Figs. 15–17. Sometimes only the helices are drawn, omitting the end lines of the threads. The effect is better on the triangular thread, Fig. 16, than on the square thread.

Finally, Fig. 15 represents the helices as all equal, parallel, and straight, and included between two parallel lines.

Fig. 17 is a screw bolt, that is, a bolt or short rod threaded to receive a nut at one end, and headed at the other.

Uniform System of Screws.

112. The extent to which screws enter into the composition of machines, either as fastenings or communicators of motion, and the distances from the place of manufacture to which machines

are often transported, and at which they must be repaired, make it very desirable that, at least for screws used for fastenings, there should be a uniform system of threads and nuts.

Screws used for communicating motion may be subject to so many special conditions as to make the use of an invariable series of them impossible.

Such system as that just mentioned is used in England, and it is of constantly increasing importance that the like should be employed in this country. The following carefully matured system, proposed in 1864 by William Sellers of Philadelphia,* is therefore given as a contribution to this desirable result. It consists of the following notation, formulas, and a table of sizes, from which a few examples are here taken. They relate only to triangular-threaded screws.

D = external diameter of screw.
a = constant subtrahend, 2.909.
b = constant divisor, 16.64.
c = D expressed in 16ths of an inch, plus 10.
d = internal diameter, or that at the bottom of the threads. $= \frac{D\ 1.299}{n}$
p = the pitch, meaning, in this table, the distance between the threads. $= \frac{\sqrt{c} - a}{b}$
n = number of threads per inch, the nearest whole number to $\frac{1}{p}$
w = width of the flat top and bottom, that is, of the outer and inner edges of threads. $= \frac{p}{8}$
l = least diameter of finished nuts and bolt-heads = perpendicular between opposite sides, or diameter of inscribed circle. $\frac{3}{2}$ D + $\frac{1}{16}$ inch
h = long diameter of hexagonal nuts, or bolt-head = diameter of circumscribing circle. $= l \times 1.155$
s = Do. of square nuts or bolt-heads. $= l \times 1.414$
t = thickness of finished nut or bolt-head. $= D - \frac{1}{16}$ in.

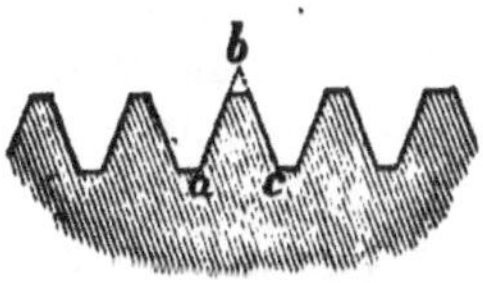

FIG. 48.

The threads are to be truncated as at Fig. 48, to give increased strength both to the thread and the bolt, and the angle abc is fixed at 60°, that being much easier to verify than the English one of 55°, besides giving a more substantial thread.

The following table presents a few examples, and the dimensions of only finished nuts and bolt-heads. All the dimensions are in inches and fractions of an inch.

* Essay on a System of Screw-Threads and Nuts.

PROPORTIONS OF SCREWS, THREADS, NUTS, AND BOLT-HEADS.

D.	*n*	*d*	*w*	*l*	*t*	*h*	*s*
$\frac{1}{4}$	20	.185	.0062	$\frac{7}{16}$	$\frac{3}{16}$	.505	.618
$\frac{1}{2}$	13	.400	.0096	$\frac{13}{16}$	$\frac{7}{16}$	.938	1.161
$\frac{3}{4}$	11	.507	.0113	1	$\frac{9}{16}$	1.155	1.414
$\frac{7}{8}$	9	.731	.0138	$1\frac{3}{8}$	$\frac{11}{16}$	1.588	1.944
1	8	.837	.0156	$1\frac{9}{16}$	$\frac{15}{16}$	1.805	2.209
$1\frac{1}{4}$	7	1.065	.0178	$1\frac{15}{16}$	$1\frac{3}{16}$	2.238	2.740
$1\frac{1}{2}$	6	1.284	.0208	$2\frac{5}{16}$	$1\frac{7}{16}$	2.671	3.270
2	$4\frac{1}{2}$	1.712	.0277	$3\frac{1}{16}$	$1\frac{15}{16}$	3.537	4.330
$2\frac{1}{2}$	4	2.176	.0312	$3\frac{13}{16}$	$2\frac{7}{16}$	4.403	5.391

EXAMPLE XXXIX.

Endless Screws and Spiral Gear.

Description.—An axis revolving in fixed supports, and having a screw thread cut upon its circumference, is an endless screw. Because such a screw makes no advance in the direction of its axis, it will advance or move any yielding piece on which its thread can act. One complete revolution of the screw will advance the point upon which it acts a distance equal to the pitch of the screw. If, then, the thread engages with a wheel whose teeth are so set as to be tangent to the thread, when in contact with it, the screw will give a slow rotation to the wheel.

Pl. XXXI., Figs. 1 and 2, shows such an arrangement. OO′K—O″K′ is a wheel actuated by the screw HAB—B′L.

In this case the screw thread is formed in the usual manner; therefore, by taking a section, *c*′HB, through the centre of the body of the wheel and the axis of the screw, we shall have the equivalent plane *rack* driving a spur wheel. This problem is here solved by the system of involute teeth (IVth Sol.), but with the pitch line, LL′, or exterior element of the outer cylindrical surface of the screw as the generating line. The involutes of LL′, generated by the unwinding of LL′ from itself, will be straight and perpendicular to LL′ as at *ac*, Fig. 4, and the corresponding involute teeth of the wheel will be involutes of its base circle, O—O′K, tangent to LL′, as at *ab*, Fig. 4.

Construction.—The screw being constructed as usual, and as shown, the inclined teeth of the wheel are thus drawn. First lay out the circular elevation, O—O′K, of the front of the wheel as just described, with involute teeth. Then develop any convenient arc, as AB—A′B′, of an outer helix of the screw, into the tangent plane AD—A′*a*′D′, at AD—A′D′, by making AD = AB, and projecting D at D′, on *a*′D′ drawn through B′, and perpendicular to the axis *a*′A′. Then D′A′ is a tangent to the helix at A′; and, by the properties of tangents to a helix, has the same inclination to a plane perpendicular to the axis, A′H′, that the helix has; moreover, this inclination is constant.

The teeth of the wheel will now be inclined to the straight elements of its cylindrical rim by the angle *a*′D′A′, the complement of the inclination *a*′A′D, of the thread to the screw axis A′H′.

We now turn aside to rehearse, for convenience, the construction of a helix, A′B′, Fig. 3, from its plan, AB, and development, A″B″, which is straight, and found by unrolling into a plane any convenient portion of the vertical cylinder ABC—A′C′B′, on which the helix lies; so that here A″C″ = A3C. Then any point, as 1, of A′B′, is the intersection of the projecting line 1—1 perpendicular to A′C′, with the line 1—1 parallel to A′C″.

Returning now to Fig. 1, and proceeding likewise, make *d*D′*a*′ = 90° and A′D′C = 90°, to make CD′*d* = *a*′D′A′, and we shall have the inclination of the teeth of the wheel relative to the projection, O′O″, of its axis, but on the under side of the wheel.

Draw E″*e*″, Fig. 2, so as to have E″*c*″*c*‴ = CD′*d* = A′D′*a*′; and E″*e*″ will then be the development of a portion of one helical edge of a tooth of the wheel, analogous to A″B″ in Fig. 3. For convenience, take the lines L′*n*′, H*c*′, and K′*b*′ produced, and others at the same distance apart, as E′E″ and *e*′*e*″, to correspond to *a*′*a*″, *b*′*b*″, etc., Fig. 3. Then, by drawing E″*e*‴ perpendicular to them, and dividing *e*″*e*‴ into four equal parts, as E″*e*″ is, we have *e*″*e*‴ corresponding to AD = AB = 03 (at the left of the screw) for the screw. Hence lay off *e*″*e*‴, and its divisions at F*f*, on the tangent FG, and so that one of the divisions shall fall at *c*, the point of a tooth. Transfer these points to the circle of radius OF, as at FE, and we shall have the horizontal projection, we will call it for the moment, corresponding to A3B, Fig. 3, of the helical portion whose development is E″e″.

The vertical projection, e′E′, is now found just as A′B′ was in Fig. 3, as is seen by the lines of construction, and the use of the same letter for the same point. Owing to the very great pitch of the wheel, considered as a screw, which it now plainly is, e′n′c′b′E′ is sensibly straight, and FE is sensibly the same as F*f*. Thus we have verified, by a full construction, the propriety of making all the longitudinal lines of the tooth of a worm wheel, or spiral gear, parallel straight lines.

Projecting b′ back to b, we find *cb* for the projection of the length of the tooth on the projection OO′K. Hence, to complete that projection, make $pq = cb$, and do the same for all the teeth, and make the back curves, as from *q*, the same as the front ones, as from *p*.

Observe that, as successive teeth of the wheel come in contact with the same thread, B′L, of the screw, and in the same relative position, they are not successive portions of the same thread, but *like* portions of *different* threads; that is, the wheel, considered as a screw, has as many threads as it has teeth. Hence the other teeth on the projection, O″K′L′, are found by projecting from the other figure, O—O′K, where the teeth all appear alike.

Often the section of the screw thread is of the same form as a wheel tooth, that is, with a separate face and flank. In that case the wheel teeth would be likewise formed by an appropriate generating circle, according to the first or second solution.

Further, to give the screw tooth a larger surface of contact with the wheel teeth, the point and root lines of the wheel teeth, in other words, all parts of the face, n′b′KL′, of the wheel, are concave arcs of the radii, HB and H*m*, of the screw thread. In such a case points would be constructed on the two plane sides, K′b′ and L′n′, and in the central section, Hc′, of the wheel.

With this description of all that is peculiar to the case, the student can readily construct, on a large scale, a few such teeth.

EXAMPLE XL.

Detailed Construction of a Tooth in Spiral Gearing.

Description.—The term spiral gear is applied to a species of spur wheel in which the teeth are formed as in the wheel OO′K—O″L′K′, Pl. XXXI. A pair of such wheels, with *parallel* axes, act together with a peculiar smoothness and stillness, because a pair of teeth begin contact at one end, and their point of contact shifts continuously to their other end, with a rolling motion between the teeth, and just as one pair are about to quit each other another pair will begin contact.

Pl. XXXI., Fig. 5, shows an approximate form of this gearing, in which a wheel is formed of a series of thin plates, each of which is a spur wheel, but set a little in advance, angularly, of the next one, as indicated by the black spaces, which represent teeth.

Construction.—Pl. XXI., Fig. 6. Let O be the centre of the wheel, and O*m*, O*g*, and O*b*, the radii of the point, pitch, and root circles, O*g* being $4\frac{4}{4}$ ins. Lay out by curves with any suitably assumed radii, taken merely for illustration, the cross section of a tooth in plan, at *abcdef* (at the left), and let there be 20 teeth, and $1\frac{1}{2}$ inches pitch. Make the two radii, O*a* and O*a*, 45° apart, so that *aa* shall be one-eighth of the circumference of the pitch circle. Then let O′O″, = 8 inches, be the *ascent* of the tooth section during one-eighth of a revolution. That is, the pitch of the teeth threads is 64 inches. Then we have only to construct in the usual way, Ex. XXXVII., as shown by the projecting lines and letters of reference, the six portions of helices, beginning at the points *a*, *b*, *c*, *d*, *e*, and *f* (on the left). The equal pitch for the different radii, O*b*, O*a*, and O*c* is regarded by dividing *bb* and O′O″ into the same number of equal parts, four in this case. Divide *aa*, and O′O″, each into four equal parts, and *cc* and OO″ in like manner, also *ff*, *ee*, and *dd*. Now *bf*, the width of the tooth at the root, happens to be the fourth part of *bb*. Then, by the radii O*b* and O*f* produced, we find *gh* = one-fourth of *aa* to be laid off from *a* and *e* four times to the right; and *mn* = one-fourth of *cc*, to be laid off four times on *mn*″ from *c* and *d* (on the tooth) 1.

Having constructed the helical arcs, by projecting up the

points thus found, assume PQ as a segment of the face of the wheel and all those portions of these helices within its limits will be real lines of a tooth.

As to visibility, the left root curve is visible from b'''' down to p, where it runs out of sight; the left pitch helix is visible from a'' down to r; the two point helices are visible throughout, and so is the right-hand pitch helix, $e'e''$, while the right-hand root helix is visible from f' up to t.

113. In regard to the manufacture of worm wheels, of concave face, to act with endless screws, the difficulty of representing the teeth whose parallel sections are dissimilar, owing to their different position relative to the successive meridian sections* of the screw, is obviated by making a model *screw* of hardened steel, and notched on the edges of its threads. It is thus made into a cutter, and will itself cut the proper teeth on the concave face of the wheel, when both are revolved together with the proper relative velocity.

By a similar expedient, and by attaching a cutter to the wheel, it is possible to form an endless screw like that shown in Pl. XXI., Fig. 8, where the threads are in contact with a large number at once of the wheel teeth, and in the central plane ABC of the two bodies.

* A section made by a plane containing the axis.

CLASS IV.—REGULATORS.

114. Under the head of *regulators*, steam governors might naturally occur to the mind at once; but it must be considered that governors are not single mechanical elements in the sense of those hitherto represented; rather they are secondary machines, attached to their principals, which they govern. That they truly are machines is evident from the definition (28), since they always consist of a train of connected pieces, embracing some or all of the following, viz., band wheels, spur or bevel wheels, pistons, racks, connecting rods, screws, oscillating arms, etc.; receiving motion as a whole at some point, and communicating it through the train of pieces to another point.

Hence, as there would be little interest or use in separately describing or drawing the essential governing member alone, whether ball, or fan, etc., the further description of governors is postponed till the chapters on compound elements of ma chines.

A—Point Regulators.

Governor balls belong under this head.

B—Line Regulators.

Example XLI.

A Fly Wheel.

Description.—A fly-wheel is here reckoned as reduced to its rim, and that a heavy *line.* It is an equalizer of velocity by being an equalizer of *work*, which is its principal function. That is, when the load is largely or wholly taken off, the inertia developed in the fly wheel, by the gradual increase of velocity which follows, results in a storing up of work which is usefully given out in the maintenance of a slowly retarded velocity, when a return of the full load takes place.

The greater the difference between the extreme loads carried by the engine, compared with the power of the latter, and the less the time in which the extreme load is brought on, the heavier should be the fly wheel.

Accordingly, the heaviest fly wheels are found in rolling mills for example, where the rolls, Ex. XVII., are alternately empty, and run with very little power; and then full, and operating against a prodigious resistance, especially in case of steel rolling mills, where rolls nearly a foot thick are sometimes snapped in two.

Pl. XXIV., Fig. 1, represents a sketch on various scales; or no scale, in some parts, of a sixty-ton fly-wheel at the Bessemer Steel Works at Troy, N. Y.

It is made in ten segments, as AB and the arm E, weighing five tons each. The massive hub, and ten 3⅛ inch rods, as F′, make the total weight sixty tons or more; whereas twenty-five tons is the weight of quite a heavy wheel.

The section of the rim is 19 ins. by 20 ins., and each segment is filed smooth at the ends of the rim portion, AB, on all the parts which are shaded in the end view, *p′q′*. Adjacent segments are then strongly bound together, as at *pq*—*p′p″*, by stout links, D,D″, made of wrought iron, 2½ ins. square. The inner ends of the arms are then turned, 15½ ins. long, and 8 ins. diameter, and keyed to the hub, as shown at *s*, and *o*. The radius rods, F′, are likewise keyed into the hub, and headed as at *u*, to set into the head socket *u′*.

There being *ten* segments, no arm will be horizontal if one be vertical as at E; yet to show both a face view, *g*KL, and an edge view, *et*, of an arm, and *f*, of a radius rod, the two latter are shown in a horizontal position in plan. *mkk″* is the hollow interior of the hub, bored on the two sides as at *ab*—*a′b′* to receive the main shaft, to which it is heavily keyed.

Construction.—Three segments, or at least two, should be constructed in full in both projections, and on a scale of not less than *half an inch* to one foot. The section *p′q′* of the rim may well be made on a scale of *one inch* to one foot; and the plan should be complete, from the horizontal bearing, *g*KL, of a vertical arm, and showing the alternate 11 inch and 16⅛ inch similar bearings of the radius rods and arms. These bearings touch each other, as shown from g' to the left on the elevation. The annular space between the circles d' and f' is fluted, as at g', and to the left, by the alternate thick and thin necks, gLK and rf, or H and J, through which the arms, E, and rods, FF′, run. This cannot be very clearly shown without shading the elevation, which may be done with fine effect.

The student may also usefully add a central section, on JF, and a cross section on ll, of the hub.

C—SURFACE REGULATORS.

After fly-wheels, column five of the general table of elements of machines affords no more examples for drawing till we come to *volume* elements.

Plane throttle valves are simply like a common stove-pipe damper, a very rude contrivance.

115. *Single poppet valves* lift off of their seats, instead of sliding across them, as do *slide valves*, whether reciprocating, or oscillating, as in the valves of a Corliss engine, which are like a door, only so thick that its thickness covers the opening which it commands, and so its wide cylindrical edge slides on and off the rectangular opening in its concave cylindrical seat.

116. *Cage valves*, illustrated in my "Elementary Projection Drawing," and so called from their name, are used in locomotive pumps, and perhaps in some other similar situations, where a valve must act rapidly against great resistance.

117. *Cylindrical throttle valves*, to be illustrated by and by in a govenor, are the modern improvement over the old damper throttles, and act to open a series of retangular openings at once, which, when fully open, give an area of opening equal to that of the pipe which they command.

118. *Ball valves* are simply balls fitting a spherical seat. They were formerly used instead of cage valve, in which the valve is a cylindrical cup with a flat seat.

D—VOLUME REGULATORS.

119. *Cocks* are named from the number of passages which they control, and consist essentially of a perforated conical stem, turning about its axis in its conical seat, whose walls are

also perforated. When the perforations coincide, a passage is formed. When they do not, it is destroyed.

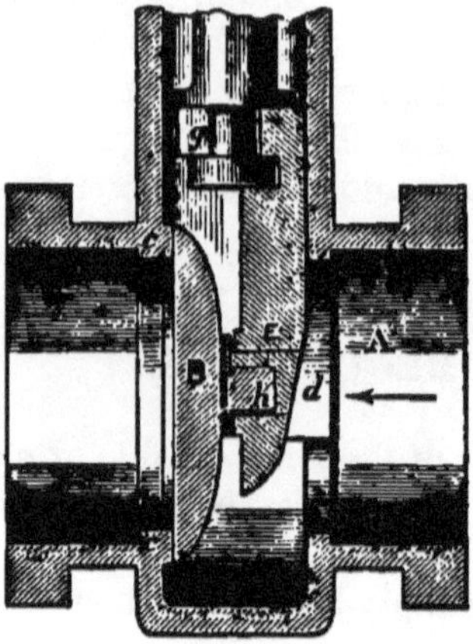

FIG. 49.

FIG. 50.

120. In *globe valves*, so-called from the partly spherical case in which the valve operates, the latter is conical, and lifts off its seat which connects two curtains, each of which cuts off half of the passage; so that the latter, when the valve is fully open, is quite circuitous and obstructed. Still the globe valve has advocates, on account of an alleged difficulty in tightly seating a flat gate without leaking, dislocation, or grinding friction in unseating.

121. *Water gates*, used also for steam and gas, are very perfect in theory, and highly esteemed in practice. Two of the best are made at Troy, New York. In one, the Ludlow valve, Figs. 49, 50, the gate is seated by the wedging action of an inclined plane at its back, with an ingenious arrangement for loosening the valve from its seat before it begins to be drawn off the same.

A vertical section of the apparatus is shown in Fig. 49. A is the valve box, through which the water flows from right to left or left to right. B is the valve or gate, ground to fit its face, *cc*. The rear of the valve box is formed with two wedge-shaped projections, *dd*, shown in perspective in Fig. 50. Between these projections and the valve, is the wedge, E, the ridge on its face pressing against the back of the valve, B. The valve stem raises and lowers the wedge, and through this, only, acts upon the valve. Upon the back of the valve is formed a lug, *h*, which enters, loosely, a circular recess in the wedge, and by this arrangement the wedge is permitted to move upward a certain distance, thus loosening the valve before it begins to rise. Whatever the position of the apparatus, the wedge cannot pinch the gate before the valve opening is covered.

FIG. 51.

In the Brown valve, Fig. 51, the valve, V, is fastened by a brace, *a*, acting somewhat after the manner of a toggle joint, rising and falling with the valve, in standing guides, see the figures. It acts with very little, if any, injurious sliding under pressure between the valve and the seat, owing to the roller form of the ends of the brace.

In all of the last three contrivances, the effectual packing of the valve stem, which should work in a tight collar, especially for steam and gas, is a matter still deserving of attention.

All such devices as these, although practically very important, are hardly elements of machines in such a sense as to afford examples for full illustration, unless, as in the following cases, on account of their ingenuity, novelty, or grand importance.

EXAMPLE XLII.

Chambered or 'D' Locomotive slide valves; Plain and Anti-friction.

Description.—The common slide valve, Pl. IV., Fig. 2, is best understood in connection with the cylinder. T,T′,T″ is its hollow interior, which is sufficient to cover the exhaust port, *bd*, and either one of the steam ports, as *ce*.

From the shape of the section, T″, of the valve, it is often called a D valve. *cd*—*c″d″*—*c′d′pq* is the rectangular *yoke*, surrounding the body of the valve, and forged to the *valve stem*, *s*, by which the valve is actuated. *baa″*—*a′* is one of two stiffening ribs on top of the valve, shown in plan in dotted lines, the valve being there upside down.

The valve is shown raised from its seat, *gt*, Fig. 1, and out of position relative to the piston. Supposing the latter to be just about to begin its stroke to the left, the valve should be far enough to the left to have opened the steam port, *ce*, already, about one-eighth of an inch. This distance is called the *lead* of the valve, and it serves to admit live steam a little before the beginning of the stroke; which *cushions* the piston, and relieves it from the jerking strain of a very sudden change of motion.

The difference between the width of the lip or face, *fh*, of the valve, and the port, *ce*, is called the *lap* of the valve. Its amount, in this case 1 inch, determines the point in the piston stroke at which steam is cut off; as will be more fully explained by and by.

The minute recesses, as *m* and *n*, which break joints with each other, serve to secure a lubrication of the valve seat by the steam.

In contrast with the common slide valve, the friction of which is very great, Pl. XXIV., Fig. 3, represents a frictionless slide valve, which is believed to accomplish the purpose of saving the wear on the valve seats, eccentrics and other gear, cause by the excessive friction inherent in all sliding valves.

This valve is not of the balanced valve variety, which, locomotives, has been thought to be impracticable, but, as seen the figure, it works by changing a rubbing to a rolling friction. When the valve is in operation, it is so suspended from the axles of the rolls R,R, by means of saddle plates SS, that it works just in steam tight contact with the seat AA, without any appreciable friction as the valve moves back and forth; for the pressure comes upon the axles, BB, through the saddle plates, and so causes the rollers to roll upon the ways CC. In this movement of the rollers and axles, there is, however, no rubbing friction, as not only is the friction of the rollers upon the ways of the rolling kind, but so is the friction of the axles on the saddle plates, and hence there is no appreciable resistance whatever to the movement of the valve created by the pressure of steam on the back of it. The rolls, axles, ways, and saddle plates, are of hardened steel, and subject to a crushing, not a rubbing force. It is said, that after a test of several years in different engines, valves so applied, do not show the least indication of wear.

These valves are in use in locomotives on a number of the leading railroads of the United States, and can be put on any locomotive in a few hours' time.

Construction.—As there is a considerable difference in the size of the ports in different engines, the student can readily assume these, and then draw the valve, either in plan and elevations, or in isometrical or oblique projection, from suitable measurements.

Example XLIII.

Tremain's Balanced Piston Valve.

Description.—The valve of the steam engine has probably been the subject of more thought than any other piece of

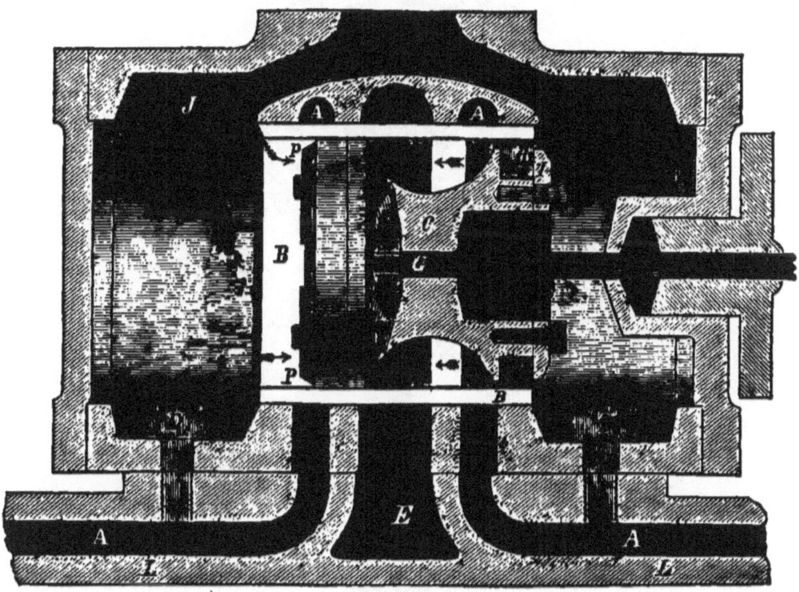

Fig. 52.

mechanism of equal size and simplicity. The great amount of power absorbed in working it, the strength and weight consequently required in all the parts connected with it, the constant wear and liability to break down, the delay and expense of repairs, the difficulty of reversing or working the valves of large engines by hand, are sufficiently well known. The object has been to relieve the common slide valve of the pressure of steam, and many ingenious contrivances have been invented for this purpose. This has been accomplished by the invention here illustrated. This balanced slide valve is of the piston variety, and its claims to superiority are of a novel character, and such as to attract the attention of engineers and owners of steam engines.

First, it is a perfectly balanced valve; it requires no adjustment; it is simple and not liable to get out of order. The cost of its application to engines now used is small, while for

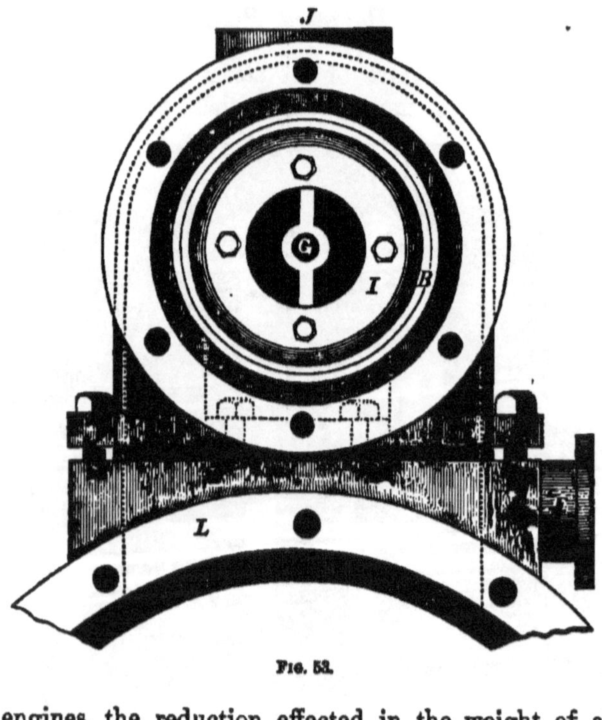

FIG. 52.

new engines, the reduction effected in the weight of all the valve-gear makes it much cheaper than the common valve. It is applicable to either high or low pressure engines, or when single valves at each end of the cylinder are used. By it a much longer port is obtained than by the common slide valve. In a steam chest eight inches in diameter, the circumference being over twenty-five inches, the steam port will be twenty-one

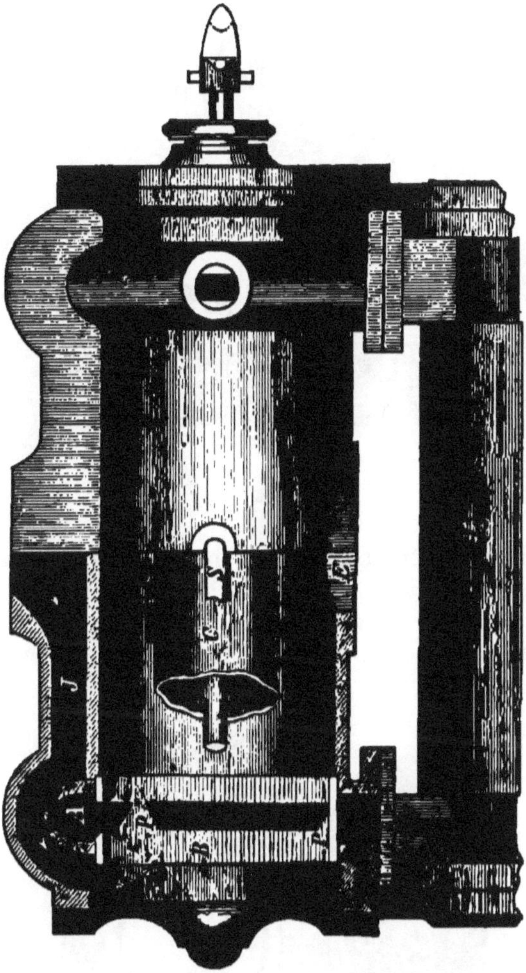

Fig. 54.

Fig. 55.

inches. This is of very great advantage in clearing the cylinder of steam after it has done its work.

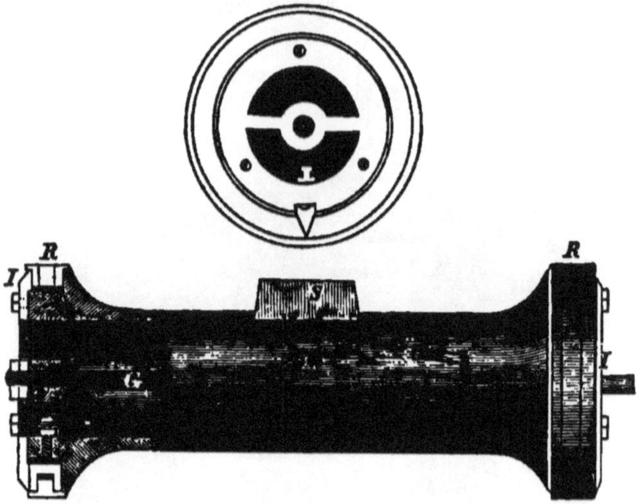

FIG. 56.

Fig. 52 is a longitudinal section, showing the application to old engines now in use. The same letters refer to like parts. Fig. 53 is an end view of Fig. 52, showing the bolts, D,D, in dotted lines. Fig. 54 shows the manner of bolting the chest on,

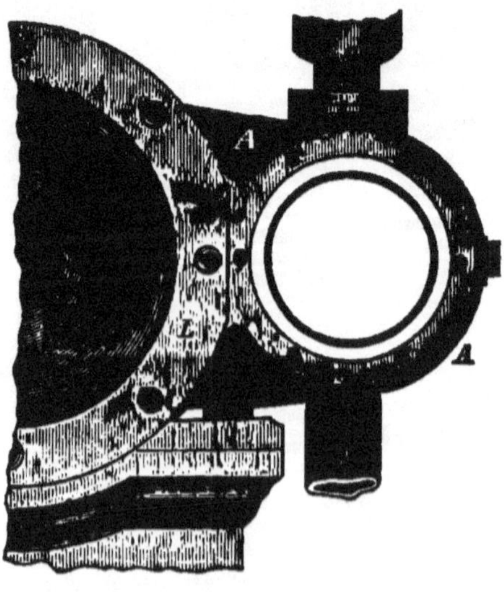

FIG. 57.

when it is not convenient to cast it and the cylinder together. Fig. 55, valve heads connected by common pipe when the cylinder is long. Fig. 56, valve showing guide, S, and the way the rings, R, are set out; these rings, R, being in three parallel parts, prevent leakage of steam by breaking joints; while the manner in which they are set out is adapted to prevent leakage by their being compressed inward by the steam. Fig. 57 is an end view with steam chest cast on the cylinder.

The valve is placed in the cylindrical steam chest, which has two grooves, AA, encircling it, which are in communication with the steam ports, A, which lead into the cylinder, L. Steam is admitted through the side pipe, J. The grooves, A,A, are covered by the ring, B, which forms the valve seat and contains the ports, *p*,*p*, through which steam passes into and out of the cylinder, L. E is the opening for exhausting the steam from the cylinder. The valve, C, is hollow, and is secured upon the stem, G. The rings, R, are secured by the follower, I, which is recessed into the ring and will not come in contact with the seat, B. DD are bolts by which the chest is fastened to engines previously using the common slide valve.

Construction.—Let the figures be arranged, with the different elevations of the same thing on the same level.

Example XLIV.

Balanced Poppet Valves.

Description.—In marine engines of comparatively short and quick stroke, as propeller engines, slide valves, like those of locomotives, but of proportionally larger size, are frequently used; while for engines of long and slow stroke, like beam engines generally, poppet valves are used.

A poppet valve lifts off from its seat, and a balanced poppet-valve, Pl. XXV., Fig. 1, where one-half of a valve is shown at AHB, has two seats, CD and *cd*. These, like the valve, are circular in plan, the figure being a vertical section; and as steam enters from the steam pipe S, as shown by the arrows, the pressure *downward* of the steam at CD resists the opening of the valve, while the upward pressure at *cd* assists its opening. Hence, by making the diameter of the upper opening one inch greater than that of the lower one, the valve is lifted only against the downward pressure on a ring of half an inch in width, and, in this case, 23¾ inches outside diameter.

E*f*IJ is a vertical section of the steam chest of a marine beam engine built by the Novelty Works for the Pacific Mail Co., and with a cylinder 105 inches diameter, and 12 feet stroke. S is the steam pipe (see also Pl. V., Figs. 5, 6), bolted at EF to the vertical pipe leading to a similar chest at the top of the cylinder. Steam flowing into the spaces, KK, rushes through the openings made by the lifting of the steam valve AHB; through the steam port, LL, into the bottom of the cylinder, and forces up the piston.

At the same time, the upper exhaust valve, corresponding to *ahb*, lifts, and sets free the steam which effected the previous stroke, and which then escapes through the exhaust pipe. This pipe is bolted on at *ef*, and the steam thus passes on through G and the exhaust port, MM, to the condenser.

Again, when the upward stroke is just about to end, the upper steam valve, corresponding to AHB, lifts and admits steam to "cushion" the piston at the end of its upward stroke, and drive it down again, while the lower exhaust valve, *ahb*, opens and steam flows out of LL into G, and through M to the condenser. The partitions NN divide the steam, from the exhaust chamber. It is thus seen that live and exhaust steam can never meet unless, as in LL, by the two valves AHB and *ahb*, being both open at once for a moment.

The valves are circular, as seen in plan, and therefore sufficiently shown in the plate, by a half vertical section of each. Each valve is keyed, as at *kk'*, to a vertical valve stem. The latter are connected by short horizontal arms to the vertical lifting rods, as V, Fig. 4, in front of the cylinder. These rods are lifted by wipers, W, keyed to an oscillating rock-shaft, R, which is operated by an eccentric on the main shaft; whose rod, E, takes hold of the free end of a rocker-arm, A, keyed to the rock-shaft. The wipers, W, act upon *toes*, T, keyed to the lifting-rods, V.

The form of the upper surface of the wiper and its angular position on the rock-shaft, together with the position of the eccentric, determine the height and rapidity of the lift of the valve, and the points in the piston stroke at which it will open and close. By opening the steam valves just before the end of a piston stroke, the piston is cushioned, so as to ease the shock of the sudden reversal of the motion of its heavy mass. By closing them at or soon after the middle of the stroke, the benefit of the expansive use of the steam is obtained.

To avoid back pressure on the advancing face of the piston, the exhaust valves are larger, or are lifted higher, and held open longer than the steam valves; as may be seen by watching the different motions of their two lifting-rods on any river-boat engine. This result is effected by making their toes shorter, while their wipers begin to act at a point nearer the rock-shaft than is the case with the steam-valve wipers.

To enter minutely into this, and closely kindred subjects, with their theory, might occupy a small volume; hence this description closes with the following data from actual practice.

First, in the Pacific Mail Steamers.

Steam pressure in boiler......................18 lbs.
Mean pressure in cylinder......................Depends on point of cut-off.
Diameter of cylinder......................105 ins.
Length of stroke......................12 feet.
Steam cut-off at......................2½ to 6 feet.
Steam valve opening at beginning of stroke=lead=$\frac{5}{16}$ ins.
Height to which steam valve is lifted=7½ ins.
Exhaust valve opening at beginning of stroke=exhaust lead or *release*=2½ ins.
Height to which it is lifted=7½ ins.

The following data are from the engines of the remarkable shore steamers "Providence" and "Bristol," and were furnished by Mr. Thomas Main, Eng., their designer for the Messrs. Roach, their builders:—

1.	Steam pressure in boiler, above atmosphere..	21 lbs.
2.	Diameter of cylinder......................	110 ins.
	Of side pipes......................	30 ins.
3.	Length of stroke......................	12 ft.
4.	Usual point of cut-off......................	5 ft.
5.	Mean pressure in cylinder, about............	25 lbs.
6.	Lead of steam (poppet) valve	$\frac{3}{16}$ to ¼ ins.
7.	Total lift of steam valve......................	5½ ins.
8.	Diameters of do......................	20 and 21 ins.
9.	Lead of exhaust valves......................	2½ ins.
10.	Total lift of do......................	7 ins.
11.	Diameters of do......................	21 and 22 ins.
12.	*Three* boilers, each	35 ft. long by 12 ft. 5 ins. diar
13.	Grate surface......................	510 sq. ft.
14.	Fire surface	..............13,850 sq. ft.

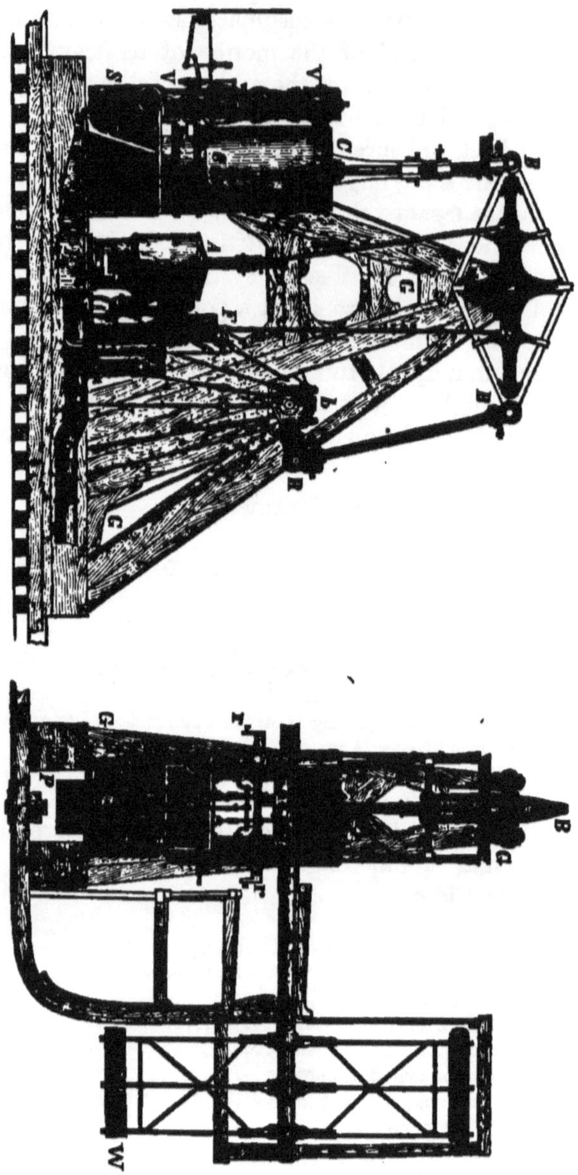

Fig. 58.

15. Surface condenser. Surface.................. 4,500 sq. ft.

16 Paddle-wheels, 38 ft. 8 ins. diameter, 12 ft. face. Paddles "stepped" thus to prevent jar.

Steamship	"Providence."
Date	April 30th, 1869.
Which end of cylinder	Both.
Revolutions per minute	$17\frac{3}{4}$.
Pressure of steam in boilers, in lbs	21.
Point of cut-off	5 ft.
Position of throttle-valve	Open.
Vacuum per gauge, in inches	26.
Temperature of hot well, Fahrenheit	122°.

Scale of indicator, 16 lbs. to an inch.

Construction.—The student can show the whole of each valve and a plan of one, and the scale for the steam-chest being reduced from $\frac{1}{10}$ to $\frac{1}{12}$, or even to $\frac{1}{16}$, the valves may be shown separately on a scale of from one-*fourth* to one-*eighth*, with enlarged sections at the seat, as in Figs. 2 and 3.

122. The general view, Fig. 58, will make the above example more intelligible.

G, is the heavy supporting frame of the whole engine, and called the gallows frame; C, is the steam cylinder; B, the working beam; BR, the connecting rod; R*b*, the crank, turning the main shaft, in the pillow-block *b*. A, is the air-pump; F, a force-pump for supplying water to the surface condenser, S (Pl. V.), through passages in the bed-plate, P (Pl. VIII.). W, is one of the paddle-wheels; V, is the steam, or valve-chest. *r* and *r* are the rocker-arms, see A, Pl. XXV., Fig. 4, the pins of which are engaged by the eccentric hooks. The rock-shaft, *rr*, is in two parts, which meet at a common central bearing. One eccentric is for the steam, and the other for the exhaust valves. The former has a large throw so as to actuate its rocker through a large arc, and quickly. A long and adjustable wiper on its rock-shaft, acting on the toe, T, Pl. XXV., only during a part of its arc, raises that toe with the steam valve quickly, and closes it at a point in the stroke, depending on the angular position at which the wiper is set on the rock-shaft.

123. The following abstracts from observations at sea are interesting, as showing the relation of cut-off to cylinder and boiler pressures:

1°. Steamship "Montana."

Which engine, main.	Which end of cyl., top and bot.	Rev'ns ⅌ min., 9.
Throttle, wide.	Steam, 16 lbs.	Vacuum, 27 ins.
Sea water tem're, 62°.	Discharge water tem're, 70°.	Feed water tem're, 120°.
Engine room tem're, 70°.	Cut-off, 3 ft.	Coal ⅌ hour, 2,536 lbs.

Initial cylinder pressure, top, 13.8 lbs., bottom, 13.9 lbs.

2°. Steamship "Montana."

Which engine, main.	Which end of cyl., top and bot.	Rev'ns ⅌ min., 9.
Throttle, wide.	Steam, 20 lbs.	Vacuum, 27 ins.
Sea water tem're, 80°.	Discharge water tem're, 90°.	Feed water tem're, 120°.
Engine room tem're, 86°.	Cut-off, 2 ft. 6 ins.	Coal ⅌ hour, 2,614 lbs.

Initial cylinder pressure, top, 17.1 lbs., bottom, 17.5 lbs.

3°. Steamship "Montana."

Which engine, main.	Which end of cyl., top and bot.	Rev'ns ⅌ min., 7.8.
Throttle, $\frac{1}{3}$ open.	Steam, 21 lbs.	Vacuum, 27 ins.
Sea water tem're, 88°.	Discharge water tem're, 94°.	Feed water tem're, 120°.
Engine room tem're, 86°.	Cut-off, 3 ft.	Coal ⅌ hour, 2,800.

Strong head-wind and sea.
Initial cylinder pressure, top, 18.9 lbs., bottom, 19.3 lbs.

4°. Steamship "Montana."

Which engine, main.	Which end of cyl., top and bot.	Rev'ns ⅌ min., 8.
Throttle, wide.	Steam, 21 lbs.	Vacuum, 27 ins.
Sea water tem're, 84°.	Discharge water tem're, 90°.	Feed water tem're, 120°.
Engine rooom tem're, 86°.	Cut-off, 3 ft. 6 ins.	Coal ⅌ hour, 2,925.

Initial cylinder pressure, top, 20.5 lbs., bottom, 21 lbs.

The increasing difference between boiler and cylinder pressure as the cut-off takes place earlier, indicates the increased wire drawing of the steam, as the steam valves are less lifted and quicker closed.

The difference between the sea-water temperature and the discharge-water temperature shows how much the sea water is heated in passing through the condenser tubes.

The increase of fuel required to maintain the boiler pressure as the cut-off is later, is also noticeable.

Example XLV.

Richardson's Locomotive and Lock-up Safety Valve.

Description.—Fig. 59 represents the form of this valve, as applied to locomotives, on a scale of one-half.

AA, represents a section of an ordinary dome cap or plate, with the projecting lip or flange *a*, which may be cast on, and form a part of the valve seat B; or it may be formed by letting the bush, B, down into the dome cap.

BB, is a section of a brass bush, on which is formed the valve seat, KK, which should be bored with a spherical rose cutter, having a $1\frac{3}{4}$-inch radius.

CC, is a section of a four-winged valve with chamber c'' formed by a lip projecting $\frac{1}{32}$th of an inch beyond and dropping $\frac{1}{32}$th of an inch below the top of the bush B.

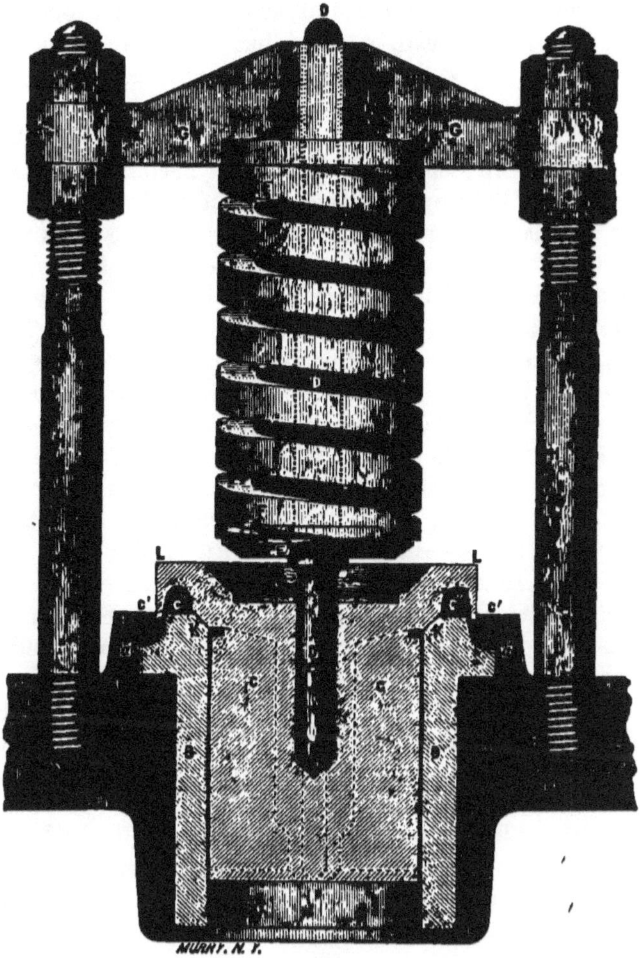

FIG. 59.

DD, is a steel spindle, with the plate E shrunk on, and turned true. That portion of E indicated by the dotted line should fit closely within the spring. The spindle should bear on its point, *d*, and be $\frac{1}{16}$th of an inch loose in the bore. The upper end of the spindle passes through the cross-head, G, $\frac{1}{16}$th of an inch loose, which keeps it in a central position, at the same time allowing it to move freely up and down.

FF, is a helical spring, supported by the plate E, and bearing against the cross-head G. This spring is formed of a bar of

⅝-inch cast steel, 40½ inches long, with the ends tapered 3 inches, and when coiled, faced off at both ends and tempered.

GG, is the cross-head, made of hard brass, and adjusted by the nuts II on the studs HH, which are screwed into the dome cap AA.

JJ, is a section of a washer fitting closely to the spindle, but moving freely thereon, for the purpose of keeping the sparks or cinders from filling up the space surrounding it.

The cut may be regarded as a working drawing for a 2¼-inch valve (the locomotive size), in which the spring is represented as uncompressed. For 100 pounds pressure the spring should be screwed down ⅝ths of an inch, and proportionately for any other pressure. In setting this valve, the steam gauge should be *known* to be accurate, and the connecting pipe *clear*. The cross-bar, GG, may be of any length which circumstances shall require; but the hole through which the spindle passes should line *accurately* with the bore in the bush. Should the pressure be reduced too much before the valve closes, increase the outlet from *c''* to *c'* by turning out the lip of the valve a very little; which relieves the upward presure in *c''*.

It is now evident that, as soon as the steam pressure exceeds what is intended, it will start the valve from its seat, and then it has a surface of the larger diameter nearly equal to LL, to act upon. Thus the additional force necessary to overcome the increased resistance of the spring as it is lifted is obtained, and a free escape for the surplus steam is secured.

Fig. 60 represents the lock-up valve for stationary and steamboat boilers. The same is shown to scale in Pl. XXIV., Fig. 4, where the plan shows half of the valve case cover, CC', and the whole of the nut cover, D', removed. The elevation is half in vertical section. Like letters refer to like parts on all the figures.

A,A' is the inlet from the boiler. B is the valve, winged, as at *b,b'*, with four wings, and resting on its seat at the top of the bush *c,c'*, and enclosing the spindle *d,d'*. S,S' is a thin screw, by which the spring E,E' is compressed to the intended pressure. Steam, lifting the valve by compressing the spring from below, escapes into the space F,F', and thence through an outlet pipe, carried from GG' to any convenient point of discharge. A padlock being locked into the staple *aa'*, cuts off access to the nuts *nn'*, which hold down the valve case cover

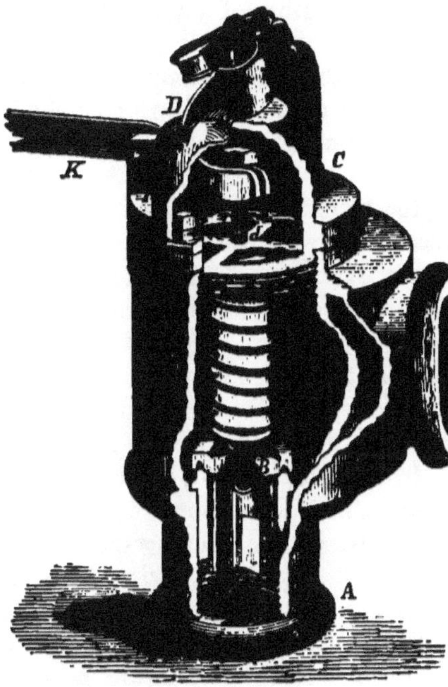

FIG. 60.

CC′, and thus prevents tampering with the valve. A hand lever, K, allows the valve to be lifted by hand, if need be, to ascertain whether it has become stuck to its seat.

Construction.—From Fig. 59, a plan view can be made, by assuming only a few of the measurements. And by careful comparison of the two projections of Pl. XXIV., Fig. 4, with Fig. 60 to serve in place of a model, an elevation or section looking in the direction of the arrow, *p*, may be made.

The lock-up valve is made of various sizes; hence, as Pl. XXIV., Fig. 4, is drawn to scale, to show its *proportions*, measurements may be assigned to it, according to any suitable scale, so as to make the student's drawing larger or smaller than this one.

EXAMPLE XLVI.

A Double-Beat Pump Valve.

Description.—Pl. XXX., Fig. 2, represents such a valve Before describing the valve itself, its place in the pump should be understood, and the nature of its action and relation to adjacent parts.

The magnitude of the great pumping engines forbids the use of any such valves as are found in common hand pumps; and the necessary size of single valves would make them cumbrous.

For example, the cylinder of Brooklyn Pumping Engine No. 3, is of 85 inches diameter, with 10 feet stroke. The pump is directly under the cylinder, of 51½ inches diameter, and 10 feet stroke. As the bucket of this pump ascends, it lifts the water above it, and the pump is filled by water rushing up through a group of 20 foot-valves, commanding inlets of 11½ inches diameter each; and all contained in a chamber of 89½ inches diameter. On the down stroke, a valve of 44 inches diameter opens in the bucket itself, together with 13 others of the size before given, and placed around an annular space surrounding the pump-barrel, and opening into it.

Now, to diminish the work of lifting these valves in the water, they are made as in the figure, where the annular seats, E and F, are called *beats*. There being two of these, the valve is called a double-beat valve. The figure is a plan and sectional elevation, and the parts shown in section line are of uniform section all around the valve. Whence it is plain that the valve is lifted against the vertical pressure of an annular column of water of the *horizontal* width from *c* to *e*. The bridges, H, slide up and down on the stem, G, as a guide. The unshaded parts are six radial arms, or wings, to stiffen the valve wall I. Thus the water escapes, as shown by the arrows *e* and *f*.

The seats E and F may be of wood, or other like material, to avoid heavy concussion in closing the valves.

Though a partial digression, it may be added that this double acting pump is built upon what is called the fly-wheel system. That is, at the opposite end of the working beam—which is 31 feet long, and weighs 30 tons—is a 26 foot fly-wheel in 10 segments, and weighing 36 tons; on a shaft 20 inches diameter. And it is reported, after official trial, that the work done is the raising of 72,000,000 pounds, 1 foot for every 100 pounds of coal consumed, which is equivalent to 81,000,000 pounds raised 1 foot by 112 pounds of coal, the standard of fuel weight given in some English authorities.

This engine is reported as doing about double the duty in proportion to the fuel, that is done by engines Nos. 1 and 2, each having a force pump at each end of the beam, and no fly-wheel. The celebrated Cornish engines are without fly-wheels, and have a steam cylinder at one end of the beam and the pump at the other, with the pump-end of the beam shorter than the other.

We cannot enter into the discussion of the relative merits of these and other forms of pumping engines, but must refer the reader to the Jour. Franklin Institute for 1868–69–70, Bourne's works on the steam engine, etc.

It seems probable that the marvellous duty of 100 to 120 *million* pounds raised 1 foot, per 100 pounds of coal burned, attributed to the Cornish engines, may be partly, at least, owing to the fact that their construction, with suitable boilers also, has been made a specialty for many years, under every stimulus of necessity for economical results, that could well exist.

Construction.—As the many dotted circles in the plan are hardly intelligible before a careful tracing out of their vertical projections, the student should, as a study, draw this valve on a scale of one-*fourth*, or even one-*half*, and then add the *tangent* projecting lines of every circle of the plan.

Example XLVII.

The Cornish Equilibrium Valve.

Description.—This is a steam valve, used on the Cornish engines already referred to. Its conical seats are *mn* and *op*, on which rest the corresponding edges MN and OP, of the valve when that is seated. Otherwise, the figure explains itself.

Construction.—The horizontal section shows in section lines the radial supports of the valve wall. See also the directions in the last problem.

Example XLVIII.

Giffard's Injector.

Description.—This, as well as the last device described, is rather an instrument (30) than a machine, as its moving parts are separately adjustable, yet as it is not *used* separately, but as an accessory to the steam engine, we make a place for it.

Giffard's Injector, Pl. XXIV., Fig. 5, is a contrivance for making the steam power in a boiler, feed the boiler with water, by means of the work developed by the rush of the steam towards the vacuum constantly tending to form where the steam is condensed by contact with the cold water supply; that is, ultimately, by the conversion of the *heat* of the steam into its equivalent in mechanical force at that point.

We will, before explaining the action of the injector, give a general description of the apparatus as improved and made by William Sellers & Co. It is in two parts, joined at *ee* by bolts *dd*. AA is the steam inlet, separated from the water inlet K by the partition *ff*. BB, a screwed rod, operated by the winch *a*, and which, by rising and lowering through the nut *n*, adjusts the opening at C, and hence the flow of steam to form the jet. B is hollow and perforated, so that a little steam will flow through it, even when closed down. D is a tube, combined with a piston, FF, which slides in the space F*b*, as actuated by the overflow of excess of water at E. The delivery tube, G, is attached to, and moves with FD, the piston tube. H is the waste valve operating in the small space at *c*, and allows any overflow from G to escape. This valve is sometimes opened laterally by hand, acting on a screw stem through H. J is the foot valve, here shown wide open; and which, when closed, prevents the return of water from the boiler, when the injector is not working. L is the outlet to the boiler. The pipes connecting at A, K, and L should be of the same diameter as those apertures, and as short and straight as possible. Steam is let on by a cock in the pipe entering at A; and there should be a regulating cock in the water pipe entering at K, in case the water flows in under pressure as from a level above the injector, instead of being lifted from below, as is often done.

Operation.—Screw down the plug B to its lowest point, when steam issuing through the small perforation in it, before described, and from its point, will act to produce a vacuum about D, which will draw water in, as by "suction," at K, and force it along to H and out at I, the valve J being closed by the boiler pressure. The screw plug B being then drawn outward, the flow of steam will increase until the force of the combined steam and water current opens J, and proceeds on into the boiler. If now the water supply be too *great*, there will be an overflow at E, which will accumulate in NN, and drive back the piston tube FFD, and thus narrow the water entrance; see the arrow at D, and properly reduce the water supply. But if the water supply be relatively too *small*, the freer rush of steam through the nozzle AC, will produce a partial vacuum around C, into which, water flowing more forcibly, will drive back FFB, and enter more abundantly. This self-regulating feature is of great value by dispensing with frequent regulations, by hand, of the relative steam and water supplies which are oftener required, the more variable is the boiler pressure.

Theory.—This can be most clearly apprehended in a general way, by reference to an analogous device, in which the steam

FIG. 61.

current from the boiler is represented by a visible solid, shot out by the agency of an elastic medium. See Fig. 61, where A and B are two closed air vessels, fixed on a common support D, and connected by a tube C; and thus charged alike by a condensing syringe attached to either of them. In one end of A, is the valve *e*, opening inward; while B is an air pistol, discharging a ball through a tube, T, when the air pressure is directed against the ball by the springing open of a cock. The ball thus discharged, and striking the valve *e*, will open it, and enter the chamber A.

To understand this, we have only to consider the difference between stationary *pressure* and the *accumulated force*, represented by a moving mass.

Pressure, as 100 pounds to 1 square inch, against an unyielding resistance, and continually neutralized by it, is a unit, as compared with the *living force* developed by the motion of a heavy mass, and representing the sum of the units due to the repeated action of this pressure in producing more and more motion at each instant while the pressure acts.

Thus, the force exerted by the ball upon the valve *e*, represents the sum of all the effects produced upon the ball by the pressure, in all the instants while it is passing through T; while the resistance of the valve *e* is due to the equal pressure exerted on it in only the single instant in which it opens.

Or, in other words, before all parts concerned can come to rest, there must be an equation of *works.* *Work* is the product of weight into space passed over; and the work represented by the motion of the ball, must be given out before the ball can stop. But without motion there can be no work; hence, when the ball encounters the valve, held down by a pressure, less than will resist the whole indenting effect of the ball upon an unyielding mass, motion will be imparted to the valve.

Returning now to the injector, the valve J, unyielding in a direction *from* the boiler, receives the static pressure of the water from the boiler, and is analogous to the valve *e*, above. The issuing steam is analogous to the ball shot from T, and no less so, though of the same form of matter as the agent which propels it, viz., other steam in the boiler, discharging itself by its own elastic force. This discharged steam, reducing its velocity by mixture, in condensation, with many times its weight of water, will still force open the valve J, so long as the quantity of water taken along with it, gives a velocity and living force to the entering jet of combined steam and water, greater than the living force of a jet of water alone issuing from the boiler. And this is just what really occurs.

A point of difference between the injector and the air pistol, and in favor of the former, is that the dilatation of the freely escaping steam, which is further increased by the tendency to a vacuum constantly existing by reason of condensation at its contact with the water, increases its velocity, and hence its living force. The *living force* of a unit of weight of steam at its issue, represents a quantity of *work*, measured by the *weight* of that steam, falling through the *height* to which its particles, considered as projected atoms, would ascend by reason of their velocity of issue from the steam nozzle. And, as before, so long as the additional weight of water drawn into the steam jet does not reduce the velocity, and consequently the living force of the mingled jet to less than that of water alone of the same temperature issuing directly from the boiler, the former will prevail and enter the boiler.

If the steam pressure in the boiler be increased, the steam jet will have a greater velocity, but still more, an increased weight relative to the water portion of the jet, comparing units of each.

Hence it will take up a *less* number of times its own weight of water before reducing the living force of the feed jet to less than that of a water jet alone from the boiler. That is, the *lower* the boiler pressure the *more* effective *relatively* will be the injector. Thus, an injector which will deliver 200 cubic feet of water per hour under a steam pressure of 50 lbs. per square inch, will deliver but 264 cubic feet at 100 lbs. per square inch, and 328 cubic feet at 150 lbs. per square inch.

If, however, the water supplied at K were too hot to permit the condensation of the steam, the injector would cease to act; and, accordingly, the hotter the feed water, the less will be thrown into the boiler. Combining this with the preceding result, greater *heat* of feed water can be purchased by the sacrifice of *quantity;* and this can be done more fully, the less the steam pressure. Hence, directions for using the injector give the maximum advisable temperature for different boiler pressures. Thus, at a steam pressure of 30 lbs. per square inch, the temperature of the feed water may be 130° F.; at 100 lbs. pressure, the feed may be at 110° F., etc.

A full physico-mechanical theory of the injector, treated analytically, and with numerical computations, is given from M. Combes in the Journal of the Franklin Institute for May, 1860, and an extended general explanation and description of its several applications, in the same Journal for July, August, and September, 1868. To these, from which the foregoing was partly taken, the reader is referred for particulars which belong more to physical mechanics than to this work.

Construction.—Pl. XXIV., Fig. 5, shows a section through the axis of the injector, which is here supposed to be horizontal, as is usual in practice. As nearly every section perpendicular to the axis, would show only circles, any number of cross sections can be made by the student.

The figure was prepared from a beautiful model presented by the makers, Wm. Sellers & Co., and differing from the working form only in having a quarter of the case cut out so that the interior was exposed, without taking the instrument to pieces.

The injector is made of various sizes. By taking the diameter of each of the openings A, K, and L as 1 inch, it will serve as a scale for the drawing, in the present example.

CLASS V.—MODULATORS.

124. MODULATORS (39) serve to discontinue motions that would otherwise go on; to maintain motions that would otherwise stop even though the motive power were not withdrawn; to change the relative directions of motions; or the *ratio* of their velocities; and that, suddenly, or gradually.

The term, *modulators*, is preferred to *modifiers*, on account of its less common use and consequent greater precision. If to modify be to adapt a general law to a special case, then to modulate may be to impress a determinate law of change, or a change according to a determinate law, upon a given form of action.

125. Many, if not most, modulators are, like many regulators, *compound organs* or *sub-machines*, and few, if any simple modulators afford valuable drawing exercises, after what have already been presented. This entire class of organs may therefore here be passed over with a few remarks upon the examples mentioned in the column of modulators in the Table I.

A—Point Modulators.

126. An *idler pulley* is merely a small wheel, or roller, mounted in a swinging frame, or in movable bearings; so that it can be pressed against a belt, to tighten it, and give it a firmer hold upon a pair of band wheels, or to throw them in and out of gear if need be.

B—Line Modulators.

127. An *escapement* is a purely mechanical means for converting an oscillating into a rotary motion in one direction by alternately engaging and disengaging peculiarly adjusted arms, with the teeth of a peculiarly designed wheel. It also serves for regularly intermitting, for the purpose of restraining, a motion that might otherwise soon run down. To be of much use or interest as a subject of study, it must be shown in connection with the parts adjacent to it, and is therefore deferred till the chapter on compound organs shall be read.

128. *Band shifters* are very generally moved by an adjust-

able projecting piece called a dog, which is clamped to some moving part of a machine, as the table of an iron planer, so as to operate a jointed lever which carries the shifting arm.

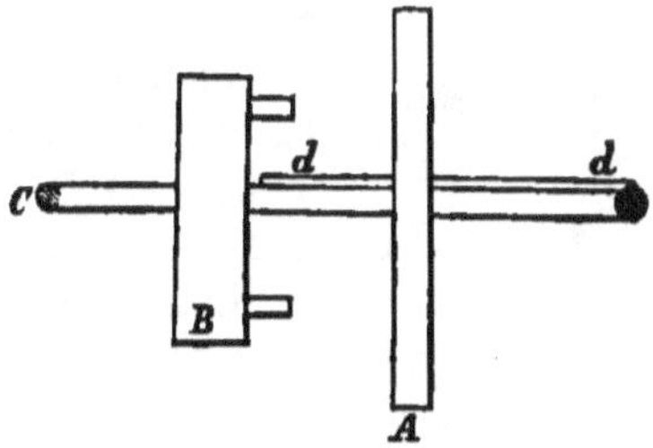

FIG. 62.

129. *Clutches* generally are devices for coupling or uncoupling at pleasure the successive pieces in a line of shafting, or of causing the shaft to communicate its motion or not.

Fig. 62 represents a pin clutch, in which the disk A, is keyed to the shaft, C, by the long feather, *dd*, and is pierced with holes, into which the pins of the disk, B, enter. Now if B revolve loosely on the shaft, or if it be keyed to a shaft, which is divided, between the disks, the shaft C, with A and B, will all revolve together by pressing up A against B till the pins enter the holes in A.

130. *Simple slide rests* are merely stationary supports, adjustable to any position for the cutting tool held by the operator of a hand lathe.

C—SURFACE MODULATORS.

a—Plane Modulators.

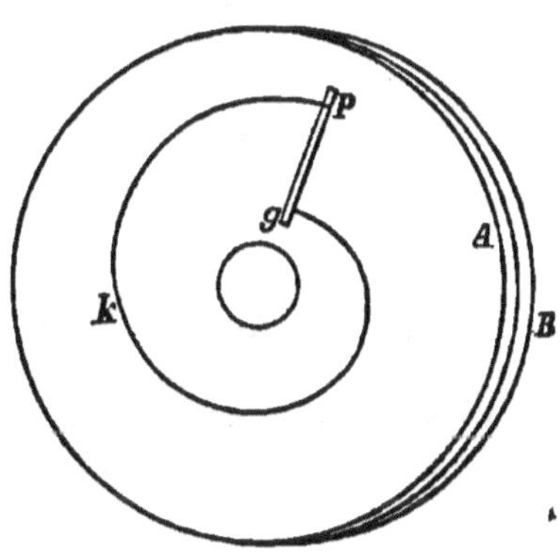

FIG. 63.

131. *Variable crank.* Fig. 63 shows what may be called a plane variable crank. A and B are two disks. A contains the radial slot, *pq*, and B the spiral slot, indicated by a single line, *pkq*. If now, a crank-pin, *p*, be free to move in both slots, it will follow both, and give a variable angular velocity to A, which may be varied by revolving B in the same way at a less speed, in the opposite way, or not at all.

b—Developable Modulators.

FIG. 64.

132. *Speed pulleys* (Fig. 64) are familiar objects in all machine shops, for altering the speed of any given machine according to the work to be done upon it, by shifting the band, *b*, from one to another of the pulleys, all of which revolve on the same shaft. Such band pulleys are arranged in two sets with the largest of one set opposite to, and working with, the smallest of the other, and with the sum of the diameters of any pair that act together constant.

The following theorem and problem express the main points of interest relative to speed pulleys.

THEOREM XXI.

If the band be crossed, it will be equally tight on every pair of opposite pulleys.

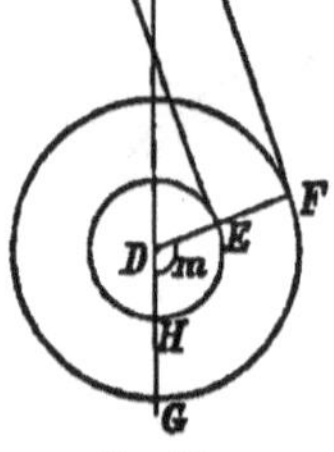

FIG. 65.

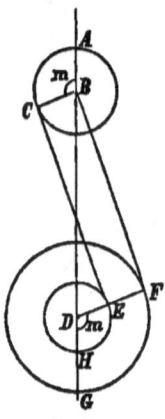

Let BC and DE, Fig. 65, be the radii of a pair of pulleys; and CE, a common tangent, a portion of the crossed band. Make BF parallel to CE, and hence tangent to the circle $DF = DE + CB$.

Now since similar arcs are as their radii, and measure the same angle at the centre, they may be represented by that angle, multiplied by their respective radii. Also note that $ABC = FDG$, and we have

$$HE + EC + CA = EC + DH \cdot m + BC \cdot m$$
$$= EC + DF \cdot m$$
$$= BF + FG.$$

That is, the half length, and hence the whole length, of the band is constant for any pair of pulleys whose added diameters equals DF. Hence, as stated, the band will be of uniform tightness.

Problem X.

To form a set of speed pulleys to give a series of velocity ratios in geometrical progression.

Let the greatest and least diameters of the pulleys be 6 inches and 15 inches. As both sets are alike, the extremes of the series of velocity ratios will be reciprocals of each other, viz.:

$$\frac{6}{15} \text{ and } \frac{15}{6} \text{ or } \frac{2}{5} \text{ and } \frac{5}{2}.$$

Now let there be three intermediate ratios; that is, terms to this series.

The property of a logarithmic spiral, that its equidistant radii are in geometrical progression, enables us to find the required terms graphically.

To construct the spiral, make a figure as follows, Fig. 66,

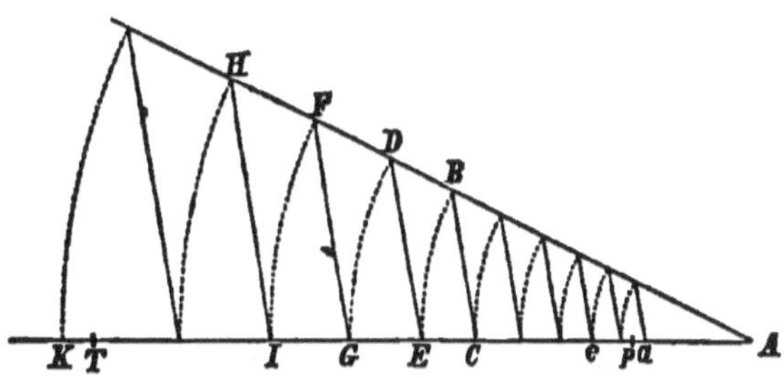

Fig. 66.

where AC : AB : : AE : AD : : AG : AF, etc.;
or AC : AE : : AE : AG : : AG : AI, etc.

Thus AC, AE, AG, etc., are in geometrical progression. Now describe a circle with any convenient radius A*b*, Fig. 67, divide it into any number of equal parts, *bc*, etc. The smaller these parts and the larger the radius A*b*, the more accurate will be the results to be obtained.

Then, beginning on A*b*, for example, make A*a* = A*a*, Fig. 66; AC = AC, Fig. 66; AG = AG, Fig. 66, etc. To avoid confusion, not all the radii from AC to A*b*, Fig. 67, are shown, on which the successive distances from AC to A*a*, Fig. 66, are

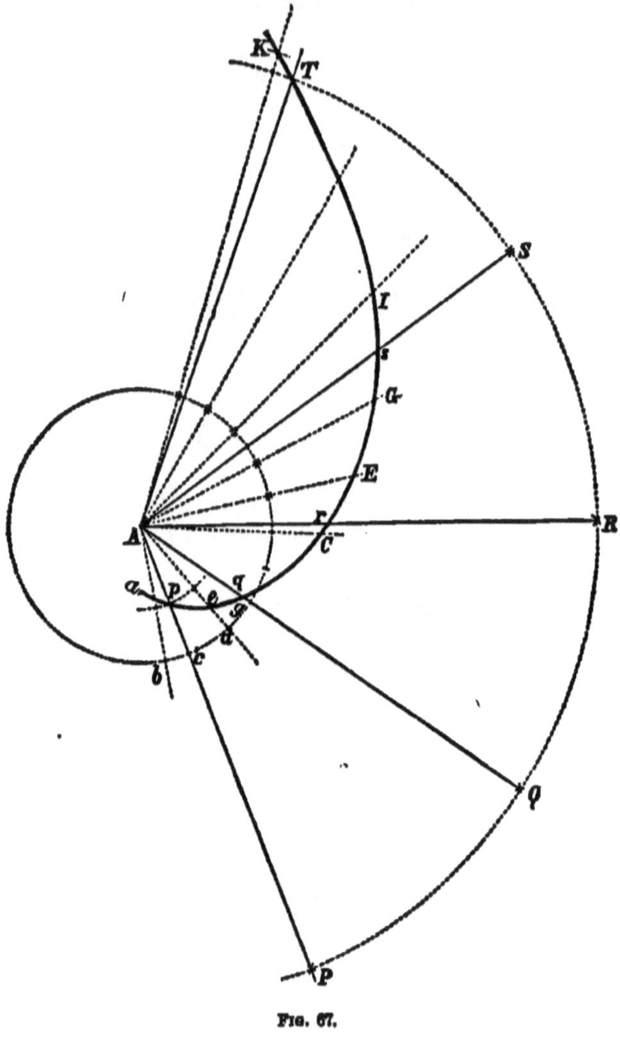

FIG. 67.

laid off. The curve through *a*, *e*, C, E . . . . K will be the required auxiliary spiral.

Now, in the given example, the extreme ratios are $\frac{6}{15}$ and $\frac{15}{6}$, or $\frac{2}{5}$ and $2\frac{1}{2}$. Then take these distances as radii on any convenient scale (a scale of inches was used in this construction), and A as a centre, and describe arcs, intersecting the spiral at *p* and T. Finally, divide PT, where P is on A*p* produced, into four equal parts, and draw the radials AQ, AR, etc., through the points of division, and Aq, Ar, and As will be the three desired intermediate terms of the progression, of which A$p = \frac{2}{5}$, and AT $= 2\frac{1}{2}$, are the extremes.

These distances are the values of the *ratios* of the velocities of the successive pairs of pulleys of the required set, and the sum of the diameters of these pairs is constant. We have then to divide a given line into two parts having a given ratio. How shall this ratio be expressed? *Six* and *fifteen* are the diameters of the extreme pulleys which act together, and to express their ratio, and thus that of their velocities, by a proportion, we have

$$15 : 6 :: 1 : \tfrac{6}{15}, \text{ or } \tfrac{2}{5},$$

and a like proportion would be found for each pair of opposite pulleys. Hence let MN, Fig. 68, *represent* the full size of the constant *sum* of the diameters, $= 21$ ins. That is, in the actual case supposed, MN would be 21 ins. We shall then, as shown

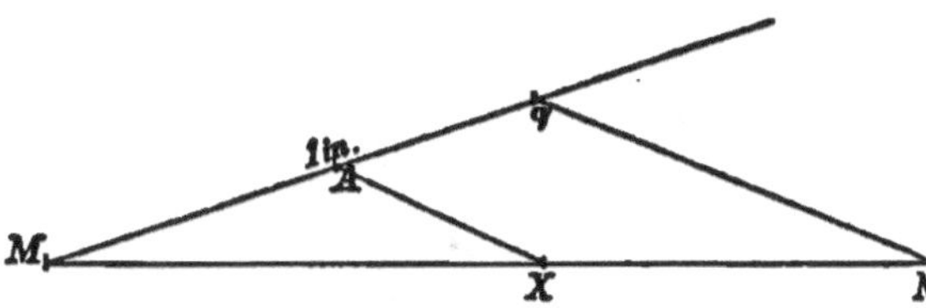

FIG. 68.

in the figure, divide it first into two parts whose ratio shall be Aq, that is two parts, MX and NX, which shall be to each other as 1 and Aq, 1 being 1 unit of the same scale (inches in Fig. 67), from which A$p = \frac{2}{5}$, and AT $= \frac{5}{2}$, were taken. The parts of MN will be the diameters of one pair of pulleys. Next divide MN into two parts, which shall be to each other as 1 and Ar, etc.

By constructing all these figures of large size and in fine lines, on very heavy smooth paper, and with a *foot* instead of an inch for the unit, MA, Fig. 68, of the scale on which Ap, Ar, etc., Fig. 67, are laid off, the results will doubtless be practically as accurate as if found by computation.

133. *Cone pulleys*, Fig. 69, are another device for adjusting velocity ratio; but, by imperceptibly small variations instead of definitely differing ones, as in the use of speed pulleys.

134. *Dead pulleys* revolve loosely on their shafts, so that when

a band is shifted to them the machine driven by that shaft stops.

FIG. 69.

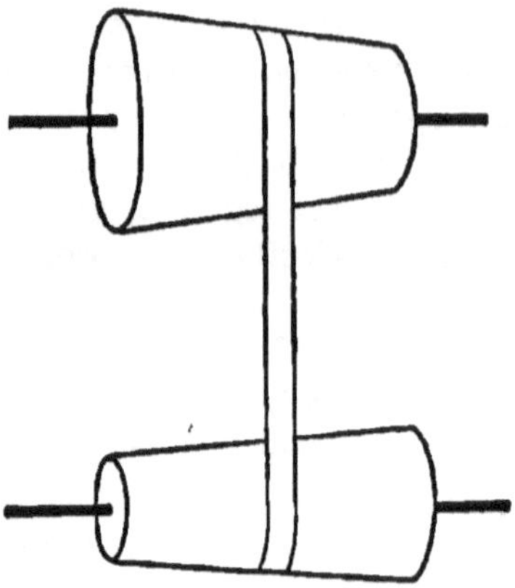

135. *Sectoral motions*, Fig. 70, afford variable velocity ratios by toothed sectors, arranged in parallel planes, so that any two, which act together, are in the same plane and the sum of the radii is constant and equal to OQ, the distance between the parallel axes of motion.

A clearer idea of such motions may be had by analyzing Fig. 70. Let the revolution be in the sense of the arrows, and let the *equal* quadrants M and M′ begin contact at A.

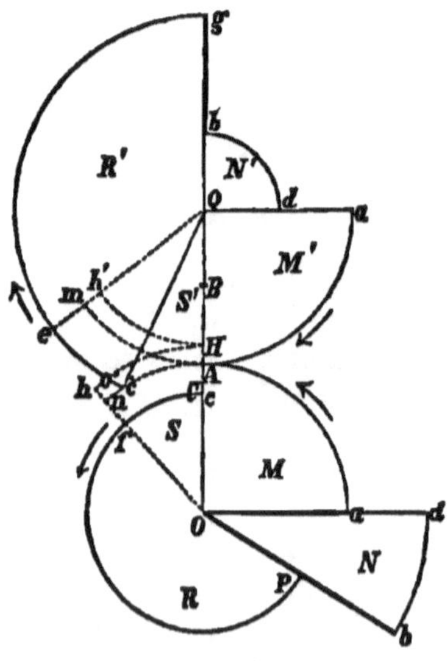

FIG. 70.

Then, after a quarter revolution, *a* and *a* will be together at A. Then the quadrant N′ engages with N, and after another quarter revolution of Q, *b* and *b* will unite at B. Arc *db* = *db* on N′ and the velocity ratio is as Q*d* to O*d*. Then R and R′ act together, with the velocity ratio $\frac{Qc'}{Oc}$ till c and c' coincide at C. O has now made a complete revolution, but Q has made less than one revolution by the angle c'QA, gc' being equal to pfc. To bring the radii QA and OA of the quadrants M and M′ together again at A, the wheel Q alone must revolve through the angle CQc'. For this, an imaginary sector, of radius QO, must act with a like sector of radius = 0 at O. Otherwise, cut off R′ by the radius Qe, and R, by the arc $cf = ec'$, and then divide QO into segments inversely as the arcs Am and An, for the radii of a fourth pair of sectors, Qh'H, and OhH, marked S′ and S, which will cause M and M′ to begin contact again after one complete and simultaneous revolution of each wheel.

136. *Elliptic gears*, when used merely to make a motion quick in some parts and slow in others, and not arbitrarily, but for a special reason, are modulators. They are simply equal and similar toothed elliptic wheels, with teeth formed by a constant generating circle as in circular gearing. But their centres of motion are foci, at a distance apart equal to the transverse axis, and the point of contact will then be on the line of centres. Elliptic gears are sometimes seen in slotting machines, for planing such parts as the insides of connecting rod straps, where the cutting tool, instead of the piece to be planed, travels back and forth. Hence, it is a valuable saving of time to have the tool move faster on its back stroke when it is idle; as the planer table does when *it* travels under a fixed tool. The driving ellipse revolves uniformly, then its longer radii will act with the shorter ones of the follower, giving the latter a high angular velocity, during the retreat of the tool, and contrariwise during the advance of the tool.

c—Warped Modulators.

137. *The helicoidal clutch*, Fig. 71, is merely one in which the coinciding edges, as ab, may be helices, and boundaries of right helicoidal surfaces. They come together easily, and if B be the driving shaft revolving as shown by the arrow, it has fair bearing surfaces, as ac, at right angles to the direction of motion.

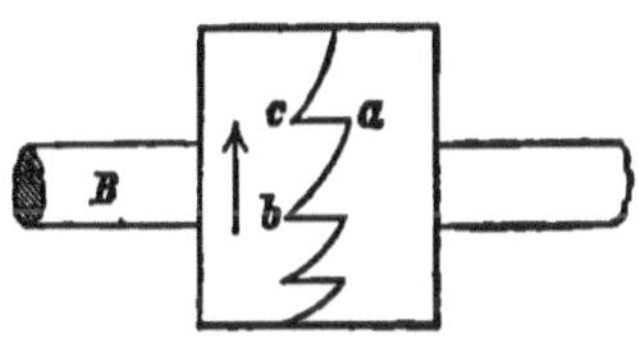

FIG. 71.

d—Double Curved Modulators.

138. *Double curved speed pulleys*, Fig. 72, may have various forms, according to the particular conditions, not necessary to discuss here, which may be imposed upon them. Thus, equal lateral shifts of the belt may be required to produce equal *differences*, or equal *ratios* of successive velocity ratio. But an analytical discussion of such data is a useless refinement, since, unless the band be reduced to a perfectly inextensible and unslipping *line*, the results obtained would not hold in practice.

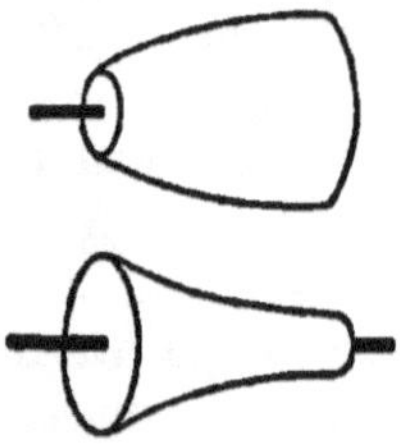

FIG. 72.

CLASS VI.—OPERATORS.

139. *Operators* (39) are those parts in any machine which act directly on the raw material, or to accomplish the work to be done; as the piston or plunger in a pump; the cutter in a wood or iron planer; the shuttle in a loom, etc. They are really tools, operated by a machine instead of by hand; and here it may be noted that when machines are used in making other machines, which in turn are employed in the manufacture or preparation of articles generally used, such as clothes, newspapers, etc., the former are now often called tools (31).

Though few of the examples in the column of *operators* in the General Table need illustration, on the plates, yet several are of interest to mention briefly.

A—Point Operators.

Example XLIX.

Movable Saw Teeth.

140. *Movable saw teeth*, a recent invention, may here be mentioned first, by reason of the magnitude of the lumber manufacture.

Fig. 73 represents a tooth for sawing small logs. Above the line AB it is tempered; the rest is soft enough to bend without breaking. This allows the throat A, Fig. 74, to be made larger than in solid saws, where the whole saw is tempered. A larger clearance is thus afforded for the shavings and dust cut away by the teeth. The perforations allow the teeth to retain their form as they wear out. The teeth are made wider at the point, P, so that the whole surface of the saw will not press against the log.

Fig. 74 shows a fragment of a saw, showing how the teeth are inserted in the plate.

A shows a tooth inserted in a section or piece of the saw plate E. D the rivet, one-half in the saw plate and one-half in the tooth. B, the incision in the saw plate. ha· V rib made on the plate to fit the corresponding groove in the tooth C. The tooth is inserted in the following manner: Place the groove in the back or convex part of the tooth on the V that is fitted to receive it, driving the end which has one-half of the rivet-hole in

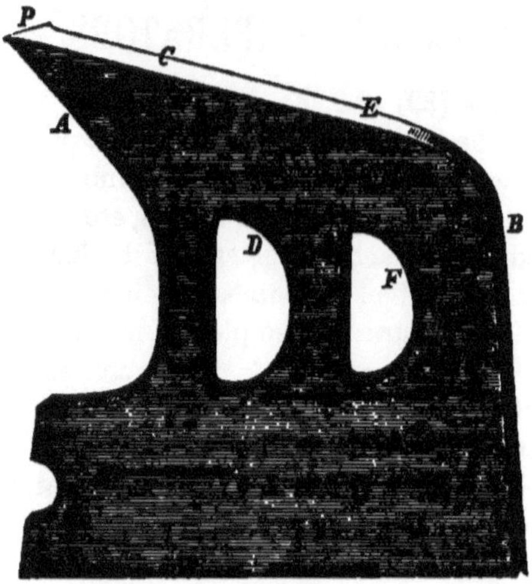

FIG. 73.

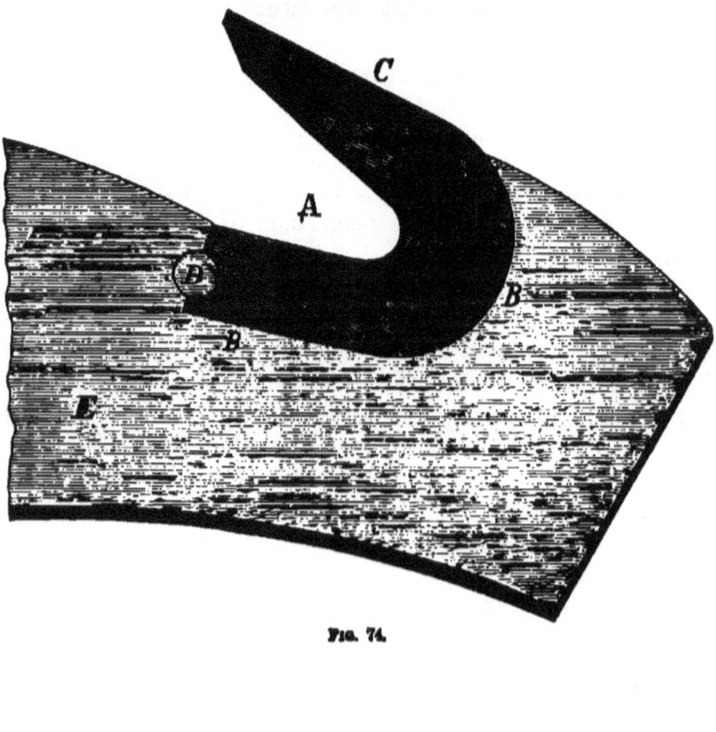

FIG. 74.

it into the saw plate from the side, then put in the rivet and head it up to fill the countersink, and cut or file it off smooth with the plate, when the tooth will be held as firmly in the saw as a solid tooth. The teeth are made thicker at the point than the saw plate, so they can be easily spread to give them the required set, and new teeth can be inserted in case an old one breaks or is torn out, which seldom happens.

FIG. 75.

Fig. 75 shows an arc of one of these saws in operation, and with the under side of the tooth, AB, tangent to a circle, CB, of three-fourths of the radius of the saw, the latter radius being estimated from the points of the teeth.

141. Figs. 76 and 77 represent a recent improvement called

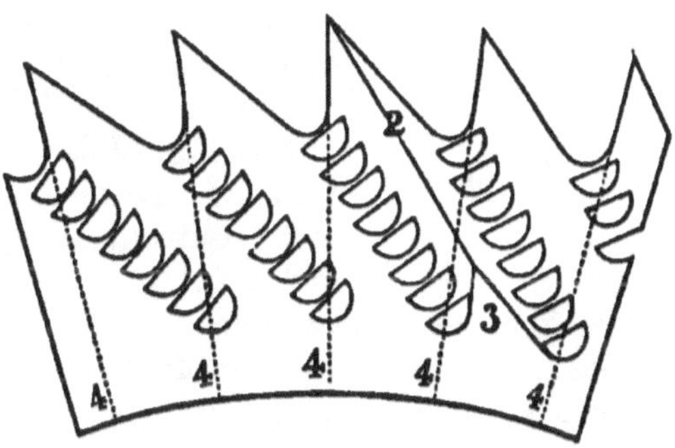

FIG. 76 represents a segment of a cross-cutting Circular Saw, 10 inches in diameter.

perforated saws. They are said to possess many advantages, such as the saving of frequent "gumming," that is cutting or

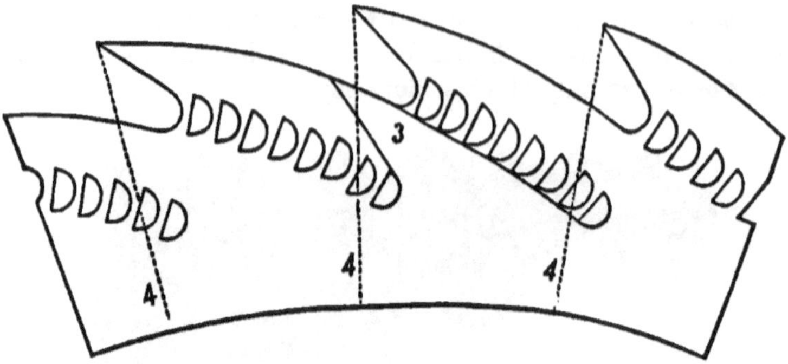

FIG. 77 represents a segment of a Circular Splitting Saw, 16 inches in diameter.

filing out the throats of teeth as they wear away; prevention of expansion of the rim, and fracture at the inner angle of such a pointed throat as may be left by filing. Line 2, Fig. 76, shows the line of wear, and 3 the last tooth before the saw is worn out. The labor of trimming the inner edges left by breaking out the bars between the perforations is small.

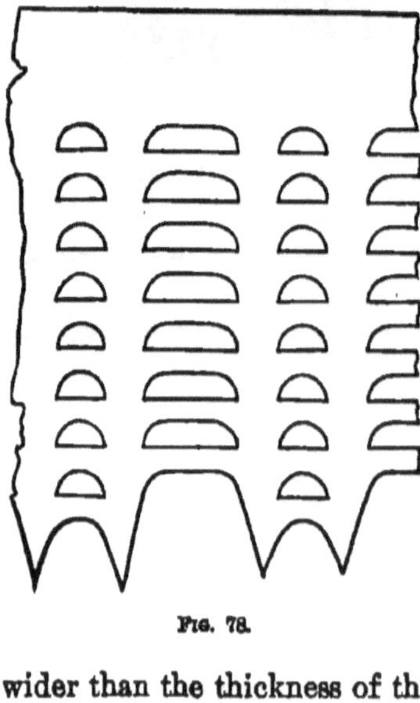

FIG. 78.

Fig. 78 represents a new patent form of cross-cut saws; the short or clearing teeth are about $\frac{1}{16}$ in. shorter than the long or cutting teeth, and act to clear the kerf of saw dust.

142. Fig. 79 shows the usual form of saw teeth, with the teeth bent alternately to right and left, to make the kerf wider than the thickness of the saw plate.

Movable-toothed circular saws are made of all diameters from 8 to 88 inches, and solid perforated ones from 4 to 72 inches diameter.

With a saw properly hung in a true vertical plane, and firm on its axis, or mandrel, and with the log carriage running steadily true on a firm bed, 9,000 feet per minute is pronounced a proper speed for the rim of a circular saw.

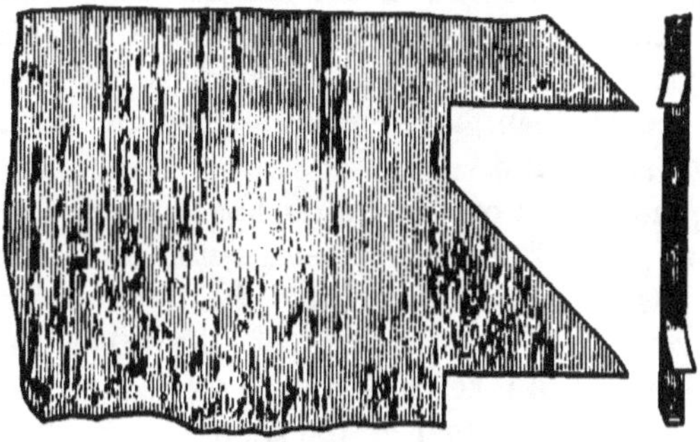

FIG. 79.

EXAMPLE L.

Lyall's Positive Motion Shuttle.

Description.—By a "*positive motion*" is meant one in which the moving piece is taken hold of by that which moves it, so that it *must* move unless there be an obstruction, sufficient to break the connection between the pieces. This is the character of most mechanical movements. On the other hand, the motion of a fly-wheel after all steam has been shut off, that of a valve closed by a weight, or the *natural* flow of water through an open pipe, are passive motions. In other words, *positive* or necessary motions take place because they *must*, while *passive* or *free* motions continue because nothing actively *prevents*.

Now the motion of the shuttle of a loom is of the latter class. Any one who has seen the operation of either hand or power weaving, will remember that the shuttle is thrown through between the warp threads by hand, or by a rod called a picker staff, whose upper or free end gives the shuttle a quick thrust, and then leaves it to find its way across the loom.

[The remainder of this description is condensed from the "Scientific American," and the "American Artisan."]

Notwithstanding the persistence with which these methods of actuating the shuttle have continued, there have always existed serious difficulties, which it was desirable to obviate,

Without entering too minutely into details, we will specify a few important defects, that the important advantages which the device under consideration is destined to accomplish may be understood.

First, the distance to which the shuttle can be thrown with certainty, either by the hand, or by the use of the picker staff, is limited; and the difficulty of weaving wide goods is consequently greater than that of making medium or narrow textures of the same kind.

Second, the motion of the shuttle, having no positive relation to the other parts of the loom, the operator has no control over it during the time it is traversing the distance between the shuttle boxes; and the motions of the other parts, if by accident they should take place a little too soon from any cause, are liable to clash with that of the shuttle. To illustrate this, suppose the shuttle, impelled by too feeble a stroke, to pause in its passage. In a power loom of the ordinary construction the lay, or vibrating frame which drives home the transverse thread to its place, would then make its beat, and either drive the shuttle through the warps, making an extensive breakage, or would spring the dents of the reed, or it would do both.

In weaving fine goods, the bending of the dents cannot be wholly repaired. They cannot be again perfectly straightened without taking the piece out of the loom, and if the piece is woven to the end with such a defect in the reed, a slack woven streak will appear through the entire remainder of the tissue. In order that the shuttle may traverse with certainty, a regular speed must also be maintained, below which it is impossible to work a power loom with success.

Third, the shuttle reaches the shuttle box after its flight in either direction, and comes to rest before the lay makes its beat. An adjustment so perfect that, at this point, the threads of the weft shall be firmly drawn up against the exterior threads of the warp opposite the shuttle, is neccessary to make a perfect selvedge. This perfect adjustment is difficult of attainment, so much so that the character of the selvedge on a piece of linen or silk goods is one of the criterions by which the quality of the fabric is determined.

To remedy these defects entirely, a motion radically different was required. The problem may be enunciated as follows:—

Required to produce a positive and uniform motion in a

shuttle, by means of an external appliance moving exteriorly to the sheds of the warp *without positive connection between the shuttle and the motor through which it receives its motion.* A

FIG. 80.

problem, which the majority of mechanics would have pronounced impossible, had not its possibility been demonstrated by this invention. But the problem is further complicated by

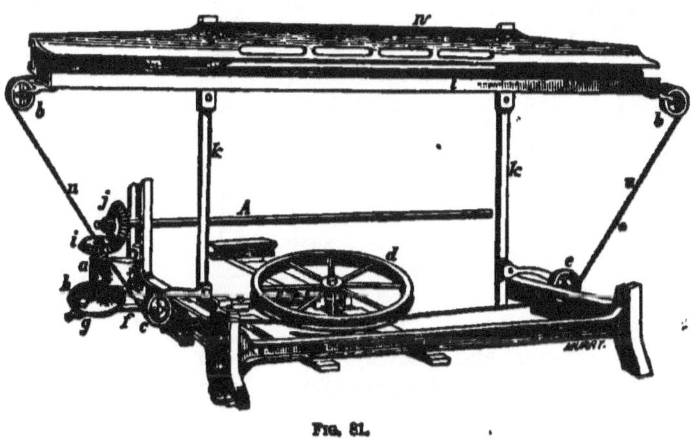

FIG. 81.

another condition which is omitted in the general enunciation; namely, no lateral motion must be imparted to the threads of the warp.

The ingenious method by which these conditions are fulfilled is shown by the aid of the accompanying figures, in which Fig. 80 represents the loom complete; Fig. 81 represents all the mechanism of the positive shuttle motion—the parts not necessary to illustrate this being omitted; Fig. 82 is a front view of the shuttle and its driver; Fig. 83 represents, in transverse

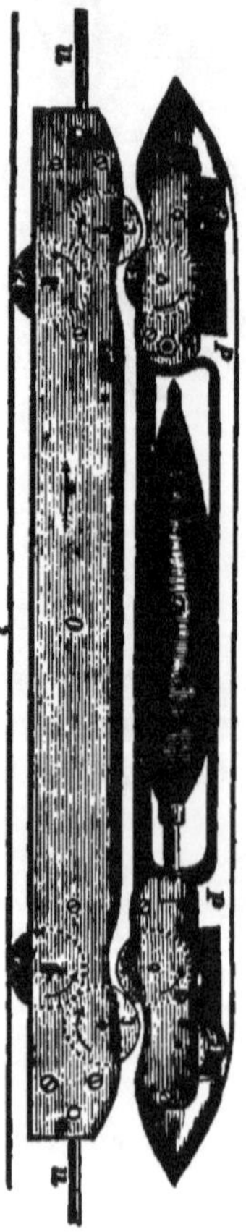

Fig. 82.

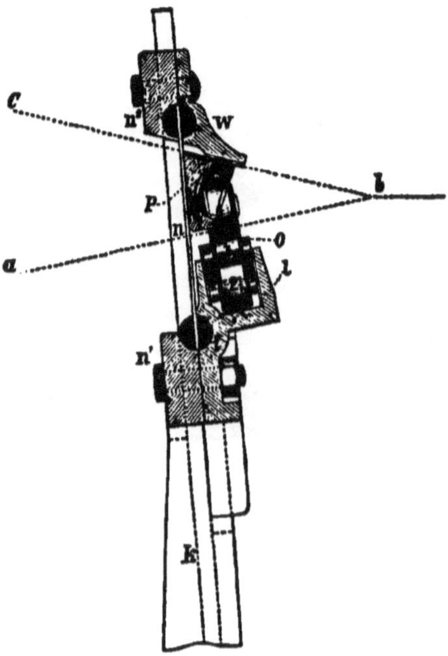

Fig. 83.

section, the lay-reed and raceway with the shuttle and its carriage; and Fig. 84 is a diagram illustrating the action of the shuttle-carriage and shuttle upon and in the warp. In Figs. 82

and 83 it will be seen that the shuttle *p*, is furnished with two rollers, 4, which are supported—the lower part, *ab*, of the shed of the warp intervening while the shuttle passes through the shed—upon two rollers, 3, in the carriage. The carriage has two lower rollers, 2, which run upon the bottom of the lowei rail of the race-way, *l*, and the shuttle has two upper rollers, 5,

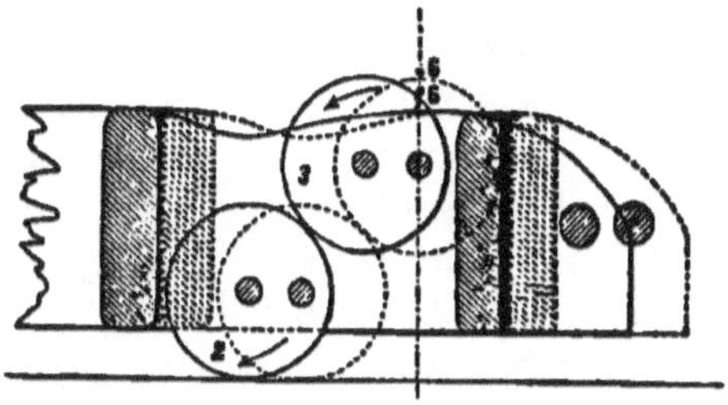

FIG. 84.

which run against the bevelled under side of the upper rail, *w*, of the race-way, which keeps the shuttle in place in front of the reed, *n*.

The lay is carried by the swords, *k*, in the usual manner, and the movement may be produced either by a crank-motion or by a cam; but the inventor perfers the cam, as it enables the movements of the lay and shuttle to be better timed. The shuttle carriage is connected at each end with a band, *u*, Fig. 81, which passes over rollers *b*, at each end of the lay; thence downward and under two rollers, *c*, attached to the lower parts of the swords, and around a horizontal pulley, *d*, near the floor. Attached to this horizontal pulley, there is a pinion which gears with a horizontal sliding-rack, *e*, which receives a reciprocating motion through a pitman, *f*, from a crank-wrist, *g*, carried by a disk, *h*, on the lower end of a vertical shaft working in a stationary box, *a*, attached to the outside of the loom framing. On the upper end of the vertical shaft there is a bevel-gear, *i*, gearing with and deriving motion from a bevel-gear, *j*, on one end of the shaft, A, which carries the cams for operating the lay and those for producing the harness motion, by which alternate warp threads, as *ab* and *cb*, Fig. 83, are alternately raised and depressed, so as to interlace with the cross-thread or woof. The reciprocating motion given to the rack, *e*, by the bevel-gears, *j*, *i*, vertical shaft crank-wrist, *g*, and pitman, *f*, produces, by its action on the pinion, an alternate or reciprocating rotary motion of the pulley, *d*, which by alternately winding and unwinding the band, *u*, on opposite sides, causes the band to move the shuttle carriage back and forth along the lay and under the warp, thereby causing the carriage to carry the shuttle back and forth through the open shed of the warp.

One important feature of the motion is that the carriage carries the shuttle over the intervening lower shed of the warp without the friction which is produced by the fly-shuttle and which tends to break the warp. The manner in which this is effected is illustrated by the diagram, Fig. 84. The carriage has not even the slightest tendency to produce any lateral displacement of the warp yarns. The lower rollers, 2, of the carriage are caused to derive a rotary motion, like that of the wheels of a road carriage, in their passage along the bottom of the raceway, and this motion is imparted, by contact, to the upper rollers, 3, in the opposite direction, as indicated by the arrows on the rollers in Fig. 84, since rollers 2 are in slotted bearings, so that the weight of O is borne by rollers 3. The shuttle is supposed in this figure to be moving to the right. The dots, 6, represent one of the threads of the warp yarn in two positions. The roller, 3, of the carrier first strikes the yarn in the lower position; and as it moves along with the carriage, the latter moving to the right, the upper part of the roller, 3, in contact with the yarn moves just as fast to the left, and so does not tend to carry the warp with it, but merely lifts up the latter to the higher position. The rotary motion of the rollers, 3, is transmitted through the warp to the lower rollers, 4, of the shuttle as the threads of the warp are successively passed between the rollers 3 and 4, with a scarcely perceptible rolling but no rubbing motion.

In the above-described operation, the shuttle being acted upon and controlled by a direct and continuous connection with the motive power, its action is absolutely positive, and is produced with very little power, and without any sudden jerk. The crank-wrist, *g*, is so arranged as to gradually overcome the inertia of the shuttle at starting from one side or the other of the loom, producing an accelerated motion as far as the centre of the lay, and afterwards a gradually slower movement—gradually checking the momentum of the shuttle as it approaches the other side. One great advantage derived from this is that the weft is not subject to sudden pulls in starting, and another is that the shuttle does not rebound, and a tight and even selvedge is sure to be produced. Another advantage of the positive shuttle-motion is that it enables goods of any width to be woven. It is as easy for this motion to carry the shuttle through the widest as through the narrowest warp, and so it enables the widest goods to be produced at the same cost per square yard as the narrowest.

Construction.—These shuttles are of various sizes. The one shown in Figs. 82 and 83 is on a scale of 3 inches to 1 foot. These figures may therefore be drawn to scale by the student.

B—Line Operators.

143. Among linear operators may be placed all cutting edges, whether straight or curved. Among the latter are the helical edges of hay cutters, where the cutting action is in a radial direction, and those of the Ruggle's slate trimming machine where the cutting of a revolving helical edge takes place tangentially to the imaginary cylinder on which the edge is traced.

C—SURFACE OPERATORS.

a—Plane Operators.

Example LI.

Air Pump Bucket of a Marine Engine.

Description.—Pl. XXVII., Fig. 1. represents a vertical section of the bucket, on a scale of 1½ inches to 1 foot. As the bucket descends in the air pump, the water of condensation below, from the well K—A′K′, Pl. VIII., Fig. 1, of the bed plate, lifts the rubber valve through the gridiron bottom, and fills the bucket above the valve. Then, as the bucket rises, being open above, the water in it and the external air close the valve and the water is emptied into the hot well over the pump, by lifting the floating cover of the air pump, which is tightly held down by the external air during the descending stroke.

This pump thus removes water, as well as air, from the con denser, but on the latter account, which is the most important one, it is called an air pump.

Construction.—All the principal parts being circular in plan, a half plan may be added by the student with sufficient accuracy, and the scale may be further reduced to one-*twelfth* or one-*sixteenth*.

b—Developable Operators.

These, in the form mainly of cylindrical or conical bending or shaping rolls, as used by sheet and plate metal workers, or in rolling mills, do not need detailed illustration.

c—Warped Operators.

The Screw Propeller.

Preliminary Remarks.

143. Here we have probably the most important example of operators, in respect to the magnitude, and the scientific, commercial, and social interest of its applications.

The adequate treatment of it, alone, would be enough for a volume, and must detain us longer than that of any other example, though only the elements of its formation and action, treated geometrically, can here be given.

144. Steam vessels have for a long time been propelled by the reaction of water against the radial floats of wheels, one on each side, and usually revolving on the same shaft; placed transversely across the vessel. Only a small arc of the circumference is submerged at once, and to a small depth. Such wheels are called *paddle wheels*.

But, of late years, wheels with arms or blades oblique to the wheel axis, placed at the stern of the vessel, entirely submerged, and revolving in a plane perpendicular to the path of the vessel, have been increasingly employed. These are *screw propellers*. They, and the hulls propelled by them, have been already so far perfected in respect to mutual adaptation, that the days of paddle wheels have been considered as numbered, except for use in waters so shallow that submerged propellers would be inadmissible.

145. The screw propeller acts upon the water in which it is submerged, so that the reaction of the water shall impel the vessel to which the propeller is attached, in the direction of its keel. The vessel is supposed to be free to move under the action of any force which is sufficient to overcome the relatively small resistance of the water, by friction and inertia, to cleavage by the passage of the vessel through it. Finally, as the water in which the screw acts is confined by the all surrounding water, its reaction becomes effective as an impelling force. So much, only, at present, for the general idea of a propeller.

146. The study of the screw propeller is encompassed with much needless obscurity, arising from want of due express recognition of the few simple geometrical facts and relations which belong to it, or are connected with it.

The treatment of it will therefore next be cleared; *first*, to those who have not kept themselves familiar with descriptive geometry, by a brief rehearsal of the great divisions of *surface* and *lines*, and a reference of the propeller surface to its proper geometrical position; and, *second*, to theoretical students by a popular description of it, and precise definitions of its several parts.

Introductory geometrical Principles.

147. The forming of a line, or a surface, by the movement of something simpler, is called its *generation*. The moving magnitude is called the *generatrix;* any fixed *point*, or *line*, which guides this motion is a *directrix;* and a similarly fixed *surface*, is called a *director*.

Now, any line, or surface, may be defined in two ways, *first* by a *description of it*, consisting of the enunciation of any one or more geometrical properties possessed by it; or, *second*, by the method of generating it. We will here define lines and surfaces in both ways.

148. Any *line* is *generated* by the motion of a point. It *consists* of a series of points having but one direction at any one of its points, and but one dimension. Every *surface* is generated by the motion of some line of constant or variable form. It has extension in but two directions at any one of its points. The different positions of the generatrix are *elements* of the line or surface. If *straight lines*, they are called *straight* or rectilinear *elements*. If two elements have none between them, they are *consecutive*.

149. All *lines* are

STRAIGHT, or
CURVED.

A *straight line* is one which has but one direction. It is *generated* by a point which moves without change of direction towards a fixed point, the directrix.

A *curve* changes its direction continually.

150. All CURVES are either

those of two dimensions = *plane curves*, or
" " three " = *curves of double curvature.*

Plane curves are also called *curves of single curvature*, and may well be called *curves of two dimensions.* They are such that all the points of each lie in the same plane; as circles, ellipses, cycloids, letter S, and figure 8 curves, etc.

151. *Curves of double curvature*, may well be called *curves of three dimensions;* to distinguish them, first, from compound curves, like polycentral arch curves; or reverse curves, like the letter S, which are naturally enough said to be of double curvature; also, second, from *double curved surfaces*, with which they are also confounded from similarity of name. The points of a curve of double curvature cannot all lie in the same plane, as in case of the handrail to circular stairs, for example.

The Helix.

152. The only curve of double curvature, which it concerns us here to mention, is the *Helix*, which we will now describe.

If a point simply revolves about a fixed axis, it will generate a circle perpendicular to that axis. If it moves only parallel to that axis, it will generate a straight line parallel to the axis. Combining both motions, the point, by both revolving and ascending—supposing the axis to be vertical—will describe *some form of helix.* Further, if both motions be uniform, the curve will be the *common helix.*

153. The ascension of the helix, while making one revolution about the axis, is its *pitch.*

The *amount* of pitch for one revolution, may thus be the same for all different radial distances of the generatrix from the axis in different helices. But in this case, the less the radius the steeper, evidently, will be the pitch, as circular stairs are steeper at their central part than at their circumference. The pitch may be as great, or as small, as we please to make it, for one revolution. If great, it is termed *coarse*. If small, it is called *fine*.

The more general forms of helix may have a variable *radius*, as well as pitch. Such would lie on conical, spherical, or other surfaces, while the common helix evidently lies on the surface of a cylinder, having the same axis as the helix.

154. All SURFACES are either

RULED, or

DOUBLE CURVED.

A *ruled surface* is any one on which *a straight line can be drawn* in one or more directions. It is *generated* by the movement of a *straight* line.

A *double curved surface*, is one on which no straight line can be drawn, as a sphere, or an egg. It can, therefore, be *generated* only by some curve.

155. All ruled surfaces are either *plane*, or

single curved.

A *plane surface* is one on which a straight line can be drawn in *any* direction. It is *generated* by a straight line, moving parallel to itself upon another straight line; by a straight line moving in any manner upon any two parallel or intersecting straight lines; or by the revolution of a straight line AB, about another, AC, perpendicular to it.

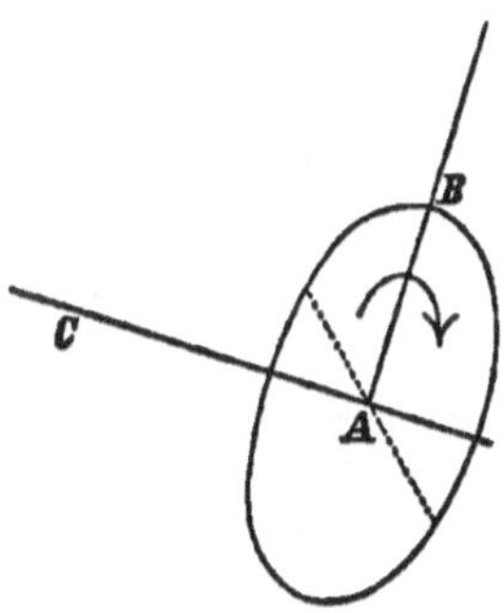

FIG. 88.

156. A *single curved surface*, is such that through *any* point on it, at least *one* straight line can be drawn; as on a cylinder, or cone. It is, therefore, *generated* by a straight line; moving so that its points *generally*, describe curves.

157. All single curved surfaces are either

developable, or

warped.

A *developable surface* is one which, like a cylinder, or cone, can be rolled out upon a plane, so that at each moment it will have a line, or *element* of contact with the plane. Such a surface is, therefore, *generated* by a line whose consecutive positions are *parallel*, as in a cylinder, or *intersect*, as in a cone. That is, *they are in the same plane.*

158. A *warped surface* is one which can have no element of contact with a plane, which, therefore, cannot be made to lie in a plane except by rending its consecutive elements from their proper relative positions. It is, therefore, *generated* by moving a straight line so that its consecutive positions shall not be in the same plane.

159. The plastering under circular stairs; the working surfaces of either square or triangular screw threads, and of propeller blades, are warped. So are those of locomotive cow catchers, railway snow ploughs, Jonval turbine water wheel buckets; and the surface, very easily illustrated by every one for himself, formed by revolving an oblique line, AB, Fig. 86, about a vertical axis, Dd, not in the same plane with AB.

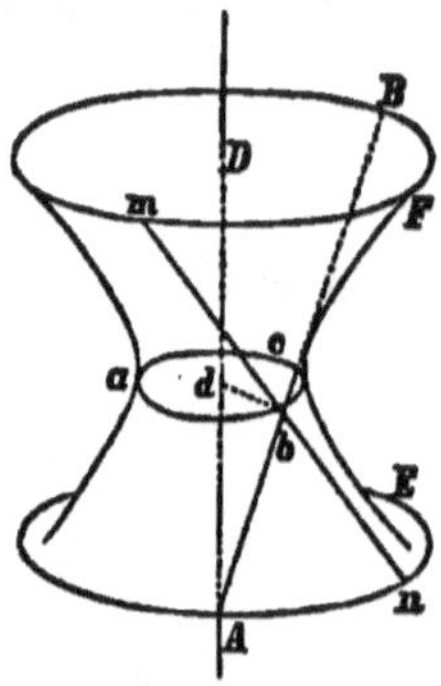

FIG. 86.

This surface is called a *hyperboloid of two united nappes*, because, by examination, any section of it, as EF, made by a plane containing its axis, is found to be a *hyperbola*; of which Dd is the conjugate axis. Dd being vertical, all horizontal sections of the surface would be circles. The least one, *abc*, of these, is called the *gorge*. If its radius *bd* became zero, the hyperboloid would become a cone of two equal nappes, of which *d* would be the common vertex. In geometry these two nappes are considered as making but one complete cone.

160. In the cow catcher, Fig. 87, the horizontal bars AC, *ab*, etc., are *elements*, the vertical bar AB, over the rail, and the oblique bar CD, over the centre of the track are *directrices*, and AC, etc., being all parallel to the ground, the latter is the *director*; in this case, a *plane director*.

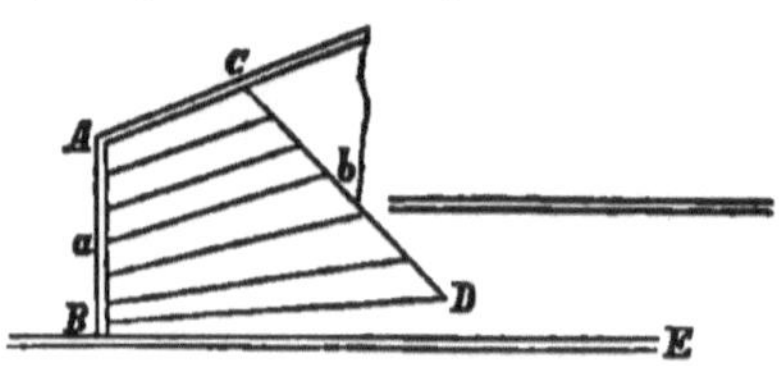

FIG. 87.

This surface is called a *hyperbolic paraboloid*, because all its curved sections are parabolas or hyperbolas.

161. *All* the elements of the *hyperboloid*, Fig. 86, make the

same angle with *any* plane perpendicular to its axis D*d*. Hence such a plane is a *director*, this *simple uniform* relation being of use in constructing the surface. Hence, also, the surface has *two sets of elements*, as AB, and *mn*, inclined equally, but in opposite directions, to the director.

Likewise, in Fig. 87, AC and BD could have been taken as *directrices*, and AB as the generatrix, moving upon AC and BD, and keeping parallel to the vertical plane ABE. None but these two warped surfaces have this property of having two sets of elements. We have dwelt upon them, on account of their simplicity, in order to lead the mind to a conception of warped surfaces generally.

162. Proceeding, now, to the screw surface. If an indefinite line, AB, Fig. 88, revolve, *horizontally*, that is, with no ascension,

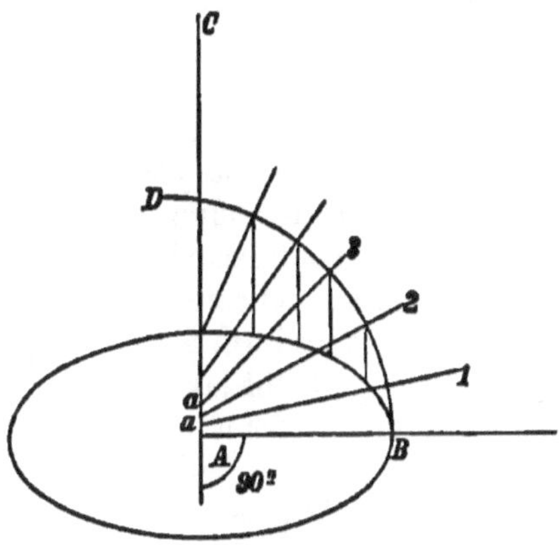

FIG. 88.

about the vertical AC, it will generate a horizontal plane.

If it only move upward, parallel to itself, it will generate a vertical plane.

But if it combine both motions, *both being also uniform*, it will generate "a spiral surface," "a twisted plane," geometrically termed a *right helicoid*, or, more specifically, the *common* right helicone. It is called a *helicoid* because the path of *any* point, as B, of the generatrix, is a *helix* BD (152). It is called a *right* helicoid because all the elements AB, *a*1, etc., are perpendicular to the axis AC. This is the surface of the under side of circular stairs; and the edges of their steps are detached elements of a similar surface. It is also the working surface of square threaded screw, and of the common screw propeller.

163. The height on AC, attained in one complete revolution of AB, is the *pitch* of the helicoid, or screw. The *pitch* is the same for all points of AB, since all ascend equally as stated above. But the circular part of the combined motions of each point is greater and greater as its radial distance from the axis AC increases; so that the "steepness" of the helicoid is greatest at the axis, where it is vertical, and is less and less the further we recede from the axis; just as circular stairs are steepest near the central post about which they wind, and flattest at their outer circumference, as already explained in describing a helix (152).

164. Merely for completeness, we add, Fig. 89, that if we substitute for BA (162) a line, as BD or BE, to revolve uniformly around the axis AC, while it also moves uniformly upward, it will generate an *oblique helicoid*, which is the surface of a triangular threaded screw thread. Thus if FG revolve about AC, the two being parallel, it will generate a cylinder, and the portion B*de*, by both revolving and ascending uniformly, will generate the thread wound upon that cylinder. The *common* helicoids, both right and oblique, thus have uniform pitch.

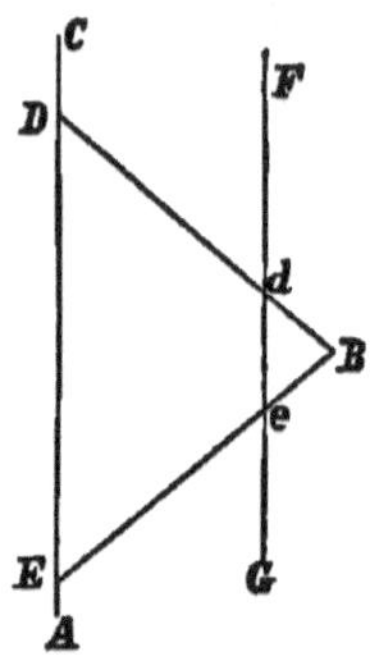

FIG. 89.

165. Returning to Fig. 88, the axis AC, and the helix described by any point of AB, are the two *directrices* of the right helicoid, and as all the elements are perpendicular to AC, they will all be parallel to a plane which is perpendicular to AC.

Such a plane is, then, the *plane director* of the helicoid. When AC is vertical, this plane will be horizontal.

If the generatrix *ascend* faster and faster, while it *revolves* uniformly, the helicoid will have an "expanding pitch." Such a helicoid is one in a more general sense than one having a uniform pitch.

166. From the foregoing geometrical definitions, we will proceed according to (146), with a more practical description of the screw propeller, as actually used.

A screw propeller is *essentially* the same in *form* and *action* as any other screw. See Pl. XXIII. It differs, *first*, in having a much greater pitch in proportion to its diameter; *second*, in being much shorter, so that each of its threads, called *arms* or *blades*, is but a small part of one complete convolution of a thread about the axis, and *third*, in having but *one* helicoidal face to its blades, whereas the threads of a common screw are alike on both sides, to enable them to work in solid nuts.

167. When a screw, working in a stationary nut, acts upon any movable object, both advance, and that equally. In screw propulsion, water, being a yielding medium, effectually takes the place of the nut only to certain extent, by virtue of its weight, incompressibilty, and inertia. So that a screw propeller, revolving in water, actuates a vessel in the same *manner*, though not to the same extent, as a common screw, working in a fixed nut, moves an object placed before it.

Slip.

168. The difference of advance in the two cases (167) is the *slip* of the screw, or, more definitely, the *backward* slip, and may be defined as the distance which a cylinder of water, of the same diameter as the screw, has been pushed backward while propelling the vessel forward. This slip, usually expressed as a per cent. of the pitch, is then called the coefficient of slip.

Thus, suppose a screw vessel to pass over a measured mile, with 200 revolutions of a screw of 30 ft. pitch. The advance of such a screw in a solid nut being 30 ft. for each revolution, would be 6,000 ft. in 200 revolutions. But $176 \times 30 = 5{,}280 = 1$ mile. Hence, in the 200 revolutions, there has been a slip of 24 revolutions $= 720$ ft. That is, the co-efficient of slip$= \frac{720}{6000}$ $= 12$ per cent. of the advance due to the pitch.

169. That radial edge of a blade which first enters the water is the *forward* or *leading* edge. That which last enters is the *after*, *following*, or *trailing* edge.

The central support of the blades is the *hub* or *boss*.

Lateral Slip.

170. In the case of a screw, working in a fixed nut, there is not only no motion of the nut in the direction of the axis of the

screw, but there is no rotation of the nut. But in the case of a screw propeller, the yielding character of the water or fluid nut, in which the screw works, occasions not only a backward slip, but a *lateral* one. That is, the friction of the blades against the water, and their lateral pressure also, communicate a measure of rotary motion to the column of water in which the screw acts.

Thus a centrifugal force is developed which results in a certain amount of lateral dispersion, or tendency to it, in the cylinder of water acted upon by the screw.

171. Screws have been made with the blades curved *backward* to confine the water, and propel it rearward with undimished diameter, equal to that of the screw. But other makers have curved the blades *forward* to secure other supposed desirable results, so that, on the whole, a simple screw is considered by others still, as good a form as any for a propeller.

These modifications will not, therefore, be discussed at present.

Negative Slip.

172. Let S, Fig. 90, be a screw working in a nut, N. Let this nut be capable of sliding backward or forward on the car-

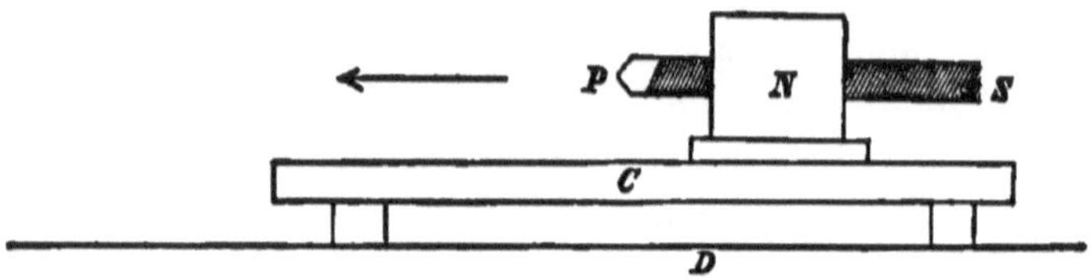

FIG. 90.

riage, C, and let C, likewise, be arranged to slide on the fixed bed, D. Now—

First, let all parts be stationary except the screw, which suppose to have a pitch of 10 ft. One revolution of it will then advance its point, P, 10 ft. This corresponds to the case of a screw *vessel moving in smooth water and with no slip*, under the action of the screw alone, a case probably never fully realized.

Second, while S, in making one revolution, advances 10 ft., let N move backward 2 ft. The point, P, will then have advanced in space, Fig. 91, but (+10)−2 ft. =8 ft. This corresponds to the usual case, where

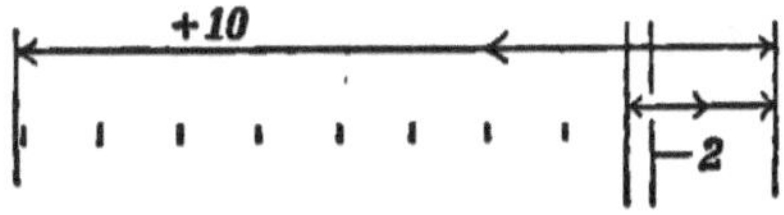

FIG. 91.

the retreat of the nut represents the distance which the water is pressed or displaced backward, in a horizontal cylindrical column of a diameter about equal to that of the screw; while the latter makes one revolution. This 2 ft. then represents the slip; in this case 20 per cent.

Third, the last conditions being still retained, let the carriage C move forward 3 ft. Then the point P will advance in space 10 ft., due to one revolution of the screw, minus 2 ft. due to the retreat of the nut, plus 3 ft., Fig. 92, due to the simultaneous

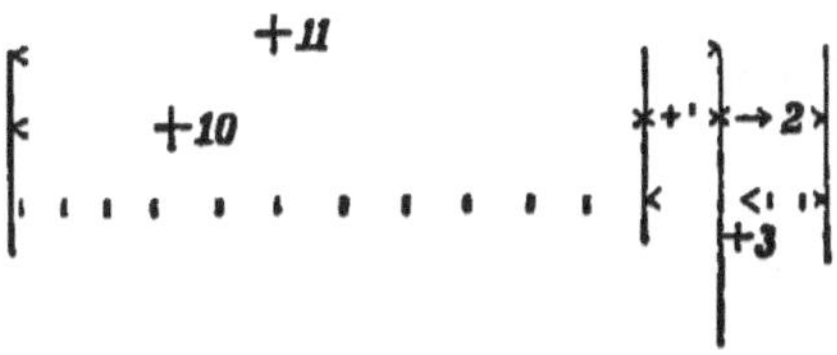

FIG. 92.

advance of the carriage containing the nut $= 10 - 2 + 3 = 11$ ft.

173. This corresponds to the phenomenon, actually observed, of an advance of the vessel, in one revolution of the screw, greater than the pitch of the screw, *i. e.*, greater than the advance of the screw would be, through a solid nut, in one revolution.

This excess is called *negative slip*. And, as in the above model, it would require the same force to turn the screw, whether the carriage moved or not, so a vessel revealing negative slip is not really dragging her screw, but is propelled by it. Negative slip may arise from a vessel's moving in a current which itself moves forward faster than the screw presses the water backward. The carriage, therefore, in the first two cases represents still water, and in the third case either a current, as just described, or a local current created by the friction of the vessel and carried along with it, or by the rush of water from the rearward into the partial vacuum at the stern, when the latter is quite blunt, as some think, or a favoring wind capable of propelling the vessel faster than the screw repels the water, or, finally, *any* cause which advances the ship *more* than the slip retards it.

In the case supposed of the blunt stern, the screw, instead of literally pushing a column of water rearward, is itself pushed along by water rushing against it in the direction of the vessel's motion. *Lateral slip* also tends to the production of a partial vacuum behind the screw, which would further invite this inflow in the line of the axis.

174. The slip, *i. e.*, common, or backward, varies up to 45 per cent., according to the *model* of the vessel; the *ratio* of its immersed section to the area of the circle described by the outermost point of the screw blade; the pitch and length of the screw, etc. Hence quite a strong favoring wind, acting on sails, or a current, may not always overcome the slip but may only reduce it.

Thus, if a screw of 10 feet pitch, while revolving once, advance a vessel but 6 feet in the same time, the slip is 40 per cent. But if a breeze or current would advance the ship 2½ feet in the same time the total advance would be 8½ feet, and the apparent slip 1½ feet, or only 15 per cent. of the pitch.

We thus see the importance of these extraneous conditions of wind and current being the same during the progress of experiments on different vessels and screws.

It is stated to be true, within usual limits, that as the pitch of a series of screws is increased in geometrial progression, their slip will increase in arithmetical progression, for all *lengths* of screw, other things being the same.

But, interesting as the subject is, we must stop somewhere, and therefore refer to the extended work of Bourne on the screw propeller, for further details based on the numberless tabulated experimental results given by him.

Irregular Screws.

175. These are screws, the acting surfaces of whose blades are other than common right helicoids. They are of two general classes, *first*, those of *expanding* pitch, and *second*, those of *variable* pitch.

Screws of merely *expanding* pitch have fixed blades, as in the plain or common screw. Those of *variable* pitch have movable, that is, adjustable blades, whose pitch may be uniform or expanding.

176. Expanding pitch is either *simple* or *compound*. *Simple expanding pitch*, again, is of two kinds. *First*, if it increase in a direction parallel to the axis, as in case of a cylindrical staircase, becoming steeper as it ascends, it is an *axially expanding pitch*. Such pitch increases from the entering to the trailing edge of the blade (169), that is, it is greatest at the latter line. *Second*, to illustrate by the circular stairs again, suppose the edges of the steps to be curved upward or downward, the height of *each step* would be greater or less, respectively, the further it was from the axis; but let all the steps be of equal height, measured at the same distance from the axis, the screw, of similar form, would have a uniform pitch at any one distance from the axis, but the pitch at different distances from the axis would be different. This is *radially expanding pitch*. The generatrix might also be a *straight* line, with radially expanding pitch, moving upward, for example, so that any point of it out of the axis would have a greater pitch than at the axis.

177. *Compound expanding pitch* is a combination of both axial and radial expanding pitch in the same screw. This is illustrated by the stairs, by conceiving the heights of the steps at the axis to increase (or diminish), while the heights at the circumference should increase (or diminish) faster than at the axis. Then the helix, at any given distance from the axis, would be steeper and steeper (or the reverse) as it ascended, and the helix at the circumference would have the greatest (or least) pitch, though the greatest circular path swept over by it might make it less *steep* in both cases than at points nearer the axis.

We will now illustrate in a simple manner the several successive elements of the formation of such forms of helicoidal surface as are employed in the designing of screw propellers, and shall then proceed to the representation of these organs in their practical working forms.

PROBLEM XI.

To Construct a Helix of Uniform Pitch and Radius.

Let the axis of the helix be vertical. Then O—O′O″, Pl. XXVI., Fig. 1, will represent such an axis; and the circle with radius O*a*, will be, at once, the horizontal projection of the helix and of the vertical cylinder with axis O—O′O″, upon which it lies.

The construction is based immediately upon the definition (152). Thus, the division of the circle, *adgk*, into equal parts, indicates the uniform rotary movement of the generating point, *a*—*a*′, around the axis; and the division of O′O″ into equal parts indicates the uniformity of the axial motion.

In the figure, O′O″ is taken as the pitch of the helix, by dividing it into the *same number* of equal parts as are found on the circle *adgk*.

Observe that the equal parts here spoken of, are made so for convenience. All that is necessary is that the parts on O′O″ should be *proportional* to the homologous ones on *adgk*. Also O′O″, the pitch, is of any length, taken at pleasure.

This being understood, let the helix leave the horizontal plane at *a*, in a forward and upward direction. Then *a* will be projected at *a*′, on the ground line; *b*, at *b*′, in the horizontal plane, B*b*′, etc.; as is obvious on inspection of the figure.

The foremost point, *dd*′, of the helix appears in vertical projection on the vertical projection of the axis; and so does the hindmost one, *jj*′.

By joining the points *a*′*b*′*c*′......*g*′....*a*″, as shown, we get the vertical projection of the helix, whose horizontal projection is the circle of radius O*a*.

Remarks.—*a*—The figure is very fully lettered, to assist in apprehending the real position of the several points of the helix.

b—Had the generatrix, *aa*′, proceeded *backward* and upward, *bb*′ would have appeared at *lb*′.

178.—The *drawing* of any irregular surface, as a ship's sides, is often called "laying down its lines."

Every *surface* is composed wholly of *lines* of some kind. *The* lines, distinctively, of a given surface, are those which are most clearly characteristic of it, that is, those which most readily enable the mind to conceive of it from a diagram.

A screw propeller is mainly bounded by a cylinder, having the same axis as the propeller, and by two parallel planes perpendicular to that axis. We will therefore, as a preliminary construction, give the projections of a common right helicoid; first, as represented by its *straight elements;* second, by its *helical* lines.

Note.—All the figures of helices and helicoids, on Pl. XXVI., should be drawn by the student at least twice as large as there shown, and with a full descriptive title for each.

Problem XII.

To construct the projections of so much of a common right helicoid, as is generated by the radius of a vertical cylinder.

Let O—O′O″, Pl. XXVI., Fig. 1, be the axis, and the circle of radius O*a*, the plan of the given cylinder.

The circle O*a*, and the axis, are each equally divided, and the helical directrix (147) *acfk*—*a′c′f′k′a″*, is constructed as in the last problem. Then by the definition of the right helicoid, the horizontal lines *a*O—*a′*O′; *b*O—*b′*B′—*c*O—*c′*C′ - - - - - - *a*O—*a″*O″ represent the helicoidal surface by its straight "*lines*," or elements, and for one complete convolution about the axis O—O′O.″

Remark.—To show the helical band which would be included between two concentric cylinders, draw any circle with centre O, and project its intersections with the radii O*a*, etc., produced if desired, upon the indefinite horizontal lines O′*a′*, B′*b′*, C′*c′*, etc., as before. See also Prob. XI.

Problem XIII.

To construct the projections of a common right helicoid, which is generated by the diameter of a vertical cylinder.

Let O—O′O″, Pl. XXVI., Fig. 2, be the axis, and A*a*—A′*a′*, the initial position of the diameter of the given cylinder; and let O′O″ be the pitch of the helicoid. Divide the circle of radius O*a*, and the pitch O′O″, each equally, as in the last problem. Indefinite horizontal lines, A*a*—A′*a′*; A_1a_1—$A_1'a_1'$ - - - A_8a_8—$A_8'a_8'$, etc., through the divisions on O—O′O″, will then be indefinite elements of the required, helicoid, and they will be limited as required, by considering the line A*a*—A′*a′* as the generatrix, and as revolving in this case, so that as AA′ moves upward and forward, *aa′* moves upward and backward. Hence

AA′ describes the helix $AA_4A_8A_{11}$—$A'A_3'A_6'A_9'A_{11}'$ and aa' describes the opposite helix $aa_3a_6a_{10}$—$a'a_3'a_6'a_{10}'$.

Remark.—Returning to Fig. 1: Any radius will generate a helicoid, thus if Oa—O′a′; Od—O′; Og—O′A′ and Oj—O′, four separate radii of the *base* of the given vertical cylinder be taken as generatrices, their simultaneous helical ascension will generate four helicoids, all like the single one shown in the figure.

Problem XIV.

Having given either projection of any element of a helicoid, to find its other projection.

Let *m*O and *n*O, Pl. XXVI., Fig. 1, be any intermediate elements, taken at pleasure, on a common right helicoid.

There are two ways of finding their vertical projections.

First. Project the points *m* and *n*, upon the vertical projection of the helix containing them, as at *m′* and *n′*, and the horizontal lines *m′o′* and *n′v′* will be the required vertical projections. *Conversely*, in the same manner, having *m′o′* given, project down *m′* at *m*, in the horizonal projection of the helix, and *m*O will be the required horizontal projection of *m′o.′*

Second. From the definition of the helicoid, make O′*o* a fourth proportional to the circumference *adga*; the arc *am*, and the pitch O′O,″ and *o′m′* will then be the vertical projection of O*m*.

Problem XV.

To represent a common right helicoid by its helical "lines."

Let the concentric circles of radii O*a*, Oa_1 . . . . . . Oa_{11}, Pl. XXVI., Fig. 3, be the horizontal projections of a series of concentric helices, all on the same helicoidal surface, whose generatrix is O*a*—O′*a′*. As the required surface is a right helicoid, all these points of the different helices, which are on the same radius, are at the same height above the horizontal plane, the axis O—O′O″ being vertical, as before.

Hence, having taken any distance, O′O″, for the pitch, and having divided it, and the circles of the plan, each, into the same number of equal parts, project all the points of O*a* into O′*a*′; those of O*b* into B′*b*′, etc., and join the joints as *a*′*b*′*c*′*f*′*h*′ and $a_1'd'f_1'h_1'$, etc., of the separate helices as shown. This will give the vertical projections of the helices, whose horizontal projections are the concentric circles with their centre at O. The two projections of these helices, then represent to the eye the helicoidal surface.

Here observe, carefully, that the *steepness* of any given helix is not to be confounded with its *pitch*. All the helices just described have the same *pitch*, O′O″, but their *steepness* decreases from the axis, where it is greatest, being vertical, to the helix whose radius is infinite, where it would be least, being, sensibly, = 0, for the height O′O″ is as nothing compared with a circumference of infinite radius (163).

PROBLEM XVI.

To construct the "lines" of a helicoid, made by its intersection with any plane parallel to its axis.

Let PQ, Pl. XXVI., Figs. 1 and 2, be the horizontal traces of vertical planes cutting the helicoids. Since these planes are vertical, their horizontal traces, PQ, are also the horizontal projections of all points and lines in them. Hence *d*‴*qst* is at once the horizontal projection of the required intersection in Fig. 1; and P*mr*Q, in Fig. 2. In both figures, project up the points *d*‴,*p*,*q*, etc.; *m*,*n*, etc., upon the vertical projections of the elements O*d*, O*e*, etc., and OA_1, OA_2, etc., upon which they are found.

We thus find on Fig. 1, *d*′*p*′ . . . *t*′, and on Fig. 2, *m*′*n*′*p*′*q*′, for the vertical projections of the required intersections.

The elements O*m*—*o*′*m*′ and O*n*—*v*′*n*′, Fig. 1 (See Prob. XIV.), are parallel to the plane PQ, and hence never intersect it. Likewise, the more nearly parallel an element, as O*c*—C′*c*′, is to PQ, the further from the axis will it intersect PQ. That is, those points of the curve *d*′*s*′*t*′, which are in the elements *om*′ and *v*′*n*,′ are infinitely distant from the axis. In other words, this curve will be tangent to those elements at an infinite distance; hence the latter are said to be asymtotes to the curve *d*′*s*′*t*′*u*′.

It is now evident that such lines as the one just described are quite inferior to the radial and helical ones in clearly representing the form of the helicoidal surface.

Also that if we trace any kind of line on *either* projection of the helicoid, its *other* projection would be found by noting its intersections with the elements, or helices, in the projection on which the line is traced, and by projecting these points into the *other* projections of the same elements or helices, and joining them.

Problem XVII.

Having given either projection of any point upon a helicoid, to find its other projection.

We will describe the general, and two particular methods of solving this problem.

The General Method.—Construct *any* line upon the helicoidal surface and passing through the supposed point. The required projection of this point will be on the like projection of this auxiliary line. Thus, in Pl. XXVI., Fig. 1, if the horizontal projection of the point were given anywhere on PQ, its vertical projection would be found by projecting it upward upon $d'p'q't'$.

Particular methods.—First.—Let m, Pl. XXVI., Fig. 2, be the given projection. Then, by Prob. XIV., construct the projections, Om—$o'm'$, of an *element* through it, and m' will be on $o'm'$. Likewise m', if given, would be projected upon Om, to find m.

Second.—Construct the *helix* containing the supposed point, and the required projection of the point will be on the like projection of the helix. Thus, having given k, Pl. XXVI., Fig. 3, construct the helix, $d_s f_s$—$d'f'_s$ through it, and k' will be projected from k upon $d'f'_s$.

Problem XVIII.

To develope one or more given helices.

When, as in Pl. XXVI., Fig. 4, we have among straight lines the relation

$$AM : ME :: AH : HK$$

the line AK, containing the extremities of the parallels, ME, HK, etc., is straight.

Now with this the helix agrees, for the distances *ab*, *ac*, etc., from *a*, Fig. 3, on the base of the cylinder on which the helix lies, being proportional to the heights of the corresponding points *b'*, *c'*, etc., of the helix above that base, it follows, that when the convex surface of the cylinder is unrolled, and made flat, the base will become a straight line as AG, Fig. 4, the successive heights will be parallel as at ME and HK, and the development of the helix will be straight, as at AF.

Hence, to develope a half convolution *adf—a'd'f'* of the helix, make AG equal to the semi-circle *adf*, and GF=A*f'*, and AF will be the required development.

Again: as the common helix has, by its definition, a uniform angle of inclination to the horizontal plane, the parts *ab*, *bc*, etc., being equal in horizontal projection, are equal in space also. Hence the same parts are equal in development. That is AB, BC, CD, etc., Fig. 4, are the equal developments of the equal parts *ab—a'b'*, *bc—b'c'*, etc., of the helix.

Making OA_1=NF_1,=the quadrant a_1d_1, and drawing A_1F_1, this line is the development of $a_1d_1f_1—a_1'd'f_1'$.

In a precisely similar manner, all the helices of Fig. 3 are developed at A_1F_1, etc., in Fig. 4, and all intersect at D.

Instead of laying off a quadrant each side of O, Fig. 4, a semi-circumference for each helix might have been laid off from A, making *that* the common point for all the developments.

Problem XIX.

From the circular projection, and development of a helix, to construct its spiral projection.

Let *adf*B, Pl. XXVI., Fig. 3, be the given circular projection, and AF, Fig. 4, be the given development of a half convolution of the same helix.

Divide *adf* and AF, each into the same number of equal parts. Then, by the properties of the helix already explained, it is clear that any point, as *c'*, of the vertical projection, is at the intersection of a perpendicular to the ground line A*a'*, from *c*, with a parallel to the ground line from C.

Helices are often constructed in this way, especially if the angle of inclination FAG be given. See also Pl. XXXI., Fig. 3.

PROBLEM XX.

To construct a helicoid of axial expanding pitch, by means of its helical lines.

First Illustration.—Let O—O'O'', Pl. XXVI., Fig. 5, be the axis of such a helicoid, and let OA—O'A' be the initial position of its generatrix, and let the successive ascents of this line, corresponding to the equal angular motions AB, BC, etc., be in *arithmetical* progression. Thus, on a small figure like this, let Mm=4-50ths of an inch; mo=5-50ths; or=6-50ths, etc., and draw horizontal lines through these points. Then project up A at A'; B at B', on the line through *m*; C at C', on the line through *o*, etc., until KK', on the line through M', is reached. The concentric helix, beginning at LL', is projected upon the same lines. Thus we find two helices of the same axial expanding pitch, the successive ascents of whose generatrix have equal differences. The helical band A'L'—K'L'' is thus one of axial expanding pitch, which obviously becomes steeper as it ascends.

Second Illustration.—Let the generatrix, O*a*—O'*a'*, in the same figure, 5, ascend so that its successive rises, N*n*, *nk*, *kp*, etc., shall be in *geometrical* progression. In the figure, N*n*= 4-50ths of an inch, and the ratio was *five-fourths.* Hence the spaces, from N upward, have a constant *ratio* instead of a constant *difference*, as on MM'. Drawing horizontal lines through the points *n*, *k*, etc., project *a* at *a'*, *b* at *b'* on the horizontal line through *n*; *c* at *c'* on *kc'*, etc. D' and *d'* sensibly coincide, because, in using so small distances on MM' and NN', the difference between the two progressions does not so soon appear, though above D, it soon becomes apparent, and the new helicoid *ad-g-si*—*a'g'-s'i'* is steeper at *i'*, the *eighth* point from N, than the former one was at K', the *tenth* point above M.

PROBLEM XXI.

To develope the four helices last drawn.

For the first, or arithmetical helicoid, let the vertical element at *d* of the cylinder on which lies the helix, Pl. XXVI., Fig. 5, be placed in the paper at *d'd''*, Fig. 6. Then make *d'a*=*d'*A= the arc *da.* or *d*A, Fig. 5, and let it be similarly divided on both figures. *d'c*=*dc* in Fig. 5, A*f*=A*f*, etc. Then make vertical lines at those points, and project *g'*, Fig. 5, on A*g*, at *g*, Fig. 6; *f'*, Fig. 5, on *ff'*, Fig. 6, and so on, and the curve, *ag*, will be the development of the expanding helix *ag*—*a'g'*.

In like manner are found *qdt*, the development of *qt*—*q'*D'*t'*; AG, the development of AG—A'G', and L*l*, the development of L*l*—L'*l'* in Fig. 5. The two latter helices, being on the back half of the cylinder, *d'd''* represents, for them, the vertical element at D, each way from which the cylinder is developed into the paper.

Problem XXII.

To make the projections of a helicoid of radially expanding pitch.

Pl. XXVI., Fig. 7, where, for convenience, the plan is over the elevation. Here the pitch is proportional to the distance from the axis, O—O'O'' and the initial position of the generatrix is OA—O'A', which after a helical semi-revolution, comes to the position OC—B'C', (176) where B', on CB', happens to coincide with B' as the vertical projection of B, a point in front of the axis, on the outermost helix.

Then C'J is the pitch of the outer helix, from AA', in a half revolution; G'K, that of the helix from EE'; *c'*L, that of the helix from *aa'*; *g'*M that of the helix from *ee'*, and O'B' that of the axis, considered as the inmost helix; all, in a half revolution. Then construct each of these helices, by dividing the half pitch of each into as many equal parts as are in the horizontal projection of its half revolution. Thus, C'J and ABC, are, in the figure, each divided into six equal parts.

Observe that, as shown by this example, the several helices do not cross the axis, in vertical projection, at the same point, as they do in all right helicoids, Figs. 1, 2 and 3, whether the pitch be uniform or expanding axially.

Variations from this figure, which the student should now make for himself, on a large scale, are, to make the generatrix B'C', a *constant curve*, as AB or AC, Fig. 8, all points of which shall have the *same uniform pitch*. Also to make the pitch at each point of B'C' greater or less than proportional to its distance from the axis. The *series* of distances as J*p*, *pq*, etc., upward from O', M, L, K, and J, will then each be in arithmetical or geometrical progression, as at MM′ or at NN′, in Fig 5, but with a larger or smaller common difference, or common ratio, the farther each is from the axis.

In Fig. 7, the different pitches O′B′, Mg′, Lc′, etc., are in arithmetical projection, they having equal differences.

To obtain pitches in geometrical projection let OA and OB, Pl. XXVI., Fig. 9, be two given lines. Then OC is a third proportional to OA and OB, and by a similar construction, joining AC and drawing a parallel BD, we find OD, a fourth proportional to OA, OB, and OC. Likewise we can find a fourth proportional to OB, OC, and OD, and so on, forming a geometrical progression. The curve, AD, Fig. 10, joining the outer points of these lines, when they are placed as equidistant radii, will be a logarithmic spiral; all equidistant radii of which are terms in a geometrical progression. We can then substitute such radii for JC′, KG′, etc., in Fig. 7.

If AB, Fig. 8, revolve uniformly around CB as an axis, and so that all its points move with the same uniform velocity, *parallel* to CB, it will generate a screw of *uniform pitch*, with a curved generatrix.

If all points of AB move *equally faster and faster* in the *direction of the axis*, while still revolving uniformly, AB will generate a screw with *axially expanding* pitch, but uniform radial pitch. If, again, all points of AB have an accelerated motion, but if the velocity of each point increases faster, the further it is from the axis, the screw has then a *radially expanding* pitch, and AB will be of *variable form*, becoming *more curved* as it proceeds.

In all these cases, the several concentric helices will cross the axis CB (in projection) at different points, as in Fig. 7; in the first two cases, merely because the generatrix is curved; in the last, for the additional reason that the pitch expands radially.

The student will learn more, essentially, about various propellers, by carefully constructing the last three cases, than by drawing the propellers themselves, without such separate drill on their essential differences.

Problem XXIII.

To develope the helices shown in the last problem.

In Pl. XXVI., Fig. 11, A′J = ABC in the plan, Fig. 7, and

JC = JC′ in the elevation of the latter figure. Hence CA′ is the development of the half turn ABC—A′B′C′, of the outer helix. Likewise, KE = semicircle EFG, and GK = G′K from the elevation. Then GE is the development of the helical arc EFG—E′G′. From this description the rest of the development can be drawn.

179. The preceding general problems furnish all the means necessary for showing how to construct the projections of the acting faces of the blades of screw propellers, these faces always being portions of helicoids of some kind. The term *pitch*, is applied to the propeller in the *same sense*, and with the same variations, as to the helicoids already explained.

The *backs* of the blades are the surfaces which result from giving such thicknesses to the blade at different distances from its axis, as are found to be necessary.

Problem XXIV.

To construct the projections of the acting faces of a four-bladed common screw propeller.

The axis of a propeller in its working position being horizontal, the screw would naturally be shown in two elevations. Therefore let the plan in Pl. XXVI.; Fig. 3, serve as one of two elevations for Fig. 12, and let the full circle, with radius O*m*, be the hub of the screw. Also let DBO, MO*b*, *c*O*e*, and KOE, be the four blades, as seen in end elevation. Construct, by Problem XV., the indefinite helices, beginning at *c*, *c*′ ; K,K′ ; D, D′ and M, M′ ; whose pitch=D′D′′′′. These are respectively, in side elevation *c*′*d*′*e*′K′′E′′ . . . . ; K′E′B′′*a*′′ ; D′*g*′B′*b*′′ *d*′′′D′′′′ ; and M′*b*′*d*′′*a*′′. Make D′B′, the length of the wheel, so that B, B′ and D, D′ shall be equidistant from vertical planes at *dg* and *d*′*g*′. Then, having projected, as in other cases, the concentric helices, as *nmr*, of intersection with the hub, we have *nmr*,B*g*D—*n*′*m*′*n*′′, D′*g*′B′, for the helicoidal face of one blade, *os*M*b*—*o*′*s*′M′*b*′ for that of another, etc.

If this screw revolve in the direction of the arrows, the reaction of the water, supposed to be at the right of it, would propel the vessel, whose hull would be at the left of it, in the direction of arrow N.

M″b″ is identical in *form* with M′b′, but is an arc of the same helix with D′g′B′. Hence the right-hand wheel, D″d″, correctly represents the left-hand one, D′d′, after the latter has made just a quarter revolution.

Example LII.

To represent variously limited propeller blades, with their concentric and radial sections.

Pl. XXVII.; Figs. 2–7. Fig. 2 is a common end elevation of the three screws Figs. 3, 4, and 5; in all of which the pitch is the same, for convenience of comparison.

1°.—Figs. 2 and 3 are two elevations of a common screw, where ABCD—A′B′E′C′D′ is identical in character with *f*F′G′ H′h′g′, Pl. XXIII.; Fig. 1, half of whose horizontal projection is *f*FG*g*. Only that in the latter figure, the difference, *g*G, between the radii of the inner and outer cylinders of the screw, is much less than *r*E—*r*′E′, Pl. XXVII., the difference of the radii of the hub, and the cylinder of radius OC, containing the outer helical edge, BEC—B′E′C′, of a blade.

NP—N′P′, etc., are concentric helices of the blade AC—A′C′, which is seen flatwise in both projections. Likewise KL—K′L′, FG—F′G′, etc., are concentric helices of the blade FH—F′J′H′G′, seen edgewise in Fig. 3. We see from this that it is not a very material error to make J′H′, K′L′, etc., *straight;* but it is quite wrong to make them in the same *direction* as their developments, *rq*′, K_1L_1, etc., Fig. 6, by the difference *q*′*h*″, Fig. 3 (for J′H′) between J′*h*″, the projection of J′H′ on the horizontal O′B′, and J′*q*=*rq* Fig. 6. The error of representing NP—N′P′, etc., as straight in the edge view, Fig. 3, of the wheel, is now apparent. Such lines may be *used* only as containing the points, N, N′ and P, P′ on the edges of the blade, and then no practical error will result. But to view them as representing the helices, as N*t*P—N′*t*′P is entirely wrong.

Another error consists in speaking of a certain artificial figure, next described, as the "development" of a blade, though no warped surface (158) can be developed. Such a figure is made by extending the chord KL, for example, till equal to K′L′ (K_1L_1), and drawing a curve through its new extremities and R. The conventional figure, formed by connecting the like extremities of the different chords, and which is broader than the projection, FH, of the blade, is sometimes called its "development;" and the curves having the extended chords are called the developments of the helices KRL—K'L'. Yet this figure represents nothing real, and is of no particular use.

2°. *Assumed form of the side view of a blade.*—Propeller blades may be limited otherwise than by parallel planes, as H'C' and *h''*B', Fig. 3. And, as the form and size of the projection, A'B'C'D', shows, the assumed form of the blade outline is made on that figure, and thence constructed by projection, on Fig. 2.

Fig. 4 shows a blade rounded at the outer corners and partly straight as at *f'g'*, and limited on the side by the line *d'a'*. The end elevation shown in a light line on Fig. 2, may be found by projecting points as *c'*, on given radii, over to the other projection of the same radius as at *c*. Otherwise, by projecting points, as *d'* or *h'*, on any given helix, as *d'i'g'* or *k'j'l'*, respectively, upon the other projection of that helix, as on *dig*, at *d*, or *kjl* at *h*. The points, *ff'* and *gg'* are taken on radii and projected upon *fb* and *gn*.

To construct the blade, G''H''*e''*, Fig. 4, we have only to construct its helices in the same way as in Fig. 3, and then to project points as *e'd'* or *s'* on any helix of the blade BB''DD'' upon the corresponding helix of the given blade. Thus *e'* on the outer helix *k'j'l'*, is projected at *e''*, on the similar helix *mm''*—*m'm'''*.

3°. *Propellers with blades "bent back."*—This rather obscure phrase, may mean that a blade, as GFC''', Pl. XXVII.; Fig. 5, is truncated by two cylinders, concentric or not, and perpendicular to the paper, whatever be the nature of the pitch, and the form of the generatrix. But it should, and probably does, mean that the generatrix is curved back as at O*r*E—HE''', Figs. 2 and 5.

In this case, FG and C'''D''' are simply positions corresponding with A'B' and C'D' in Fig. 3, of the curved generatrix E'''H. Hence FG and C'''D''' are not circular arcs, but curves, whose ordinates from LK could be determined by scale, at various points on a careful construction of large size.

4°. *To construct the concentric sections.*—These are made by concentric cylinders, whose common axis is perpendicular to the paper at O. Make K_1s, Fig. 6, equal to the arc, KRL, for example, Fig. 2, and sL_1 equal to O'O'', Fig. 3, the pitch of a blade; then K_1L_1 will be the development of KL—K'L'. Now make $K_1S''V''L_1$ at pleasure, for the section of the back of the blade, in the same cylindrical surface with the helix, KL--K'L'. The lines K_1s, $o'U''$, etc., are the same as the traces $h''B'$, Uw', etc., in Fig. 3. In the latter, make the lines through S' and X' at the same distance below K' and L' as those through S'' and X'' are below K_1 and L_1, Fig. 6. Then project the points S'', etc., of the back of the blade upon any line, AB, Fig. 6, parallel to K_1s, and transfer them to the arc KL in Fig. 2, giving the points there *numbered.* Finally, project the latter points; 0, upon the line A' B' at K'; 1, upon Q'S' at S'; 2, upon B'A' again, at T'S', see Fig. 6, because T'' and K_1 happen to be on the same line, parallel to AB; etc., and we have K'S'4X'L' for the vertical projection of the section of a blade, made by a cylinder whose radius is OK, and whose axis is perpendicular to the paper at O. In Fig. 6, F'''' and H'''' are the developments of sections similar to the one just described, at the hub, and at the outer edge. Their projections can be made in the way just described. F''', K_1L_1, and H''', Fig. 7, are the three sections, laid parallel to each other for convenient comparison.

$K_1L_1R_1$ shows the section on KL; first, developed into the plane K_1L_1; second, revolved about OI till its *edge*, KL, comes into the paper; third, revolved about K_1 L_1 till its whole surface is in the paper.

It is here to be noted, that as the screw surface is warped, neither these nor any other cylindrical, nor any plane or conical sections could show a series of *perpendicular* thicknesses of the blade. For the different tangent planes at points on any one helix would not be parallel to any one line, as in case of a cylinder, where all the tangent planes are parallel to the axis. Hence, the perpendicular thicknesses, which would be perpendicular to the several tangent planes, at the point of contact of each, would not lie in the same plane; or, conversely, a section on any helix, KL, or radius OI, and composed of perpendicular (normal) thicknesses at all points of either, would itself be warped, and therefore undevelopable.

5°. *To construct radial sections of a blade.*—Lay off on the tangents, as at *n*, *p*, *g*, etc., the vertical thicknesses at those points, viz., *p'p'*, Fig. 6, at *g*, Fig. 2; *n'n'*, at *nn*, etc., and the shaded figure, *nn pg*, will be the radial section on *ng*, revolved into the paper.

For the section on *r*E, we should have taken the thicknesses from Fig. 6, on the centre line, OO''O'''. *These* sections are in planes *containing* the axis.

To have made sections, as is often done, *perpendicular to the axis*, we should have taken K,T'', o'U'', etc., for the thicknesses. In a case like Fig. 5, these sections are taken on chords, as those of C'''D''', or FG, or E'''H.

The figures just described and indicated, should be constructed in full on a much larger scale. Let the radius OE, Fig. 2, for example, be *ten feet*, and let that and the figure 3, also 4, and 5, or similar ones, be made on a scale of *half* an *inch* to one *foot*.

Ideas Expressed in Modified Forms of Screws.

180. Every modification of the screw propeller, from the form of a common right helicoid, or "true screw," except some merely fanciful or arbitrary ones, has been the result of a certain determining idea, suggested or confirmed by experiment or reflection.

181. *The Idea of Axial Expanding Pitch.*—In this case, it being known that a screw, beginning to work in smooth water, soon acts to move rearward a column of water, the idea is that, as the entering element of the blade moves the water as described, the next element should, so to speak, chase it up, so as to press upon the moving water as heavily and effectually as the first element. This second element would give an increment of velocity to the already moving water, and so the third element, by a continued expansion of the pitch, would move backward faster yet, so as to catch up with the water, just as men, to push a rail-car with a uniform pressure as its speed increases, must walk faster and faster.

182. *The idea of radially expanding pitch*, and of all screws which, by having a curved generatrix, appear bent back, that is, *from* the vessel, in a side view of the latter, is, to counteract the centrifugal action of the water, and confine it to a cylindrical column, having the disc (end elevation) of the screw for its base.

Among the most curious screws of this kind, is *Holm's conchoidal screw*, having a rapidly expanding *axial* pitch, so that at the trailing edge the blade is tangent to a plane containing the axis of the screw, and therefore has, at that point, an infinite pitch. Also, at the outer circumference the edge of the blade is bent over *from* the vessel into a narrow cylindrical flange, which finally is *rounded* into the trailing edge at the corner, by a spherical or spoon-shaped surface.

183. Finally, the *opposite idea of bending the blades toward the vessel*, and of mounting them on rings, so as *to leave the central portion of the disc open*, is, to favor the rush of water from all sides into the partial vacuum which tends to exist behind, or on the after side of the screw. In the *Griffith screw*, the blades are widest at about the middle of their length, are bent towards the vessel, and are fitted by cylindrical arms into similar sockets in a large spherical hub, or "*boss*." Each can be turned on the axis of its cylindrical arm, and thus the pitch is variable.

If a screw is made of a "bent" form, as seen in Pl. XXVII., Fig. 5, merely by the movement of a straight generatrix, which makes a constant angle of less than 90° with the axis, the screw surface is simply the common oblique helicoid (164), or that of a triangular threaded screw (Pl. III., Fig. 10).

184. While preparing these pages, I am informed by an engineering friend who has made the experiment, that if saw dust be poured upon a screw model, while the latter revolves rapidly in a lathe, it will rather be drawn towards the axis of the screw than dispersed by the centrifugal force, developed by the rotation.

The result just stated may be explained as follows: The rearward discharge, by the screw, of a cylinder of water creates a constant tendency to a vacuum at the position of the screw. This tendency, being constant, is as effectual as a sensible vacuum, in inducing a constant rush of water from all sides to the spot where the screw works. This centripetal rush of water is believed to prevail over its centrifugal tendency, so that some, instead of bending the blades *aft*, or *from* the vessel, to confine the water radially, and so prevent its lateral dispersion, and discharge it rearward in an undiminished cylinder, have bent them *forward* as already explained, or *towards* the vessel, in order to favor the inward rush of surrounding water at the base of the water cylinder acted upon by the screw. The Griffith screw, as said before, is thus formed.

Historical Note.

185. Sufficiently patient investigation into the history of every perfected mechanical device would probably show that it

had its origin in some bodily movement. The screw propeller is no exception. In swimming, the hand may be disposed, relatively to the water, somewhat in the screw form, as well as in that of a flat paddle; and when intelligence seeks some external aid to reinforce and extend the hand, she fashions an oar, as used in sculling; where it is constantly submerged and is virtually a one-bladed reciprocating screw, alternately right-handed and left-handed; while oars wrought from the sides of a boat in the usual manner are the mechanical rudiments of paddle wheels.

186. Robert Hooke, a celebrated English mechanician (1635—1703) believed that the ancient galleys were propelled by oars held nearly vertical, always submerged and subjected to a sculling motion. This then was rudimentary screw propulsion, and, reciprocally, a screw propeller is a "continuous sculling machine."

Screw propulsion is said to have been known to the Chinese for ages. In water, it is but the companion in a new office, of the "Archimedes' Screw" for raising water, and the counterpart of the immemorial windmill in air.

These remarks serve to show how shadowy and dimly defined is the origin of the invention of the screw propeller. The point selected for beginning its history must therefore be somewhat arbitrary. This history cannot here be given. Suffice it to say, that, beginning with Hooke, Bourne rehearses about two hundred and seventy-five forms or modifications of inventions relating to propulsion by the screw or by analogous devices.

The principle of axially expanding pitch (176) was first patented in America, and applied to mill water wheels by Clark Wilson, of New Hampshire, in 1830.*

It had been patented in France and applied to propellers in 1824, and was patented in England in 1832.

The overhanging, or "bent back" form was first patented by Ebenezer Beard in Connecticut, in 1841, and in England by the "Earl of Dundonald," in 1843.

Of course among so many designers as have been indicated, many reinvented the works of others, many merely treated details and accessory features, such as various relations of the propeller to the hull of the vessel, and to the rudder; and means for raising the propeller. Others sought to produce pro-

* Bourne, p. 42.

pellers, which should be adjustable either in length of blade or in the inclination of the surface of the blade to the axis, by revolving it about a perpendicular to the axis.

Others, emulous of ducks and fishes, devised fish-tail and other flexible propellers of hinged flaps, or of steel frames covered with gutta-percha, which should adjust themselves to the action of the water. Attempts have also been made to steer partly or wholly by screws, either by having two independent screws, one on each side of the stern, which could be revolved at unequal velocities, or by deflecting the entire screw, in case of a single propeller, so that it might revolve in any plane oblique to the shaft.

Example LIII.

The Screw of the "Dunderberg."

Description.—Pl. XXVIII., Fig. 1, represents an end view of two blades, as seen in facing the stern of the vessel, and an edge view of the hub, two blades and part of another. This screw is "bent back" 15 inches. The direction of the revolution being as shown by the arrow x,x'; ME′—M′E″ is the entering edge, and ND′—N′D″ the following edge. At the former the pitch is 27 feet. On RF′—R′F″ it is 28½ feet; and at ND′—N′D″, it is 30 feet. 3′ : 9″ = 45″ is the pitch of one blade, = one-eighth of the total extreme pitch of 30 feet, and 100″ is its effective length ST.

Thus this is a screw of axially expanding pitch, bent back, by having a curved generatrix RT—R′T′.

Construction.—This is pretty nearly evident from the figure, by the letters of reference, after the preceding example. Fig. 2 shows a good method of laying out the cylindrical blade sections (Ex. LII. 4°) in case of axial expanding pitch.

DC, for example, is the length of the arc, 1—1, of the end view, MTN, of the blade, and AD is the *pitch* of the blade, then AC is the development of the helical arc 1—1. But this development is curved; and at A,B, and C, should be tangent to straight developments of helices having pitches of 27, 28½ and 30 feet. Hence make AX = twice the pitch, 3′ : 9″, of one blade = 7′ : 6″; $Xa = yb = zc$ = twice the arc 1—1 of the end view. Ay = twice the pitch of one blade at RT, = 7′ : 2″ nearly;

and Az = twice the pitch of one blade at NE′ = 6′ : 9″. Then Aa, Ab, and Ac will be the required auxiliary developments. The curved development, AC, is now drawn so as to be tangent, at A, to Aa; at B, to a line parallel to Ab; and at C, to a line parallel to Ac, as required.

The dotted figure Am Tn is an example of the conventional "development" of a blade, mentioned in (Ex. L; 1°).

187. Referring the reader again to works on propellers alone, for fuller details, the subject is dismissed with the following data of celebrated examples, from which the student can construct additional figures for himself.

1. Screw of the U. S. N. Steamer ONEIDA.

Diameter, as seen in end elevation, 12′ : 9″.

Length, parallel to the axis, 22″.

Generatrix, curved, with a radius of 15′ : 10⅛″.

Pitch, expanding axially, from 18′ to 20′.

Thickness of blades at the hub, 5¾″, greatest.

" " " " periphery, ⅝″.

Radial section of greatest thickness, ⅜ of the width of blade from the entering edge.

Hub, greatest diameter, 22″.

" bore 10½″, tapering to 9½″.

No. of blades, 4

2. Screw of the U. S. N. FRIGATE "FRANKLIN."

Diameter, 19′.

Length, 41″ at the hub. 44″ at periphery.

Generatrix, curved, versed sine = 6″ to a chord of 19′.

Pitch, expands axially from 26′ to 30′.

Hub, centre diameter, 38″, Fig. 93.

" forward " 30¾″.

" rear " 26″.

" (hollow) internal diameter, 20″.

" length of body, 41″.

" total length, 83″.

Shaft diameter, 8″.

Greatest perpendicular thickness of blade at hub, 9½″.

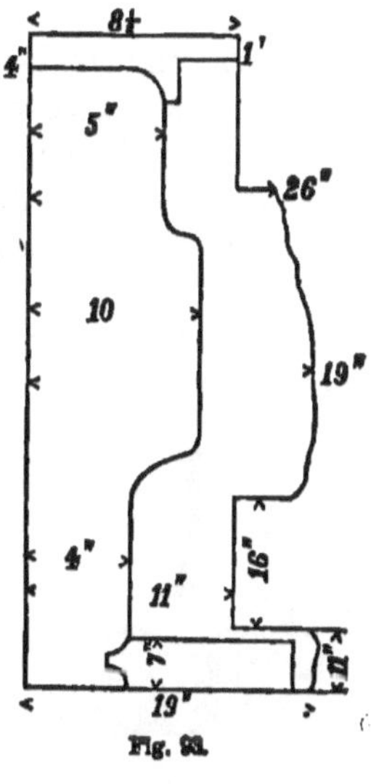

Fig. 93.

Greatest perpendicular thickness of blade at hub, perpendicular to axis, $9\frac{3}{4}$.

Thickness at periphery, $\frac{3}{4}$.

Radial sections, or templets, in planes nearly perpendicular to the axis, that is taken on, or parallel to the chord, as E‴H, Pl. XXVII., Fig. 5, of the generatrix, and *perpendicular to the paper*. The edges of these templets are curved. The centre one is of various widths, from $11\frac{1}{8}''$ at the hub to $7\frac{1}{8}''$ at the periphery (Ex. L; 5°).

D—VOLUME OPERATORS.

Example LIV.

Andrews's Centrifugal Pump.

Description. Plate XIII., Figs 6 and 7.—A is the base of the pump, cast in one piece with the case C, and strengthened by brackets *aa*. To the chamber C, by flanges *bb′*, is attached the conducting-case, composed of two parts, DD′, united by flanges *dd′*, forming a conic spiral discharge-passage, *g* and E, gradually enlarging to its outlet. F is the stuffing-box, through which passes the cast-steel driving-shaft G, having a series of grooves turned in its surface at J, which are accurately fitted in a Babbitt-metal box in the standard H, and its cap *h*, counteracting all tendency to end-thrust or vibration. I is the bed-plate, having cast upon it the standard H and brackets, to which the pump is secured by the flanges *dd′* and base A. The base A also forms a flange, to which is bolted the bend, Q, with suction-pipe B attached (shown broken off), with a foot-valve (Fig. 96) at its lower end.

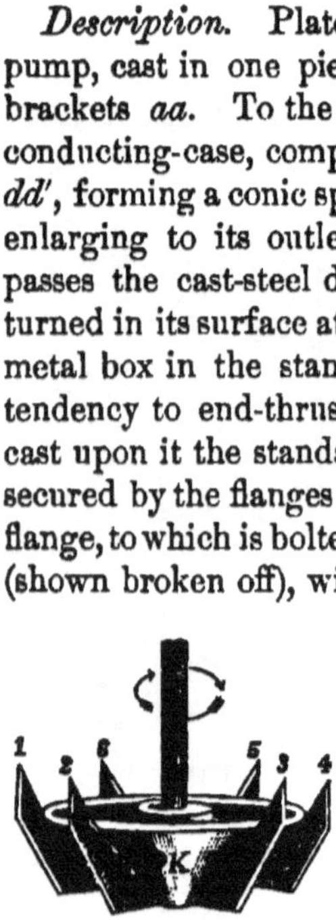

Fig. 94.

To a flange on the discharge-orifice, are attached pipes for conveying the water wherever required.

KK is the disk secured upon the shaft G, having wings, Fig. 94, 1, 2, 3, 4, 5, 6 (see Fig. 7), upon its periphery, closely fitting the space between it and chamber C, within which they revolve without touching. Their discharge-ends, *ee*, extend beyond K, close to case D′, without touching it, and terminate on a line parallel to the shaft G. L is the hub cast with and connected by flanges *lll* (see Fig. 95) to chamber C, forming spiral introduction-passages.

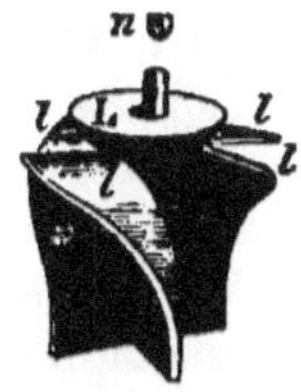

FIG. 95.

In the end of the shaft G is a steel button, *n*, with a convex face, which revolves in contact with the convex end of the step N, secured in the hub L, supporting the shaft and disk when run vertically. Motion is communicated to the disk by a belt upon the pulley P.

Operation.—The pump and pipes first being filled with water, rapid motion is given to the disk K, when the centrifugal force imparted to the water between the wings causes it to flow through the passages *g* and E, to the outlet; a vacuum being thereby created between the wings, causes the water to rise through the pipe B, to keep up the supply.

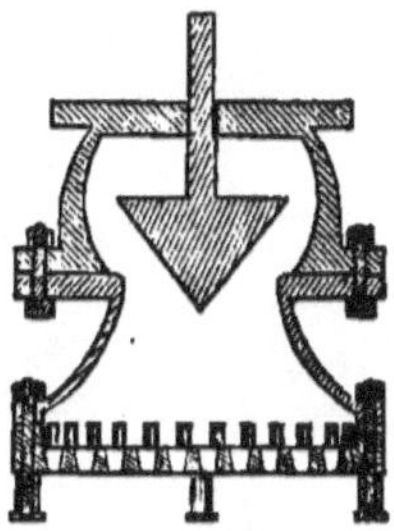

FIG. 96.

By means of the spiral passages around the hub L, the water from the suction-pipe is turned gradually from a direct forward course and delivered to the propelling wings in the line of their action; thence, through the spiral passages *g* and E, it is again, by an easy, gradual curve, brought back to a straight course, upon reaching the outlet.

The wings on the disk K, passing beyond its outer edge, create and maintain a vacuum between it and the case D, and prevent sand, dirt, etc., from coming into contact with the shaft. The step N is in like manner protected from dirt, enabling the pump constantly to discharge a large proportion of sand, gravel, etc., without injury to any of its parts. There being no valves in action (the foot valve remaining open while the pump is in motion, and used only to retain the charge when at rest), and no wearing parts, except the shaft in its bearings, which is perfectly protected from dirt, the *friction* is *reduced* to the *smallest possible* fraction, enabling the pump to run for *years* without repairs.

The power, lost in piston pumps in overcoming the momentum and inertia of the water at each stroke, is saved by this pump; also the large amount of power lost in changing the currents of water at right and other angles—all changes of direc-

16

tion being made by easy gentle curves—enabling it to perform the same work with much less power and a greatly decreased velocity.

Construction.—This can be made sufficiently well, since these pumps are of various sizes, by taking the diameter of P=12″, and of B=6″, for a scale.

An end elevation can be added, by making the *vertical* plane of the outlet flange, tangent to the flange *dd′*, at its foremost point.

BOOK SECOND.

COMPOUND ELEMENTS OR SUB-MACHINES.

187. As soon as, by leaving single mechanical organs, we come to connected series, or trains of pieces, some new and peculiar principles relative to the drawing of the latter are required.

Not only the *form*, *dimensions*, and *location* of each piece must be known, but the *character of its motion*, must be understood also, as a condition for knowing the *nature of the motion which it will communicate* to whatever is actuated by it. From all this the final result can be derived, which is the ability to assign to each piece that one of its successive positions which corresponds to a *given* position of some *one* part, taken as a standard, to which the others are referred.

188. Thus, not to take the too hackneyed illustration of the crank and piston, first imagine two spur wheels, A and B, in contact at T, and let the diameter of A be double that of B, and let their arms, AT and BT, be on the line of centres. Then, as equal lengths of arc on each must pass the point T in the same time, a revolution of 45° of AT will produce a revolution of 90° in B.

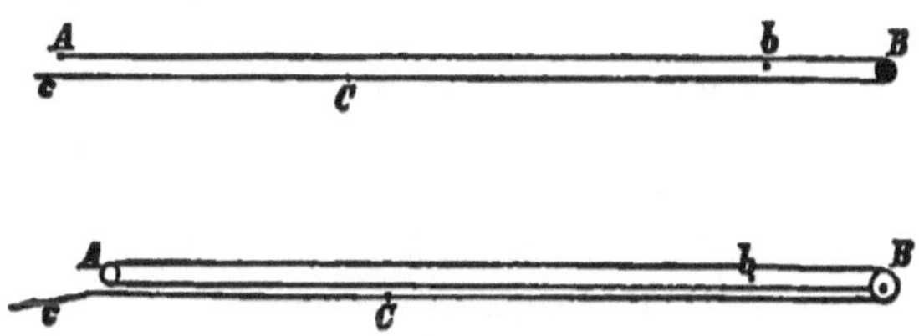

FIG. 97.

189. Again, to take an illustration from pulley work, Fig. 97. Let A be a fixed point or pulley, from which a cord proceeds to the movable point, B, and around it to C. By drawing

the cord till B moves to *b*, we shall have, by reason of the invariable length of the string:

AB + BC = A*b* + *bc*, or by decomposing,

A*b* + *b*B + B*b* + *b*C = A*b* + *b*C + C*c*, in which, by cancelling, C*c* = 2B*b*. That is, if the free end, C, of the cord be moved *a given distance*, the movable point will move *half as far*.

In the second case, A still being the fixed pulley, but the cord being fastened to the movable one, B, we have:

BA + AB + BC = *b*A + A*b* + *bc*, or, by decomposing,

B*b* + *b*A + A*b* + *b*B + B*b* + *b*C = *b*A + A*b* + *b*C + C*c*, from which, by cancelling, C*c* = 3B*b*.

And generally the place of a movable pulley will be at a distance from its initial position equal to *one-n*th of the space described by the free end of the cord, where *n* the number of cords at the movable pulley.

190. As most examples of sub-machines consist of fewer parts than any whole machines possessing much interest, it would be only a needless anticipation of the subsequent examples to give further illustrations here of the *general idea* which should be constantly in mind in constructing the following examples, which is: *Where* is *each* piece, and *what* is its motion, at any given stage of the motion of a *given* piece?

Now, when the given piece is the *operator*, the answer to this twofold question is a practical invention, consisting of a train of pieces for making the operator accomplish a desired result.

The careful study, then, of the following examples, with constant attention to this radical idea, and to the kindred question: How else could the same result have been accomplished? is the study of the art of inventing, so far as that is an acquirable or teachable one.

From the foregoing principles it may be seen that a compound element does not merely consist of many pieces, having fixed relative positions, and acting as one piece, for *that* is true of cross-heads, pillow-blocks, pistons, and many other simple mechanical organs. But a compound element truly consists of several moving pieces, each of which moves in a certain way, according to the motion imparted to some one of them.

SUPPORTERS.

Example LV.

A Compound Chuck.

Description.—A *chuck* is a contrivance attached to the *mandrel*, or supporting axis of a lathe, for the purpose of holding the work.

Chucks are of various forms, according to what they have to hold, and the work to be done upon them, whether circular, or other turning, external or internal; and will be found described in any of the accessible popular treatises on turning.

Compound chucks are of three principal species :—

Scroll Chucks, Fig. 98, where the spiral tooth, C, acts to ad-

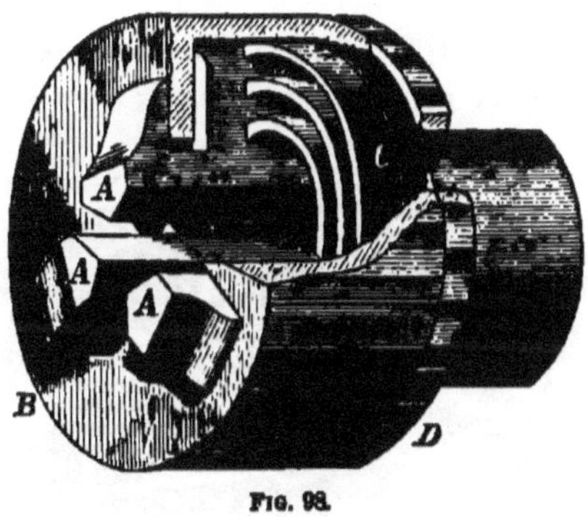

Fig. 98.

vance or retreat the jaws A, to or from the centre—

Conical or wedging Chucks, which separate the jaws or close them together, by urging them into a conical sheath in the latter case, or withdrawing them in the former, by the action of a screw; and—

Geared Chucks, which are of various forms. The following figures represent Horton's Geared Screw Chuck, which is highly spoken of by persons engaged in the finer mechanical pursuits of making small and exact mechanism.

Other chucks are very generally modifications or combinations of the foregoing. When the jaws, D, D, D, are separately operated, the chuck is said to be an independent jawed chuck.

Fig. 99 gives a view of this Chuck ready for use.

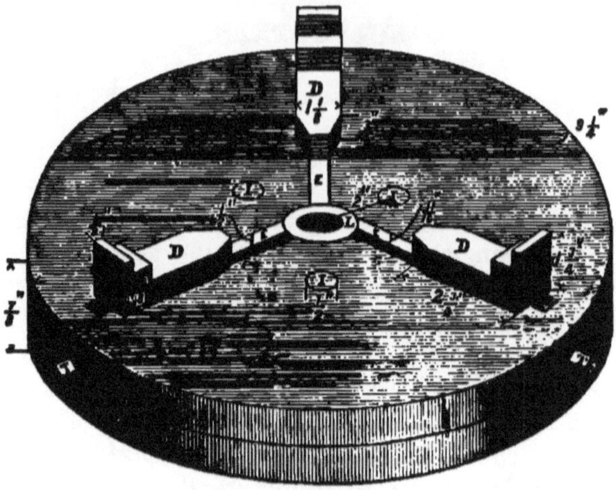

Fig. 99.

Fig. 100 represents the interior of the back plate, with the deep groove or recess containing the rack G, within which it re-

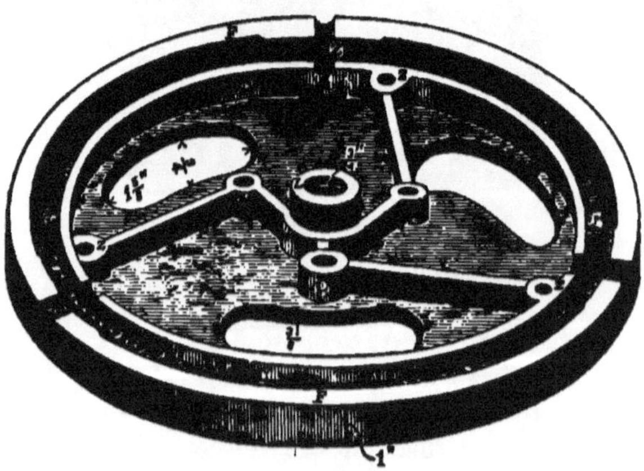

Fig. 100.

volves freely. This groove, by means of its outer and inner flanges, as shown by figures 100 and 101, forms a tight casing for the gears, thus protecting them from chips, dust or any mat- that would otherwise injure them.

Fig. 101 represents a view of the interior of the front plate showing the carrying screws, A A A, and the nut part, N, of the jaws, with the bevel pinions upon the former. The inner ends of these screws have their bearing against the hub L, Fig. 100, while

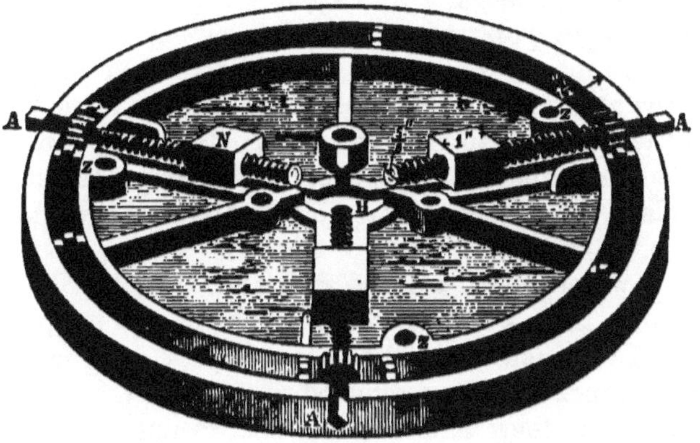

FIG. 101.

the shoulder formed by the pinion near the outer end, rests firmly against the outer rim of the groove, in which the rack G moves.

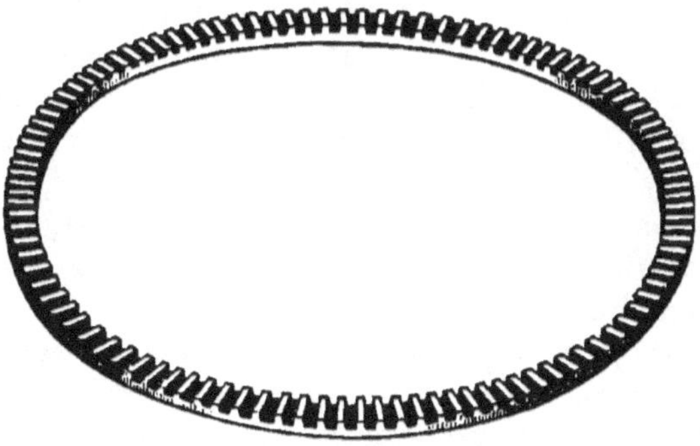

FIG. 102.

FIG. 103.

Fig. 102 is the circular rack, seen at G, Fig. 100.

Fig. 103 gives a view of the jaw, which is forged of one piece of metal. The slots for the jaws, radiating from the centre, leave the periphery entire, thus securing the greatest possible strength. By means of the rack and geared screws in this chuck, great facility is secured for moving the jaws, so as to confine any article, by applying the wrench to *any* one of the carrying screws. By doing this, all the jaws are carried forward at the same time, bringing the article to be confined at once to a centre. Then when wishing to confine the work very firmly, pass the wrench from one screw to the other, and pull upon it sufficiently to take up the back lash in the gearing. *By continuing this operation, one can fasten the work tight enough for all the purposes required.* The jaws to this chuck can always be kept very accurate. For, by means of the gearing, they may at any time be adjusted to within about one-hundred and fiftieth part of an inch of a true circle. They can also be operated independently, by simply taking the rack out of the back plate, thus allowing it to hold any kind of irregular form of work.

The jaws are case-hardened with animal coal. To do this thoroughly, takes a number of hours, and is apt to spring the jaws. To obviate this, the bite of jaws is ground true after the chuck is finally assembled for market, which makes the jaws as true as they were left after first turning.

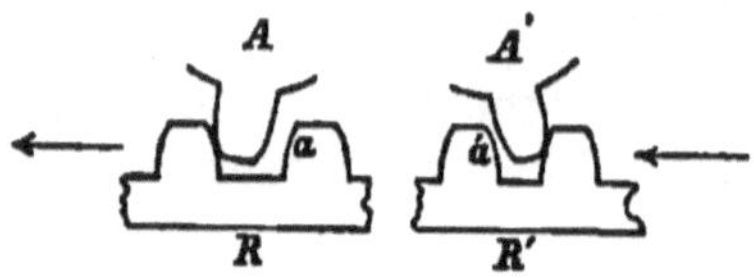

FIG. 104.

The tightening of the jaws by "taking up the back-lash" means this. When a pinion, A, Fig. 104, is turned, it draws the rack R, and *that*, as shown at R′, turns the other pinions. The *back-lash*, which is the difference between a tooth and a space, will then be at the right of tooth A, and at the left of A′. By turning A′ till it bears upon tooth a', the back-lash will be said to be taken up. It may be more, and then the back-lash will be at the left of A. Hence, as directed, all the pinions should be turned in succession.

Construction.—Figs. 99–103, with the measurements, which are sufficiently given, may serve as an example of sketches and measurements, from which projections can be drawn with sufficient accuracy to answer every purpose of a graphical exercise.

191. Other compound supporters, are *compound slide rests*, and *tool-holders*.

These are designed to give one or more motions to the point of the cutting tool in the higher power lathes, planers, shaping, and surfacing machines.

One will be found illustrated in connection with "feed motions" on a subsequent page.

COMMUNICATORS.

192. *Communicators*, are perhaps never compound, but with the small space left at our disposal, we cannot better illustrate the differences (187–190) between the drawing of *detached* and of *connected* organs, than by a few examples of groups of connected communicators.

Example LVI.

A Beam Engine Main Movement.

The skeleton figure, Pl. XXVII., Fig. 8, answers every purpose of a finished one, so far as to show how to find the positions of other parts for a given position of a given part.

Let the vertical, Bg, be the indefinite axis of the cylinder, and let the vertical, Ho, contain a diameter of the crank pin circle JII″. This *diameter* is equal to the *stroke* of the piston, and so are the *vertical chords* of the arcs described by the ends of the *beam*, whose centre is at O. Suppose the cylinder end of the beam to be connected to the piston rod by a link, B′b; attached at b to a cross head. Then make $om = ge =$ BD, the piston stroke, = JII, and describe the arcs, onm and gfe, to be described by the ends of the beam, whose length will then be nf. But now observe that *all* positions of the ends of the beam are outside of the verticals, Bg and Ho. It is desirable, however, to have the average position of the connecting rod, Hm, vertical; and so of the link B′b. Hence bisect hf at C′, and nr at I′, and draw the parallels oJ′—mH′—etc., to nf, so as to make the new vertical chords as J′H′ equal to BD, and take OI′ = OC′ for the final radius of the beam which will make the mean of the positions of the connecting rod, and of the link, vertical. Draw H′H, the lowest, and J′J the highest position of the connecting rod. To find the positions of the crank pin when the beam-end is at I′, take that point as a centre, and HH′ as a radius, and note I and I″, where the arc so described cuts the crank pin circle. Likewise the link, being of constant length, B′b = C′c = D′d, corresponding to the lowest, middle, and highest piston rod positions Bb, Cc, and Dd.

Finally, to find the position, of all other parts, due to *any* given piston position as P.

1°.—Lay off vertically from P the constant length Pp (not shown) of the piston rod.

2°.—With p as a centre and the constant length B′b, of the link, as a radius, describe an arc, cutting the beam arc, D′C′B′, in the required position, which we will call K, of the right-hand beam-end. Then KO to K′ on J′I′H′ will be the beam position.

3°.—With K′ as a centre, and H′H as a radius, describe an arc cutting the crank pin circle in a point Q, which, finally, will be the required crank pin position.

Example LVII.

Wheeler's Tumbling Beam Engine.

A skeleton view of this very curious form of steam engine in five different positions of the crank, etc., is shown on Pl. XXX., Figs. 4–8. Here, a,a are the cylinders, standing on the bed plate B. C is the crank shaft, which, with the rocker shafts, g,g, works in fixed bearings in a common frame. Rtt is the tumbling beam, no point of which is fixed, R being attached to the crank-pin R; and t and t to the piston rods. It may have various proportions, but an isosceles triangle, whose altitude BD, Fig. 105, is equal to half the base AC, is preferred.

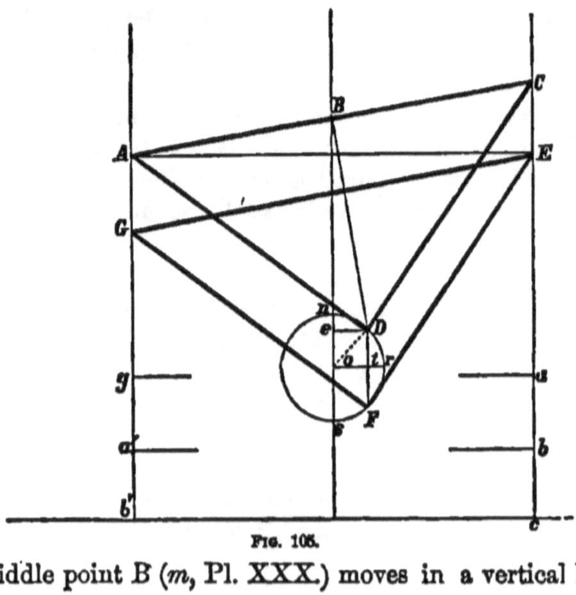

Fig. 105.

The middle point B (m, Pl. XXX.) moves in a vertical line, it being also the middle point of the link *nn*, which joins the free ends of the equal radius rods, K,K, so as to form a "*parallel motion.*" K,K turn on the centres *g,g*. P and Q are the pistons, and the arrows show the direction of their motion, at five equidistant positions, 45° apart, of the crank pin.

Starting with the given proportions of the beams, and the facts, that when one piston, *a'*, Fig. 105, is at mid-stroke, the other, *a*, is at the end of its stroke, and the point A, Fig. 105, of the beam, is on the cylinder axis A*b'*, we can determine any other pair of positions; also the peculiar property of this engine, that the piston stroke is greater than the diameter R*d*, of the crank pin circle. The amount of this excess can also be found. That it exists is obvious on mere inspection; for if the piston positions were alike, all points of the beam would move in vertical straight lines equal to the diameter R*d*. But, by reason of the unlike piston positions, the beam oscillates besides ascending, so that *t* and *t*, and thence the pistons, must have a vertical movement greater than R*d*.

To ascertain the relation of *ns*, Fig. 105, to the piston-stroke *ac*, let ACD be the beam, with B on the vertical OB, and BA = BC = BD, and D, the crank pin at the 45° point from *n*. Thus BD = ½ AC, and

$$De = Dt = \tfrac{1}{2}\,CE = \tfrac{1}{4}\,ac \text{ (compare Figs. 5 and 7.)}$$

$$\text{But } OD = \sqrt{2\,(De)^2} = \sqrt{\tfrac{1}{8}\,(ac)^2}$$

$$\text{or } 2OD = ns = \sqrt{\tfrac{1}{2}\,(ac)^2}.$$

For example, let $ac = 20''$, then *ns*, or the *throw* of the crank = 14.2; and the length, OD, of the crank = 7.1.

The points *tt*, see the Plate, do not move in straight lines, but in elongated figure 8's. Therefore, in the absence of the usual connecting rods, this engine belongs to the variety called trunk engines, in which the piston rod is a hollow cylinder, called the trunk, and links join the points *t* with the *piston*, and pass through the trunk.

193. Other novel forms of steam engine, besides rotary ones, are *Root's Square Engine*, consisting of a rectangular "cylinder" and two rectangular pistons, one within the other and moving on the two centre lines of a rectangle, while the inner one is attached directly to the crank pin, and *Hicks' Engine;* both of which are further remarkable for compactness.

Example LVIII.

An Eight-Day Clock Train.

Pl. XXXIII., Fig. 1, shows such a train. The circles in fine lines, represent the pitch circles of those wheels, which especially belong to the hour hand movement.

Each wheel is designated, 1*st*, by its radius; 2*d*, by its number of teeth, *t*; 3*d*, by the time of one revolution of it. The arrows indicate the directions of the revolutions. AB is the *pendulum*, vibrating once in each second. App' is the *escapement*, oscillating in unison with the pendulum. p and p' are its *pallets*, so adjusted that, as the pendulum swings to the left of its vertical position, tooth a of the *scape wheel*, passes the point of pallet p, but just then p' catches tooth c on its right side, so that the scape wheel moves but half its pitch at each vibration of the pendulum. Hence a' will not pass p, till the pendulum returns again to its left hand position, that is, in *two* seconds. Thus the scape wheel, having 30 teeth, revolves once in a minute, and may carry the *seconds* hand. The pitch being made the same for all the wheels, their *radii*, and thence their *times* of revolution depend on their *numbers* of teeth. These are clearly expressed on the figure. Wheel 96*t* is on the end of a barrel around which the weight cord makes sixteen turns. It can, therefore, make 16 revolutions of 12 hours each before the clock will run down. Wheels, like 64*t* and 40*t*, which revolve in the same time, may be on the same physical axis; but wheel 72*t*, which carries the *hour* hand, is on a hollow axis, through which that of 64*t* and 40*t*, carrying the *minute* hand, passes.

Other Trains.

194. Let us now examine this train so as to find some general relations, variable or not, between the train and the final results to be accomplished.

1*st*. Given a pendulum vibrating seconds, which is convenient; it is also convenient that a tooth should pass a pallet only once in two vibrations, so as to reduce the number of teeth in the scape wheel. Thirty teeth are therefore fixed for *that* wheel; and *one* minute for its revolution. The action of wheel 96*t* being more conveniently prolonged by making many turns of the cord about its *barrel*, than by making that wheel with 96 $\times 12 = 1152t$, in order to revolve it once in eight days, we give it 96 teeth and 12 hours = 720 minutes for a revolution. That done, we have

$\frac{720}{1}$ as the *velocity ratio desired*, between wheels 96*t* and 30*t*.

Now the wheels 96*t*, 64*t*, and 60*t* are successive *drivers*, and the pinions 8*t*, 8*t*, 8*t*, are their respective *followers*, and revolve in the same times as the wheels on the same axis, and we have

$$\frac{96 \times 64 \times 60}{8 \times 8 \times 8} = 12 \times 8 \times 7\tfrac{1}{2} = \frac{720}{1}$$

Whence we infer that *the velocity ratio of the extreme axes of a train is equal to the product of the number of teeth in the drivers, divided by that of the numbers in the followers*, which write $= \frac{D}{F}$

195. To show by a converse operation how simply this may be seen to be true, let the ratio $\frac{720}{1}$ be given, which for a moment we will take as the ratio of a wheel of 720 teeth working with a pinion of *one* tooth, though this would practically be impossible. Now decompose the terms of this fraction into any factors as in the example,

$\frac{12 \times 8 \times 7\frac{1}{2}}{1 \times 1 \times 1}$ and the value of the ratio is unchanged. But, *mechanically*, pinions of one tooth are impossible, while *arithmetically*, a fraction, or a ratio, is not changed by multiplying both terms by the same number, in this case by $\frac{8 \times 8 \times 8}{8 \times 8 \times 8}$, giving $\frac{96 \times 64 \times 60}{8 \times 8 \times 8}$, which are practicable numbers of teeth for the wheels and pinions of the train.

Proceeding in this manner, as the numerator may be variously decomposed, we may have

$\frac{720}{1} = \frac{12 \times 5 \times 12}{1 \times 1 \times 1} = \frac{96 \times 40 \times 96}{8 \times 8 \times 8}$, where the numerators of the last fraction will be the numbers of teeth in the *drivers*, and

8 the number of teeth in each pinion follower, the last of which carries the scape wheel on its axis.

Again, the barrel and the hour hand revolve in the same time, twelve hours. Accordingly we have for the hour hand train $\frac{96 \times 40 \times 6}{8 \times 40 \times 72} = \frac{12}{12}$.

196. These trains, and other similar trains, may be tabulated as follows, by a simple system of notation in which each wheel is expressed by its number of teeth, each pinion number is written under the number of that wheel which drives it; and those of wheels on the same physical axis are written on the same horizontal line.

We have, then, the following, from which the student can construct the drawings.

I.—Eight-Day Clock.

TRAIN.	PERIODS.
96 Barrel....................	1 h.
8——64——40....................	1 h. minute hand.
8——|——60	
40——6	
8—|—30....................	1 m. second hand.
72....................	12 h. hour hand.

II.—Twelve-Hour Clock.

TRAIN.	PERIODS.
48 Barrel——25——4	1 h.
6——45	
6——30 scape	1 m.
25 min. hand	1 h.
48 h. hand	12 h.

III.—Eight-Day Clock.

TRAIN.	PERIODS.
96 Barrel..	12 h.
8———40———10..................	1 h.
8——96	
8——|——30 scape.........	1 m.
30———10	3 h.
40....	12 h.

where for the hour hand train, $\frac{96\times10\times10}{8\times30\times40}=\frac{12}{12}$ as it should.

Eight-Day Clock.

TRAIN.	PERIODS.
108 Barrel..	810 m.
12——108———54———10	90 m.
12—100	
10—30 scape.	1 m.
36 min. h.	1 h.
80 hour h.	12 h.

where, for the hour hand train $\frac{108\times10}{12\times80}=\frac{9}{8}=\frac{810\text{ m}}{12\text{ h}}=\frac{810\text{ m}}{720\text{ m}}$.

197. In this example the pinions are larger, the wheels of the pendulum train more nearly equal, and no wheel in that train (which is the same thing as the escape-wheel train) revolves in one hour.

When, as in the *hour hand* train, the extreme axes are the same geometrical axes, there may be a difficulty in adjusting the radii, if the pitch is the same as in the escape-wheel train. It is only necessary to remember, however, that the velocity ratios are as the *numbers*, not the *sizes* of the teeth; and only those teeth which gear into each other need have the same pitch.

198. To include the pendulum, or balance wheel, in the train, consider that as a tooth, or full pitch, of the escape wheel passes

a pallet, only after *two* vibrations of the pendulum, if e be the number of escape-wheel teeth, $2\,e =$ the number of pendulum vibrations in 1 revolution of this wheel. And if the pendulum makes p vibrations per minute, then, to make $2\,e$ vibrations, $= 1$ revolution of the escape wheel, will take $\frac{2\,e}{p}$ minutes. Then, as the hour axis revolves in 1 hour, $= 60$ m., the ratio of the revolutions of that, and the escape wheel, will be

$$60 \div \frac{2\,e}{p} = \frac{60\,p}{2\,e} = \frac{30\,p}{e}.$$

199. If, then, a watch balance vibrates 300 times in a minute, and the escape wheel has 15 teeth, $\frac{\mathrm{D}}{\mathrm{F}}(194) = \frac{30 \times 300}{15} = \frac{600}{1}$ and if there be three axes in the train, each with a pinion of nine teeth, then $\frac{\mathrm{D}}{\mathrm{F}} = \frac{600 \times 9^3}{9 \times 9 \times 9} = \frac{81 \times 72 \times 75}{9 \times 9 \times 9}$, will express the train from the hour axis, to the balance wheel axis inclusive.

Similarly, a month, or 32-day clock may be designed. Lunar and annual clocks are more difficult.*

Change-Wheels.

200. Analogous principles belong to *change-wheels*, by which any given velocity ratio is given to two axes in fixed positions, as in a lathe. Thus, let the required set of velocity ratios be $\frac{1}{1}$, $\frac{1}{2}$, $\frac{1}{3}$, $\frac{1}{4}$, $\frac{2}{3}$, $\frac{3}{4}$. Since the axes are at a fixed distance apart, and since it is convenient that all the wheels should have the same pitch, the *sums* of their numbers of teeth must be constant, and must be divisible by the sum of the terms of each ratio, that is, in this example, by 2, 3, 4, 5, and 7. The number of teeth in each pair is therefore the least common multiple of these divisors $= 2 \times 2 \times 3 \times 5 \times 7 = 420$ and the teeth in the successive pairs will be as follows:

RATIOS.	NO. OF WHEEL TEETH.
$\frac{1}{1}$	210 — 210
$\frac{1}{2}$	140 — 280
$\frac{1}{3}$	105 — 315
$\frac{1}{4}$	84 — 336
$\frac{2}{3}$	168 — 252
$\frac{3}{4}$	180 — 240

* See Willis' Prin. of Mechanism.

For further information on these interesting topics, to which only an introduction can here be given, the reader must consult the larger treatises on clock and mill-work.

Example.—Let the ratios be $\frac{1}{1}$, $\frac{1}{2}$, $\frac{1}{3}$, $\frac{2}{3}$, $\frac{4}{5}$.

THE SLIDE VALVE AND ITS CONNECTIONS.

201. The slide valve, and its connections, may be considered among the foremost of sub-machines, in interest and importance; on account of the high office which it fulfils, in the working of the grandest of motors in present use, the steam engine.

The office of the valves connected with any steam engine cylinder, is to give the steam access to each side of the piston, alternately, and, at the same time, to provide an escape for the steam which has just been effecting a piston stroke.

202. Valve motions in general are classified as

1. Cock valves.
2. Poppet valves, see Ex. XLIV.
3. Slide valves. Ex. XLII.

Slide valves are the ones most used, and are those which most invite investigation. They, with their connections, may be classified as

1. Valve motions with one valve,
2. " " " two valves,

each of which is subdivided, according to the most important ground of distinction, into

1. Motions with invariable cut-off.
2. " " variable " "

203. We will now suppose the student never to have examined the subject at all, further than by simple inspection of engines in motion, or partly dissected for repair; and will therefore begin with a rudimentary example, in which only the radical features, without the various adjustments of refined practice, will be noted. See Pl. XXIX., Fig. 1.

General Description of Parts.

204. B is a fragment of the bed plate (Ex. XVI.) or general support of the engine. P is the *steam piston,* see Pl. IV.; Fig.

1, and (Exs. XX. and XXI.) which is urged backward and forward, in a rectilinear path, by the pressure of steam, acting on its opposite faces alternately.

M is a section of the *main shaft* or *crank shaft*, which is made to revolve in fixed bearings (Exs. I., II., and III.) by a suitable connection with the piston.

CC is the *steam cylinder*, within which the steam is confined, so as to take effect upon the piston. H is the front, and *h*, partly broken out, the back cylinder head. Both are bolted to the ends of the cylinder. See Ex. X.

P′ is the *piston rod*, which passes through a stuffing box, not shown, in the back head *h*, and is firmly attached to the *cross-head* O, which moves in fixed guides—not shown—and parallel with the piston rod. See Ex. VIII.

R is the *connecting rod*, partly broken to show other parts, and jointed at *o* to the cross-head and at the other end to the crank pin Q, by a joint like Ex. XXVI.

N is the *crank*, keyed, as at *k*, to the shaft, M. Supposing, then, that motion is already established in the direction of the arrow, the motion of the piston P, from its present extreme right position to its extreme left position, will turn the crank from the $+90°$ point to $-90°$, through F. This motion of the piston, from *m* to *n*, is called its *stroke*. The return stroke turns the crank and shaft from $-90°$ to $+90°$, through D. The complete result is expressed by saying that *one double stroke of the piston produces one revolution of the shaft, or an angular motion of* 360.°

205. When, as is sometimes done, the crank shaft M, is not also the main shaft, but is connected with the latter by gearing, the double stroke of the piston might produce more or less than one revolution of the main shaft.

Pile driving engines, where the piston speed is usually very great, are examples of the latter case, and some propeller engines, of the former. Since the piston has no tendency to move the crank, when the latter is in the positions MQ or MQ′, the points Q and Q′ are called *dead points;* also, *centres.*

206. How now does the steam gain access to the opposite sides of the piston alternately?

An *eccentric*, see Ex. XXIX., being suitably mounted on the crank shaft M, the *eccentric rod*, partly shown at X, where the connecting rod R s broken away, is jointed at *d* to the foot of the *hanging rocker arm a*. This arm, and its counterpart, the *standing rocker* A, oscillate together on the *rock shaft*, *r*, to which both are keyed. A is jointed at *e* to the *valve stem*, L. The latter carries the *yoke*, *yy*, which embraces the *steam valve* V. This valve is hollow, as at I, like an uncovered box turned upside down. See also Fig. 2.

T is the *steam chest*, into which "live steam" enters from the boiler, by the *steam pipe*, terminating at S. Its walls and bottom are solid with the cylinder C. Its cover is bolted on.

s and *s'* are the *steam passages*, leading from the steam chest to the opposite ends of the cylinder. In the direction of the circumference of the cylinder, or as seen on an end view, these passages extend about one-third around the cylinder.

E is the *exhaust*, and opens through the chimney or smoke-stack into the atmosphere; or, in condensing engines, into the condenser, Exs. XI. and XII. *p* and *p'* are the *steam ports*, and *u* is the exhaust port. These are three long and narrow parallel rectangular openings in the *valve seat*, *fg*, on which the valve moves. *bb*, the partitions between the steam and the exhaust ports are the *bridges*.

The valve, V, is so proportioned to the ports, and is so moved that but *one steam passage at a time* can be open into the steam chest, and but *one steam port together with the exhaust port* can be covered at once by the hollow interior, I, of the valve. It is by such an arrangement as this, that steam is admitted to the opposite ends of the cylinder alternately.

207. Let us next look at the *action* of the parts.

General Action.

As a rudimentary example, sufficient to illustrate the general action of the slide valve, let the valve be adapted to the parts as shown in Pl. XXIX., Fig. 1. At its extreme right position, let the port *p*, be wholly open to the interior I; at its extreme left position, let the same port be wholly open to the steam chest. It is, therefore, now just at mid-stroke, moving left, and ready to open the port *p*. As it opens this port, steam enters and pushes the piston to the left to the mid-stroke position GK, when the port, *p*, will be wide open, the valve at the extreme left, and ready to return. When the piston has reached the extreme left, at *nn*, the valve will be at mid-travel again, but moving to the right, and ready to open the port p' for the return piston stroke.

While the valve is opening and closing the steam port, p, to the steam chest, it is simultaneously opening and closing the port p' to the exhaust passage E, through the interior chamber I. Thus the steam used in the preceding stroke from n to m, escapes into the atmosphere, through s', the port p', and E.

208. Such, then, as to its main features, is the composition and action of a slide valve motion. What are the modifying *causes* of subordinate variations in this action?

The line from the *shaft centre* to the *crank pin centre*, properly represents the crank, and may be called the *crank arm.* The line from the *shaft centre* to the *eccentric centre*, is likewise the *eccentric arm*, and is the linear crank which is equivalent to the eccentric. Twice the length of the latter line is, in simple valve motion, exactly equal, and in valve motions generally, as in link motions, approximately equal to the *travel*, that is, the *stroke*, of the valve. Finally, the acting surfaces vv of the valve may just cover the ports, or may overlap them on *one* or on *both* sides of each port p and p'.

209. Three things, then, 1*st*, the *angle* between the crank and the eccentric arms; 2*d*, the length of the latter arm, called also the *eccentricity*, and 3*d*, the *relative sizes* and *position* of the valve *faces*, and the valve *ports*, materially affect the particulars of the distribution of the steam *to* the steam cylinder and *from* it to outer spaces, as we shall now proceed to show, taking them up from the point of view of desirable *results* to be produced.

Modifications and Adjustments.

210. The preceding rudimentary account up to (208) neglects the following particulars.

1°. The influence of the connection of the *reciprocating* motion of the cross head, with the *rotary* motion of the crank pin.

2°. The advantage of employing the steam expansively, by *cutting off* its admission before the completion of each stroke, so that the *simple elasticity* of the confined steam should effect the completion of the stroke. The point in the stroke at which admission of steam ceases is called the *cut off*.

3°. The importance of using the steam itself, in place of complications of balancing mechanism, even if such were possible, to gradually overcome the inertia of the heavy reciprocating parts and so avoid damaging shocks at the points of changing the direction of the piston stroke. This is partly accomplished by closing the exhaust before the end of the stroke. The period during which the exhaust thus remains closed is called the *compression.*

4°. The *release* of the steam behind the piston, a little before the completion of a stroke, so that it may have time to escape in part before the beginning of a new stroke.

5°. The convenience and elegance of likewise employing the steam itself to neutralize the results of certain minor and almost necessary imperfections in workmanship. This, with a further accomplishment of the third object (3°) is secured by admitting steam for a given stroke just before the completion of the preceding stroke. The opening of the steam port at the beginning of the stroke is called the *lead* of the valve.

Definitions.

211. In connection with these proposed results, and available means for producing them, the following definitions arise; which are here presented, together, for convenience of future reference, though some of them may have been given already.

1st. The distance from one extreme position of any given point of the valve to the other like position of the same point is the *travel* or stroke of the valve: The centre line, *yy*, of the valve, Pl. XXIX., Fig. 3, is a convenient line to represent the valve for the purpose of marking its travel.

2d. The excess *outward*, as *l*, Fig. 4, by which the valve face extends outward, beyond its steam port, when the valve is at midstroke, is the *outside lap* of the valve, commonly called simply the *lap.*

3d. If, as in Fig. 4, the valve being there at midstroke, its interior chamber were limited as at the inner dotted lines; the small space $=i$, would be the *inside lap.*

4th. The amount of opening of the steam port at the beginning of a stroke is the *outside lead* of the valve commonly called its *lead.* That is, *lead* admits steam to the piston *before* the latter *begins* a stroke.

5th. The amount of opening of the exhaust passage, at p' for instance, if the piston is moving from O to A, Fig. 4, when the piston has reached A, is the *inside lead.* That is, *inside* lead allows steam to escape *before* the *end* of a stroke is reached.

6th. The point at which admission of steam ceases is the *cut-off.* This term is also applied to the separate valve often used, by which the cutting off is effected.

7th. From the point of cutting off, to the opening of the exhaust passage, is the *period of expansion*, as the one from where the steam port begins to open, to the time of cut-off, is the *period of admission.*

8th. From the time the exhaust passage closes till it is re-opened for admission is the *period of compression.*

Exhaust naturally and mainly takes place *before* the piston, but *inside lead* opens an exhaust passage *behind* the piston, *before* the latter has finished its stroke.

9th. The point at which the exhaust opens is the *release ;* and is, as said, the close of the period of *expansion.*

10th. If, as in Fig. 4, the length oo' of the interior of the valve is greater than the exhaust port + the two bridges, the excess, as o, on each side, is called the *clearance.*

11th. The difference between 90° and the angle made by the eccentric arm with the crank, is the *angular advance* of the eccentric.

12th. The arc or angle of the eccentric arm motion, which would move the valve through a space equal to a given *lap*, and so as to just close a steam port by that motion, is called the *lap angle.*

13th. The angular distance of the crank-pin from Q, when the steam port begins to open, is the *lead angle.* An equal angular motion of the eccentric is *its* lead angle.

212. Taking up the above topics (210) in the order named, the connection between the cross-head, O, and crank-pin, Q, may be *indirect*, through the medium of the connecting rod, R, or direct as in Pl. XXIX., Fig. 3, where the outer extremity of the piston rod is expanded vertically into a slotted yoke ; which is simply the mechanical equivalent of a connecting rod of infinite length ; or of a *line*, perpendicular to the piston rod, and always containing the centre of the crank-pin, and therefore equal to the diameter of the crank-pin circle. In this form of connection, the motions of the piston, piston rod, yoke, and

crank-pin, in the direction of the axis of the cylinder, are simultaneous and equal.

The ends of the rocker arms might, likewise, move in small slotted yokes, attached to the valve stem, and eccentric rod; but by balancing their arcs as in Pl. XXVII., Fig. 8, or Pl. XXIX., Fig. 3, no sensible irregularity will appear in their small motion.

Theorem XXII.

In either mode of connection, the velocity of the crank-pin is uniform; and that of the piston is variable.

It is important, in order to avoid injurious shocks, especially in heavy machinery, that its parts should move with uniform velocity. But all the machines of a given assemblage derive their motion ultimately from the main shaft of the prime mover, through one or more lines of shafting, from which belts or gearing pass to the separate machines, and which revolve uniformly. Hence the main shaft should likewise revolve uniformly.

The rest of this preliminary and general theorem may now be sufficiently demonstrated from Pl. XXIX., Fig. 1.

Let the crank-pin circle be divided into eight equal parts, to represent eight equidistant positions in the uniform rotary motion of the crank-pin. Then take the constant length, Q*o*, of the connecting rod, in one pair of dividers, and the constant length, *o*P, of the piston rod in another. Then from 1, 2, 3, etc., on the crank-pin circle, as centres, describe arcs, with Q*o*, as a radius, and from their successive intersections, temporarily noted, with the centre line MP, lay off the distance *o*P; which will give the piston positions 1, 2, 3, etc., corresponding to the equidistant crank-pin positions above described; as is evident from the nature of the connection of the moving parts. Now, because the successive equal arcs 0°–1; 1–2; etc., are more and more nearly parallel to the line of direction M*p* of the piston motion the nearer they are to U, it follows that the corresponding successive advances of the piston from P, to 3, 2, etc., must be at first greater and greater; while as the crank-pin approaches the point, Q′, they must be less and less, as at 2′–3′, and 3′*n*′. And it also follows that this result must be true, in general, for both the described forms of connection.

THEOREM XXIII.

The piston positions, corresponding to crank pin positions which are equidistant from the same dead-point, are identical for either connection separately.

This very evidently results in the direct connection, Pl. XXIX., Fig. 3, from the fact that the yoke, *de*, is constantly perpendicular to the piston rod, and moves with it as one piece; so that the piston, for example, will be at *f*, whether the crank pin be at *f'* or *f''*, these points being equidistant from the same dead-point, *a* (205).

The same result follows in the indirect connection, from the constant lengths of the crank and the connecting-rod, so that, for example in Pl. XXIX., Fig. 1, triangles, as *o*J'M and *o*J''M, are always equal, and have the side *o*M common, the points J' and J'' being equidistant from Q.

213. The movement of the piston from *m* to *n*, being its stroke, the same, together with the return from *n* to *m*, is called a *double stroke*, and evidently corresponds with a complete revolution of the crank pin, beginning at Q.

Let the crank pin motion UQW, be called the front half of its motion, and WQ'U its back or rear half, and let the two divisions of the double stroke corresponding to these semicircles be called the *front and back segments* of the double stroke.

The piston motion is thus properly referred to the crank-pin motion as a standard, because the latter is uniform, Theor. XXII. And the crank-pin circle is divided thus by the diameter UW, instead of by any other diameter, because, Theor. XXIII., the piston positions in either connection are the same for crank-pin positions equidistant from the opposite extremities of UW.

THEOREM XXIV.

The segments of the double stroke are equal in the direct connection, and the front one is the greater in the indirect connection. Conversely, etc.

This proposition is sufficiently obvious from mere inspection of Pl. XXIX., Fig. 3, in case of the slotted yoke connection, since the motions of the piston, yoke, and crank-pin, are all equal in the direction of the piston motion. Thus, when the piston has advanced from A to B, the yoke has advanced from *a* to *de*, carrying the crank-pin an equal *horizontal* distance, *ab*, from *a* to *d*. That is, the half semicircle, *ad*, of the crank-pin motion, corresponds with the half-stroke, AB, of the piston.

Again, in Fig. 1, operating as in Theor. XXII., we find G′K′ to the left of the centre line, GJ, as the piston position corresponding to the crank-pin positions, U or W, which establishes the first part of the theorem.

Conversely, the segments of the crank-pin circle, corresponding to the equal segments of the double-stroke each side of GJ, are unequal in the crank connection, the forward one being the less. To show this, we have only to take the length, *o*Q, of the connecting-rod, as a radius, and *o*′, the position of *o* at mid-stroke, GJ, of the piston, as a centre, and describe the arc FMD, through M, since M*o*′=Q*o* or *o*′*o*=K*m* or K*n*; and D and F will be the crank-pin position, corresponding to the mid-stroke position of the piston.

Natural Zero Points of the Piston and Crank-pin Motions, and Segments of the Double-Stroke.

214. At first glance it seems natural to divide the double-stroke into its two single strokes, as its most simple component parts, and to place the zero points at Q and Q′, Pl. XXIX., Fig. 1.

But, as we have seen, the two segments of the single-stroke which correspond to two successive quadrants of the crank-pin circle, reckoned from Q to Q′, are unequal; and the piston positions, corresponding to equal arcs each side of MU, are unsymmetrical with GJ. And, besides this, in the earlier segment of the stroke are the lead and admission, and in its later one are the cut-off and compression and release. That is, the two segments of the single stroke are unlike, both in their piston-positions, and their characteristic events.

On the other hand, if we take the zero point of the piston motion at the position corresponding to the position of the crank-pin on MU, we shall have *lead*, *admission*, *cut-off*, *compression* and *release* on *each* side of G′K′, corresponding to the crank-pin motions, UQW, and UQ′W.

Again, as the *crank-pin motion is uniform*, let *its* path be the one to be divided into equal segments by its zero points, and let the irregular divisions be on the stroke, where the motion, also, is variable, and where the events of the two parts of each stroke are dissimilar.

We will therefore take U as the zero point of the crank-pin circle, and reckon 180° each way from it. Then let G′K′, the corresponding piston position, be the zero point of the piston motion, and let the distances from G′K′, each way to the end of the stroke and back, be the segments of the double stroke.

The steam cylinder may thus be regarded as a compound one, composed of two cylinders of unequal length, estimated in opposite directions from the section G′K′, as a common base, in the common initial, or zèro plane of both.

Distinguishing the pistons of the two forms of connection as the *crank piston* and the *yoke piston*, we have the following theorem.

Theorem XXV.

The crank piston is ahead of the yoke piston during the stroke towards the shaft, and behind it during the opposite stroke.

Since the segments *m*K and *n*K of the double-stroke, Pl. XXIX., Fig. 1, are equal in the yoke connection, the *accelerations* of the piston from *m* to K are exactly symmetrical with those from *n* to K, that is also with the *retardations* from K to *n*.

Since the like segments, P0 and *n*′0, are unequal in the crank connection, P0 being the greater, while both are traversed in the same time, the acceleration from P to 0 is more rapid than from *n*′ to 0, that is, than the retardation from 0 to *n*′.

Hence it follows that as the pistons of each connection start together, from the position P*m*, the crank piston will gain on the yoke piston till the former is at K′G′; the latter being at the same time at GK. Then the yoke piston will gain on the crank piston till both coincide at *nn*.

Therefore the crank piston is ahead of the yoke piston during the stroke *towards* the shaft, and conversely, as is sufficiently evident, is behind the yoke piston during the stroke *from* the shaft.

But it is still to be noted, that after the crank piston has gained, till it is ahead by the space from GK to G'K', the yoke piston makes up this loss and catches the former at *nn*. Then starting together at *nn*, the yoke piston gains during the back semicircle of the crank pin, till ahead by the same space it lost before, when the crank piston makes up its loss and catches the yoke piston at P. Thus the division of the double stroke adopted in (214) is further justified.

Cut Off.

215. Let that part, *v*, of the valve, which covers a steam port, be called its *lip;* and when the lip is of the same width as the steam port, and when the eccentric centre, as in the foregoing examples, is just 90° behind the crank pin, let the arrangement be called a *radical valve motion;* it being the one from which, as a base, to proceed to make all necessary modifications. The eccentric centre will be thus situated because the valve is at mid-stroke when the piston is at the end of its stroke; exactly so in Pl. XXIX., Fig. 3, and sensibly so in Fig. 1, owing to the length of the eccentric *rod*, as compared with the eccentric *arm*.

The main events in the valve, or piston, stroke are the point of *admission* of steam to produce a stroke; the point of *cutting* off; the moment of *closing the exhaust*, by which steam is pent up before the piston; and the moment of *release*, when the steam which is producing a given stroke begins to escape.

It will now be convenient to study the separate effects of *angular advance*, and of *lap* (211) upon these events.

THEOREM XXVI.

The effect of a given angular advance of the eccentric, will be to afford "admission" for a new stroke, "cut-off," "exhaust closure" and "release," all at an equal number of degrees before reaching a dead point.

Let the crank pin be at any position as +45 Pl. XXIX.; Fig. 3, and moving towards *a*. The eccentric centre will then be 90° behind it, at *h*. Let *h* now be revolved $22\frac{1}{2}$° to *k*. Then *hbk* is the angular advance. Then when, in the revolution of the main shaft, *k* has come to the diameter, *ed*, the crank pin will be at $+67\frac{1}{2}$°, that is $22\frac{1}{2}$° from the next dead point *a*.

But when k comes to ed the valve will be at mid-stroke, and moving towards the shaft, and hence admission at n begins for the next stroke; cut off at n' takes place for the present stroke, the piston being at k', found by making $k'h' = Bb = a\,A$; and exhaust closure and release occur simultaneously at t and t', respectively, at the same time with the other events.

Theorem XXVII.

The effect of a given lap, alone, corresponding to a certain number of degrees from the zero diameter, is, to postpone admission for an equal number of degrees beyond the dead point; to produce cut-off at the same number of degrees before the dead point; with release and exhaust closure at the dead point.

This theorem is best established by considering the valve as at mid-travel, Pl. XXIX., Fig. 4, where, to avoid confusing Fig. 3, the cylinder and valve are considered as simply translated to the left, with the valve placed at midstroke, and lengthened by the lap, l, at each end. When the valve is at mid-travel, and moving backward, its slotted yoke is on de, Fig. 3, and the piston at A is ready for a stroke to the left. The piston rod, issuing, as before, at C, is supposed to be connected with its yoke de by links from a long cross head, and passing along each side of the engine.

Remembering that the valve and its yoke move in opposite directions, with a rocker arm, l must be laid off to the right of de, on ba, to give the yoke position, mm', corresponding to the beginning of admission, at p, when the crank-pin will be at N, so that Nbm shall be 90°.

Thus N is as many degrees beyond a, as m is beyond de. The angle mbd is called the *lap angle.*

Again, cut-off, on the stroke from C to A, evidently took place at p', when p' was just closed by the valve, moving to the left. Then mm' was as far to the left of de, and N as far above a, as the same points now are beyond de and a, in the direction of rotation, S. Exhaust closure occurred at mid-stroke of the valve, and, there being no angular advance, this would be at the dead point of the piston. Release, also, only begins when the valve has reached mid-stroke, which is at the end of the piston stroke, that is at a dead point.

It is thus evident that no very serious evil results from lap alone except to postpone admission beyond the moment of beginning a stroke.

To avoid this effect, angular advance and lap must be combined, observing that, separately, they have opposite effects upon the time of beginning admission. Let us next examine their joint effect, as illustrated in a problem.

PROBLEM XXV.

To produce a cut-off at a given crank-pin position, without preventing proper admission, etc.

Let it be required, Pl. XXIX., Fig. 5, to cut off at a crank-pin position of 50° ; where *ac*, the stroke diameter, = AC the stroke, where *dg* is the yoke, and G the piston position at cut-off, G*g* being = BB′ =A*a*.

B′*h*, perpendicular to the crank arm, B′*d*, will be the position of the eccentric arm, for the radical valve having neither angular advance nor lap. The port, *p*′, is therefore open by the space *h*′*b*′ =*hb*, and, as the two last theorems show, it cannot be closed at the present piston position, by angular advance alone, or lap alone, without displacement of the other main events of the stroke. Now, if we advance the excentric arm B′*h* to B′*k*, 20°, or half way to the mid-stroke position, the opening *h*′*b*′ will be partially closed by the amount *h*′*c*′=*hc*, and, with the same valve, admission, by Theor. XXVI., would occur 20° too soon. If, then we complete the closure by a lap, *b*′*c*′ = *bc*, corresponding to the lap angle *n*B′*k* of 20°, the cut-off will be effected at the desired point ; and admission, which would be 20° too late, by Theor. XXVII., with lap alone, is hastened by the equal contrary effect of the angular advance.

Hence to produce cut-off at a *given crank-pin position*, set the pin at that position, give an angular advance to the excentric equal to one-half the difference *ad* between the Crank-pin position and 90°, and add to the outer edges of the valve a lap = to the perpendicular distance from the new excentric position to the mid-stroke diameter, B′*n*.

Problem XXVI.

To determine the exhaust closure and release, for the adjusted cut-off and admission.

As the addition of lap, that is outside lap, to the valve does not at all affect the positions of the inner edges of the valve, relative either to each other or to the edges of the ports, we have only to consider the effect of angular advance, alone, upon the times of exhaust closure and the opening for release, just as in the case of a valve without lap.

But, by Theor. XXVI., the effect of a given number of degrees of angular advance is, to fix the occurrence of all the main events of the stroke at an equal number of degrees before reaching a dead point. Hence, for a valve giving a certain point of cut-off, the exhaust closure at p, Fig. 3, and the opening for release at p', which are simultaneous, take place at a crank position as far from a (90°) as there are degrees of angular advance.

Theorem XXVIII.

The travel of a valve with lap is the sum of twice the lap added to twice the steam-port opening.

To establish this clearly, refer to Pl. XXIX., Fig. 4, where the valve is at mid-stroke. The total travel evidently consists of the sum of the distances traversed to the right and the left of the mid-position.

First, then, at the port, p', the valve must first travel to the right, by a space $= l$, the lap, before the port will begin to be opened, and then further in the same direction until the port p' is opened to the extent required.

Second, the like successive movements must take place from the mid-position to the extreme left in order to first begin, and then continue the opening of the port p, equally with p'.

The entire movements from mid-stroke are thus equal; each is thus the semi-travel, and their sum is the *travel of the valve*; which is sensibly equal (208) to the *throw of the eccentric*, that is to the diameter of the circle described by the centre of the eccentric.

Theorem XXIX.

Inside lap prolongs the expansion and hastens compression, while inside clearance hastens the release and postpones the beginning of compression.

In Pl. XXIX., Fig. 5, the valve b'KRF, whose interior length is H, has neither inside lap or clearance, H being equal to oo'. If the piston, G, be moving in the direction of arrow P, the valve will be moving as at P′, the crank pin and eccentric centres being at d and k, and cut-off is just occurring at b', and expansion beginning. If, then, the length of the interior opening were *less* than H, in which case there would be *inside lap*, it is clear that the port, p, would be closed sooner than it will be now, which would cause *compression, before* the piston, that is between G and A, as the piston is now moving, to begin sooner than it now will.

At the same time, p' would evidently be opened later than it will be now, which will prolong the time of *expansion*, and postpone the *release*, which takes place *behind* the piston.

On the other hand, if H were greater than it it now is, there would be inside *clearance*, and p would evidently be closed *later* than it will be now; and the *compression* would be *postponed* and *shortened.* At the same time, p' would be opened *sooner* than it will be now, and thereby *expansion* would be *abridged* and *release hastened.*

To repeat, and summarily: When there is neither inside lap nor clearance, the opening of one port and the closing of the opposite one *by the inner edges of the valve*, are simultaneous and dependent on the *angular advance* alone (Theor. XXVI.). Inside lap *hastens* the *closure* of the port *before* the piston, and thereby *hastens compression;* and *retards* the *opening* of the port *behind* the piston, and thereby *prolongs expansion* by *postponing the release.*

Inside *clearance* has just the *opposite effects.*

Continuing the summary, but relative to the *outer* valve edges for convenience of reference, we have, from Theor. XXVII.,

Outside lap postpones admission and *hastens cut-off;* but has, of itself, no effect on exhaust closure or release.

Angular advance hastens admission, cut-off, exhaust closure, and *release.*

Outside clearance, if ever there were such a thing, would have just the *opposite* effect from *outside lap*. Thus, if the extreme length of the valve, V, Pl. XXIX., Fig. 1, were *less* than the sum of the bridges and the three ports, both ports would be partly open at once at mid-stroke; steam would have access to both sides of the piston at once, and cut-off of the stroke from *n'* to P would not take place till P had proceeded some distance from P to *n'* on the next stroke.

Problem XXVII.

To determine the effect of the eccentric upon the valve motion, and to counteract it, in part.

To illustrate the nature of this effect, it will be sufficient to refer to Pl. XXIX., Fig. 1.

When the piston is at P, if the valve, V, is at mid-stroke, as it should be, the eccentric centre will be in a position analogous to D, found by taking *d* as a centre, when *de* is vertical, and *d*M as a radius, and describing an arc which will cut the circle made by the eccentric centre, in the corresponding position of that centre; *above* QQ', for the motion represented by the arrows. The motion of the eccentric and valve is thus, though on a smaller scale, a counterpart of that of the piston and crank; but the irregularity is relatively smaller, owing to the much greater length of the eccentric rod as compared with the eccentric arm, than is found in the ratio of the connecting rod to the crank.

Also, *through the intervention of the rocker*, the *smaller* segment of the double stroke of the valve is its *forward* one. That is, the common base, analogous to G'K', of the two segments is to the right of GJ.

To keep the valve at mid-stroke, as at V, Fig. 1, while the eccentric centre is 90° back of the crank pin, as on UW, we have only to lengthen the eccentric rod. It will then follow that the shorter segment of the double stroke of the valve will be to the right of GJ, and therefore the width of port opening at *p'* will slightly exceed that at *p*. That is, the port *p* may not be fully opened, or the port *p'* may be overpassed, or more than opened.

Distribution of Power.

216. As the motion of the crank pin is uniform, while the average load carried by the engine is supposed to be uniform also, the *power* applied, while the crank pin is traversing the semicircle UQW, should be equal to that applied to it on the semicircle WQ′U. A *perfect* engine therefore would seem to be one in which the *power exerted* in the sub-cylinder, K′G′P, while the piston should pass from K′G′ to P and back, should equal that expended in the smaller sub-cylinder, K′G′*n*.

Whether this could be practicably and advantageously accomplished by making the valve faces and steam ports unsymmetrical with GJ, so as to cut off nearer to *n*′, on the stroke *n*′P, than to P, on P*n*′, is a question which may be left to mechanical engineers and designers, and to the larger treatises on valve motions.

The limits of this volume only permit the question to be raised, whether, as just implied, it would be possible to make the work in K′G′P, that is the average pressure on the piston when passing from *q* to P and back, multiplied by the distance 2*q*P, equal to the work in K′G′*n*′ = the average pressure when going from *q* to *n*′ and back × 2 *qn*′.

Just this may be noticed: With a constant effort applied to to the crank pin tangentially to its circle, a less total effort will be required on the arc DQF, and hence a less average pressure on the piston, from GK to P and back, than from GK to *n*′ and back; and as the eccentric motion makes the cut-off occur *later* on the stroke from P to *n*′, owing to the *greater* irregularity of the piston motion, Theor. XXV., this diminished pressure can be obtained by the *earlier* cut-off on the stroke from *n*′ to P.

Lead.

217. Owing to wear of the bearings of the crank, and the cross head pins, and the necessity in some cases of a minute play of these pins in their bearings, to avoid too much stiffness and binding, the sum of QO and OP, Pl. XXIX., Fig. 1, is not precisely constant, in actual mechanism. During the stroke *from* the shaft, that is from *n* to *m*, Q and *o* are at their greatest distance apart, being pulled apart by the steam, acting on the left side of P. During the opposite stroke, Q and *o* are pushed together. The change takes place at the ends of a

stroke, and if not made gradually, by slowly overcoming the momentum of the moving parts, the result is an injurious, and, to the ear, disagreeable "thumping," or "pounding," upon the centres. Now *compression* tends to overcome this momentum, but if employed to the very extremity of the stroke, or a hair beyond, it injures the *admission* for the succeeding stroke. Instead, therefore, of relying only on the compression of the confined steam of the previous stroke, it is better to also *admit the "live steam"* for the next stroke, an instant before the beginning of that stroke. The opening of the port at the point of beginning a stroke is called the *lead* of the valve. The corresponding distance of the crank pin from a *dead point*, is called the lead angle (212). The lead angle may vary from 0° to 8°.

Problem XXVIII.

To provide a certain lead angle without disturbance of the cut-off.

By Theor. XXVI., *angular advance* hastens *both admission* and *cut off.*

Outside lap *retards admission* and *hastens cut off.* Hence a *reduction* of it *hastens* admission, and *retards* cut off.

If, then, we *increase* the *angular advance* of the valve, say 3°, from k, Pl. XXIX; Fig. 6, the valve will begin to open the port, p, 3° *before* the crank pin reaches the + 90° point, and the cut off will also occur 3° sooner than now, or at 47°.

But if we also *reduce* the *lap* by an amount corresponding to 3° from k, the cut-off point will retreat 3°, from 47° to 50° again, and the admission will be further hastened from 3° to 6° in all. That is, admission will begin when the crank pin is 6° from the + 90° point.

Hence to produce a given lead angle, as required, *increase* the *angular advance* by *half* that angle, and *reduce* the *lap* by the other half.

Problem XXIX.

To determine the effect of lead on exhaust closure, release, and travel.

Since, by Prob. XXVI., exhaust closure and release, when there is neither inside lap nor clearance, depend only on angular advance, which hastens both, the increase of this advance by 3° hastens the both of the events named by the same amount.

The total angular advance being now 23°, the exhaust closure and release will take place at the + 67° position of the crank pin, instead of at the + 70° position as before.

Finally, as the semi-travel equals the sum of the port opening and the lap, the reduction of the lap, made to procure lead without altering the cut off, has *reduced* the *total travel* by *twice this reduction of lap.*

Theorem XXX.

The Angular Advance, estimated from the zero radius hitherto taken, is equal to the sum of the lap and lead angles, estimated from the same point.

See Pl. XXIX., Fig. 6, giving an enlarged view of the quadrant *n* B'M of figure 5. B'M is the *semi-travel.* Here, when the angular advance is increased from *hk* (= *kn*) to *hr*, by half the lead angle, the new lap = the space L.

Now, reckoning from *n*, we have *ns* = *hr*, by making *ks* = *kr*, as *kr* is at once the *increase* of the angular advance, and the decrease of the *lap* angle, *n* B'*k*, by half the lead angle. Thus *ns* = the angular advance, = the angular measure of the sum of the *lap* L, and the lead *l*, also estimated from B'*n*.

Theorem XXXI.

When the steam port is open by the amount of the lead, the opposite port is open for exhaust by the amount of the lap and lead.

See Pl. XXIX., Fig. 4, where the valve is at mid-stroke, and supposed to be moving left. Before the steam port *p*, for the time, can open *at all*, the lap *l* must be overcome, and then the exhaust port *p'* will be open to an equal extent.

Again, when the steam port *p* is opened to *a certain amount of lead*, the exhaust port, *p'*, will be evidently opened further, by the same amount, which makes its total opening as enunciated.

Port Opening.

218. In many engines, the steam ports are alternately used for the admission and the exhaust of steam. Various considerations bear upon their proportions.

First. Steam, during admission, maintains a nearly uniform pressure, while during the exhaust a single cylinder full of steam forces itself into the atmosphere by its own elasticity, and with diminishing velocity as its tension decreases by expansion. Hence when separate ports are used both for admission and exhaust, the latter should be the larger, and when one port serves both purposes, it should be adjusted, as to size, so as to secure a free exhaust.

Second. The speed of the piston evidently affects the areas of the steam and exhaust ports, both in relation to each other, and to the piston area. When the piston speed is great, so that the piston follows up the escaping steam so rapidly as to partly push it out of the cylinder, the average tension of the escaping steam will differ less from that of the incoming steam, than in case of slow piston speed; and the steam and exhaust ports may be more nearly equal. Thus, it is stated, that for piston speeds not greater than 200 ft. per minute, the area of the exhaust may well be 0.04 of that of the piston, and that of the steam port 0.02½ of the same; while for a piston speed of 600 ft. per minute, the exhaust port area should be 0.10 of that of the piston and that of steam port 0.08 of the same. In the latter case both ports are larger, and also more nearly equal.

Third. When, as in most cases, one port serves for both admission and exhaust, the size and travel of the valve, and the positions of the ports, should be so adjusted, that the valve will open the ports *fully for exhaust*, and partially, to the due proportionate extent, for *admission.*

This, of course, cannot be done with a valve having no lap, as in Pl. XXIX., Fig. 1, since, as is evident by inspection, a full opening of p', for instance, for exhaust involves a simultaneous full opening of p for admission. But, see Pl. XXIX., Fig. 4, when the valve has moved to the left till p', for instance, is fully open for exhaust, the opening at p will be less than the width of the port by the lap, l. If this opening be too little, move the valve still further to the left by a greater throw of the eccentric, till the required opening at p is obtained, and contrariwise, for due steam opening at p'.

Fourth. Having thus the proper travel, see Theor. XXVIII., the eccentric can be set, with a throw just equal to this travel. The action of the valve, by which the port p' is more than opened, for exhaust, in order to secure due steam opening, as at p, is advantageous, since it keeps p' wide open for exhaust for more than a bare instant.

Fifth. While a port, which is large enough to afford a free exhaust, *need* not be fully opened for the admission of steam, still it will do no harm to have it so opened, and may yield some advantages, especially if the port be very long and narrow.

When the travel of the valve is quite short, the valve will move more slowly, and the ports will be opened somewhat slowly, and the steam will enter with some difficulty through the very narrow opening which it first meets. This obstruction is called *wire-drawing* the steam. It may be avoided by increasing the travel and hence the speed of the valve, so that the ports will be rapidly opened. Also by making the outer edges of the valve quarter round instead of square, as at k, Fig. 4.

Increase of valve travel, so as to give a port opening greater than the width of the steam port, gives the further advantage of *quickly* bringing the ports wide open, and of keeping them so during that part of the travel by which the port opening exceeds the width of the steam port. This excess may seem a contradiction in terms, but the *port opening* simply means the distance which the outer edge of the valve, as f, Pl. XXIX., Fig. 5, moves from the outer edge, as k', of the port to its extreme position, which may be to some point as f' beyond the inner edge o, of the port. Thus $k'o$ being the port, $k'f'$ would be the port opening.

Sixth. Exhaust port opening. Whenever an inner edge, as m, Fig. 5, of the valve, being at mid-stroke at o, moves inwards towards the centre line BO, a greater distance than the width of the bridge r, it partly closes the exhaust port E. Now, to secure a proper exhaust, the remaining unclosed portion of the port, E, must be *at least* equal to the steam port. But the total movement of m, from o to the left, is the semi-travel, and thus would bring m to m', when f travels to f', $m'f'$ being simply the extreme left position of *mf*. We thus have the expression for the width of the exhaust port,

$$E = om' + m'n - r.$$
$$= om' + ok' - r.$$

that is, E = the semi-travel + the steam port − the bridge.

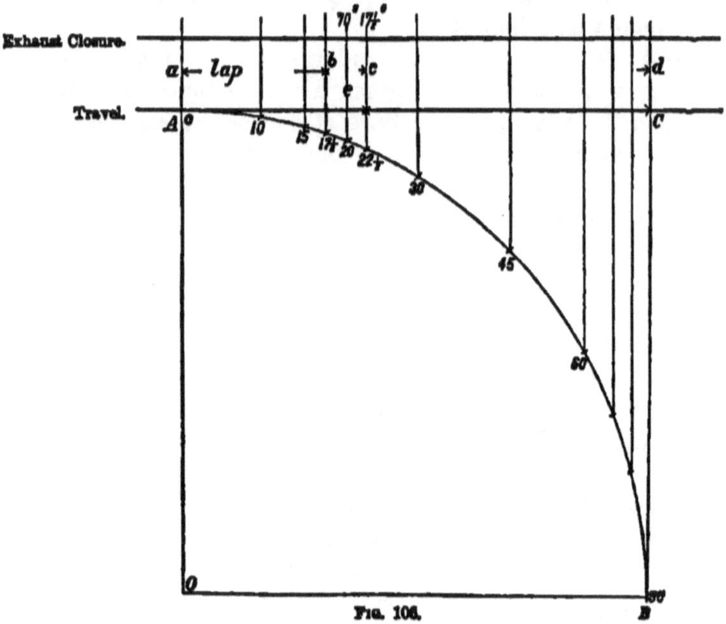

FIG. 106.

Summary of Elements.

219. The main particulars hitherto presented can be conveniently impressed upon the memory by a diagram, the construction of which is sufficiently illustrated by Fig. 106, where OB represents the crank position, and AC = OB, the *semi* valve-*travel*. OA represents the corresponding position of the eccentric arm of a valve without lap or angular advance. It is therefore the line from which (see *kn*, Pl. XXIX., Fig. 6) the lap angle is to be estimated.

Now, *first*, suppose an angular advance of 20° and a lap-angle also of 20° from OA (Prob. XXV.) giving *ae* for the *lap* and *ed* for the *port opening* (Th. XXVIII.), and cut-off at 50°, and exhaust closure at 70°.

Then, *second*, let the angular advance be increased to $22\frac{1}{2}°$, and the lap angle reduced to $17\frac{1}{2}°$ (Prob. XXVIII.), and we shall have—

ab = lap.
bc = lead.
bd = port opening.
ad = semi-travel.
exhaust closure at $67\frac{1}{2}°$, and
cut-off at 50°.

Or, if the port opening = *ed*, as before, the total travel will be reduced by twice the reduction of the lap. Notice that *bc* is not reckoned for the 5° from OA, since, in fact, the angular advance is made first from *h* to *k*, Pl. XXIX., Figs. 5, 6, to leave the *primitive lap angle* *k*B′*n*, and *then* increased to give the semi-lead angle, between the full and dotted lines *k*B′, and to reduce the *lap* (Prob. XXVIII.). By inside clearance = *be* (Th. XXIX.) the point of exhaust closure may be restored to 70°.

Upon this diagram, it is only necessary to remark:

1*st*.–Similar illustrations may be made, beginning with any other angular advance and lap angles, taken at first equal to each other.

2*d*.–Having found the lap, port opening, and lead, for any travel, the same can be found for any other travel, by the simple principle that they are all proportional to the travel. For if Fig. 106 were reduced, till AC were any given part of what it now is, all parts of the figure would be reduced in the same proportion.

220. The matter already presented enables us to determine various interesting particulars. The *piston position* for any given *crank pin position* can easily be found, as already shown. The angular advance and lap angles for given *lead* and *cut-off* being given; the travel, and the *eccentric position* for any *crank-pin position* can readily be found, and the *valve position* for any given *eccentric position* can then be determined, so that finally the valve position due to any piston position can easily be constructed.

Indeed, the student cannot now better exercise himself on this subject, than by constructing, very accurately, and on a large scale, from given data, taken from actual practice, a figure similar to Pl. XXIX., Fig. 5, but with the piston positions for all the crank positions, 10° apart, and for those at 5° apart, near the dead points; with the corresponding valve positions ranged in a vertical column, and all referred to the same vertical centre line, or plane, BO.

Another most useful way of becoming familiar with the relations of the valve edges and ports in all positions, is to make a large section, like any of the sections of the valve and its seat on Pl. XXIX., to scale from measurements (see Fig. *n*). Then cut the figure apart on the line as *fg*, Fig. 1, of the valve face, and move the valve back and forth, through equal small intervals, which will clearly show its relations to its ports.

Problem XXX.

To Reverse the Motion of an Engine.

The Drop Hook Motion.

Referring once more to Pl. XXIX., Fig. 5, with crank-pin at *d*, moving as at arrow (1), and the eccentric centre at *k*, moving the valve (rapidly) to the left; let a second eccentric be placed at k'', where $k''t = kt$. Let the rod of eccentric *k* be disconnected from the rocker $H'b''$, and let the equal rod from k'' be put into connection with $H'b''$. To do so, H′ will be moved to *s*, where $B''s = b''k''$, and the rocker will appear at ss'. Then lay off from s' the distance $h''i$, and it will give i', the new position of the valve centre (shown on b'F carried vertically upward to Fig. *n* to avoid further confusion of the lower figure). All the edges $b'l$, F, etc., of the valve, b'KR(H) F, and its ports, are projected on b'F, Fig. (*n*).

Then, finally, lay off on each side of i', half of the exterior length b'F, and interior length, H, of the valve, and we shall have its position, $b'H'F'$, when in connection with the eccentric at k''.

Before, *d* was moving as at arrow (1), the piston G was moving to the right, by the expansion of the steam behind (to the left of) it, the exhaust was open by the amount *ov*, and the engine was going *forward.*

Now, the port *p* is open by the space $F'k'$, the port p' is wide open for exhaust, G is therefore moving to the left, the crank-pin *d* is moving as at (2), and the engine is going *backward.*

Whence we conclude that, to reverse an engine, the eccentric must be adjustable upon the shaft, as from k to k''; or there must be two eccentrics, as at k and k'', each of which, separately, may be put into communication with the rocker.

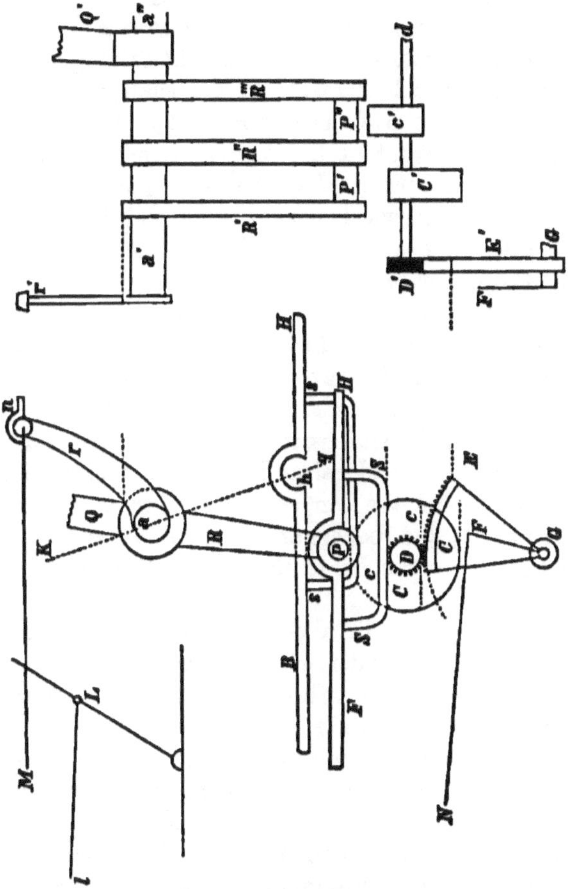

FIG. 107.

The latter method is now in general use.

Formerly the eccentrics were separately put in gear with the rocker by the old drop-hook motion still seen only on a few old engines, though generally used in this country before 1850, and often previous to 1860. Space forbids more than the rude illustration of it in Fig. 107, the study of which will greatly add to the student's appreciation of the "handiness" of the link motion, to be afterwards briefly explained.

The right-hand view is as it would appear in an end view of the engine, and the other is as appears on a front elevation of the engine.

a—*a*′*a*″ is the rocker shaft in two parts, one on each side of the engine. R,R′—R‴ the lower, and Q,Q′ the upper rocker arm. FH is the forward eccentric rod, or rod for forward motion, proceeding from an eccentric on the driving shaft, and now behind the vertical diameter of the shaft, as R is drawn well back. BH is the back eccentric hook, out of gear. Its eccentric, now, is in front of the vertical diameter of the driving shaft. The crank pin is *above* the driving shaft; the valve, see R and Q, is well forward; steam is entering the rear steam port, that is the one nearest the shaft, and the piston and the engine are moving *forward.*

To reverse the engine, draw back the arm, FG, by a lever in the cab. This will revolve the sector, E,E′, and thence give the small gear, DD′, half a revolution on its shaft, D*d*, also the same to the semi-circular cams, C—C′, and *c*—*c*′. This operation allows BH, which rests, by its stirrup *s*, on the cam, *cc*′, to fall by its own weight; and raises FH off the rocker pin, *p*,*p*′. Then, by drawing back the arm, *rr*′, by a lever in the cab, and through the rod M*n*, the rockers are revolved to the position K*h*, where the hook, *h*, settles on to the pin, *pp*″, revolved to the position *qp*″.

The valve, being then in a position to admit steam to the right-hand end of the cylinder, while the crank is still above the shaft as before, the engine will go *backward* from a state of rest, when before it would have gone *forward.*

The eccentric rods being side by side, the lower rocker, R, is double, as shown on the end view; where, to avoid confusion, the end view of the eccentric rods is omitted.

This was not a very rude contrivance, and was used on new first-class express passenger engines built as late as 1857. Still it had many inconveniences. The *hooks* (*h*) might bound off their pins, or, when raised, they might get jambed between the rockers, as R′ and R″ and fail to fall into place promptly when the cam shaft was tumbled over. But worst of all was the delay when instant reversal was required. For, to prevent the rapid oscillation of the levers in the cab, for swinging over the rock-shafts in reversing, the rod, M*n*, was lifted off the pin, *n*, by an arm, L, drawn to an upright position by a lever acting through Ll, unless the latter presented a ring to hand, at the cab. Then 1*st*, M*n*, must be settled on to its pin on the arm, *r*. 2*d*. The cam shaft must be tumbled. 3*d*. The rockers, R, must, by their motion, catch the hook intended for them. And, after all, from a scientific point of view again, there was no cut-off, at least no variable one. Engines in those days generally had little or no lap to their valves, so that the eccentrics were essentially at right angles to the crank, and so that there was no cut-off, or but very little.

Cut-offs, then, were separate valves, in separate two-ported steam (valve) chests, directly over the main valve chest, and were generally invariable cut-offs at that. This required one or two more levers, while the link motion, with but one lever, is a *variable* cut-off.

Example LIX.

A Stephenson Link Motion.

Description.—This motion is so called from the engineer who first brought it into use. It was invented by a Mr. Howe, an Englishman.

There are various other English and continental link-motions. The Gooch, or stationary link, except as to its oscillation by the eccentrics, and in which the link block rises and falls to reverse an engine. The Allan, or straight link, and others.

Pl. IX., Fig. 3, is merely a sketch and measurements from a link-motion model, not showing the valve, rocker, and eccentrics in their true relative position, but which may quite as well, for that, illustrate the operations of drawing such a model and adjusting it. L is the link, and to catch the main idea of its operation it is only necessary to conceive the forward and back eccentric rods in the last example to be attached to it as at *f* and *b*. The link block pin, *l*, is held by the lower end of the rocker arm, R. Then, by letting down the link, by the hand-lever, C, working on the *arch*, A, till the mean positions of *f* and *l* are about on O*d*, the forward eccentric, alone, will actuate the valve. But if the link be raised, till *b* and *l* are in like manner together, the back eccentric, only, operates the valve, essentially as in the drop-hook motion, only that but one motion of one lever is required to raise or lower the links on both sides of the engine. When it is set so that l is at various points between b and f, l will have the resultant motion due to the joint action of both eccentrics. The effect will be to reduce the travel, and hence, obviously, to *vary the cut-off*, by the quicker shutting again of a steam port after it begins to open.

Thus the link motion is a variable cut-off.

Construction.—There are so many parts and points to a link-motion, that, in practice, it is usual to fix some of the latter, with reference to convenience or practicability, as governed by surrounding parts of the engine.

Thus, taking O, the centre of the crank shaft, as an origin, or fixed point of reference, and Od as the horizontal centre line, containing the axis of the cylinder, fix T, the centre of the "*tumbling shaft*," by the co-ordinates $25\frac{7}{8}''$ and $6\frac{3}{16}$ths, in this example. Also, S, the centre of the rocker shaft, by the co-ordinates $34''$ and $4''$.

The fixed dimensions of the link in this case, are the radius of its central arc, the "link-arc," $34''$, the distance between the eccentric rod pins $5\frac{7}{8}''$, their distance back of the link-arc $1\frac{1}{2}$,″ and the length of the saddle $4\frac{7}{8}''$. The *radius of the link arc* must = the eccentric rod, Ff, + the $1\frac{1}{2}''$ just mentioned, to avoid shifting the centre of the travel, in different "gears."

Let the *travel* of the valve be $2\frac{1}{4}''$. For this purpose, if the rocker arms are *equal*, as in the figures, each $= 4\frac{3}{32}''$, the half throw of the eccentric, or radius of the eccentric centre, will be sensibly $1\frac{1}{8}''$, or half the travel.

When the forward gear eccentric rod, $F'f$, is in action, it will be nearly, or exactly, on the centre motion line Od, and when the valve V is at its extreme left. The point l, the *link-block pin*, which is held by the lower end of the rocking arm R, must then be at its extreme right or $34''+1\frac{1}{8}''=35\frac{1}{8}''$ from O. Hence the length of the eccentric *rod*, as Ff, from the eccentric centre to f, must be $35\frac{1}{8}-(1\frac{1}{8}''+1\frac{1}{2}'')=32\frac{1}{2}''$.

The distance from the eccentric centre F, to the outer face of the collar k, Fig. m, of the eccentric strap being $4\frac{3}{8}''$, the length kf, of the eccentric rod proper will be $28\frac{1}{8}''$, or F to $l = 34''$.

These descriptions show the relation between the eccentric rods, and the rocker shaft S. That is, when the valve is at mid-stroke, the rocker arm is vertical, usually, and the link block pin, l, must then be at the same *horizontal* distance from O, that the point S is. The circle O—FB is that described by the centres of the eccentrics, sensibly of $1\frac{1}{8}''$ radius, for a travel of $2\frac{1}{4}''$. The position of l can be found for any assumed valve position.

Let us, then, consider the seven leading positions of V and l.

First. When the valve is at mid-stroke, the rocker arms will be vertical, and any motion, either way, of the valve, will open an *exhaust* passage, or give release to the steam.

Second. If V be drawn *each way* from mid-stroke by the amount of the *lap*, the valve will be at the points where cut off takes place, and l will be at positions, which we will call $l'l'_{,,}$ Fig. (y), at the same distance on the opposite sides of its mid-position, when the rocker is vertical.

Third. If V be *further* drawn *each* way from its mid-position, by the amount of *lead* proposed, l will have a pair of positions $l'',l''_{,,}$ at the same additional distance each way from its mid-position. That is the *horizontal* distance from l to $l''=$ *lap* + *lead*.

Now the last pair of positions of the valve are, of course, those which occur when the crank-pin is at the *centre* or *dead points*. The *lead* must then be the same at one port, *or* the other, according as the crank pin is to go one way or the other. Hence the *eccentric centre positions*, corresponding to l'', and $l''_{,,}$ will be as at B′ and F′, equidistant from D′, the forward dead point of the crank pin. And $90°-F'OD'=90°-B'OD'$ will be the angular advance of the eccentrics.

This gives both eccentrics, crank, and valve in one set of simultaneous positions. If the link be dropped into place for "*full gear forward*," that is, for forward motion of the engine, with the valve moving with full stroke, l will be at its position $l''_{,,}$ Fig. (y), and the required *lead* will be seen in the amount of opening of the port p.

Fourth. Let the valve finally be drawn to each of its extreme positions, and we shall then have the corresponding extreme positions l''' and $l'''_{,}$ of the link block pin l.

To find one position of the link.

221. By (Theor. XXX.) knowing the *travel*, or diameter of O—FB, the *lap*, and the *lead*, F′ and B′, the positions of the eccentric centres, when the crank pin is at D′, can be found by making their perpendicular distance from DO = the *lap* + the *lead*, as just now indicated.

Then, from F′ and B′ as centres, with radius equal to the length to f and b, 32½″, in this case, describe arcs, which will contain the points f and b, as the link rises and falls.

When F′ and B′ are the eccentric centres, the points f and b will have one pair of positions at a perpendicular distance from Od, on each side of it, equal to half fb or $2\frac{1}{16}$″. Then draw lines parallel to Od, at this distance from it, and note where they intersect the arcs before drawn from F′ and B′ as centres. An arc, through the positions of f and b, thus found, with 32½″ radius, will have its centre, Q, (not shown) on Od, and an arc of 34″ radius, and Q as a centre, will be one position of the link arc.

Data for finding any position of the link.

222. 1*st.* F and B are always on the circle FBF′, and at a chord distance apart = F′B′, as previously found.

2*d.* f and b are at a constant distance apart.

3*d.* Ff and Bb are of constant length, for any one arrangement of the model.

4*th.* s, the saddle-pin, forms with f and b a known triangle, fsb, where fs may be less than bs, to reduce the *slip* of the block in the link, the path of b being more nearly parallel to that of l.

5*th.* T being a rigid joint, and t a flexible one, the different positions of s will all be on an arc with t as a centre, and radius ts, in this case of 7″.

6*th.* The centre from which the link arc is described is always on a perpendicular to fb, at its middle point.

With these data, the student is left to construct various positions of the link, either by intersections of lines, or by a slip of stiff paper or thin wood, cut to fit the curve of the link arc, and on which the points f, b and s are fixed.

For any position of the link, l is always the intersection of the link arc, with the arc described by l from S as a centre, with the rocker arm, R, as a radius, $4\frac{3}{32}$″, in this example.

When the valve rod is attached directly to l, slip is avoided, as Zeuner shows, by making ts and tT, each, half of a parallelogram, turning on two opposite centres of which T is one, while the side ts carries a slotted guide in which s moves horizontally in all gears, and thus without vertical motion of the link.

To adjust the Model.

223. The several adjustable arms and rods are telescopic, and fitted with clamps to adjust their length.

First. To adjust the Travel. Set the fixed measurements as given, to locate T and S, and fix the lengths, F*f*, and B*b*, and make $sl = \frac{1}{4}''$. Drop the link into the *full-gear-forward* position, and make the half-throw of each eccentric = half the proposed travel, and set both eccentrics as nearly as possible to their intended positions, as F′ and B′ for the crank at D′, according to the lap and lead required. Turn the crank till the vertical centre line of the valve coincides with VH, that of the cylinder. Then turn the crank carefully and measure the distances of the extreme positions of the valve centre from VH. If their sum varies from the travel ($2\frac{1}{4}''$) alter the half-throw of the eccentric by *half the error* till the required *amount* of travel is secured. Then if these distances are unequal, alter the *length of the valve stem, ab*, till they become equal. Thus, if the valve move further to the *right* of VH than to the left, we should *lengthen* the valve stem by *half the difference.*

The *amount* and *equalization* of the travel for one eccentric are now accomplished, but without regard to the relative position of the valve and piston. That will next be attended to.

Second. To give a certain lead to the valve. This is accomplished by the *angular* movement, only, of the eccentric. Then clamp the crank at OD′, unclamp the eccentric, F, and rotate it on its shaft till the port *p* is open the desired amount, $\frac{1}{16}''$. Then clamp it. Then, if the crank be clamped at OD″, the lead at *p′* will be the same, and for a single eccentric the adjustment will be complete.

Third. To adjust the backward gear, B*b*, *etc.*

1*st.* Test the *amount* of travel, and perfect the throw of the eccentric, B.

2*d.* Test the *equality* of travel each way from VH, and if unequal, equalize it by a slight adjustment of the length of B*b*, since to alter the valve stem would disturb the equalized travel of the forward gear.

3*d.* Clamp the crank at OD′, and rotate the back eccentric till a lead of $\frac{1}{16}''$ appears at *p*, and clamp it, when, if the crank be revolved to OD″, the same lead should appear at *p′*.

Fourth. To readjust the full gear forward. The adjustment of the gear backward will sometimes a little disturb that of the forward gear. If so, re-equalize the travel by a slight adjustment of the length of F*f*. Then reset the *lead* by a small angular movement of the eccentric, the crank being at OD′ or OD″.

Remarks.—The foregoing operations are easy, yet may become tiresome by overlooking some little practical points. Clamp each fixed part tightly. The rocker arms should be vertical at mid-stroke of the valve. See that the eccentric rods are of the proper length and do not slip. Let the points *f* and *l* be about in a horizontal line in the extreme forward position of the eccentric. If the crank overpasses a dead point where it should stop, do not *back* it up to the point, but go back some distance and then come *forward* to the point. This is to "take up the lost motion," or play between parts at all the joints.

224. When the model is finally adjusted, the valve can be set by turning the crank at the points—of beginning to open a steam port; cutting off; exhaust closure, or beginning of compression; and of release, or end of expansion, and the corresponding piston positions can be noted on the scale H for—Full gear forward; full gear backward, and any intermediate gear.

The model being adjustable in all parts, other travels may be taken. The following are specimen results: Lap = $\frac{1}{2}$″.

1°—Full Gear Forward. Travel = $1\frac{7}{8}$″.

Lead . . . . = $\frac{1}{16}$″ at each end.
Front port opening, p = $\frac{6\frac{1}{2}}{16}$″. Back port opening, p' = $\frac{7}{16}$.
Cut off at. . . . . = 17.4″ (half inches) forward stroke.
" " ". . . . . = 16.7 " " backward "

2°—Partial Gear Forward. Travel, $1\frac{1}{4}$″.

	Front.	Back.
Lead	$\frac{3}{32}$″	$\frac{3}{32}$″
Out off. . . . . ,	6.1 (half ins.)	6.4 (half ins.)
Opening	$\frac{9}{64}$	$\frac{9}{64}$

3°—Full Gear Forward. Travel, $2\frac{1}{4}$″.

	Front.	Back.
Lead	$\frac{1}{16}$	$\frac{1}{16}$
Cut off at	19 (half ins.)	18.75 (half ins.)
Opening	full.	full.

4°—Full Gear Backward.

	Front.	Back.
Lead	$\frac{1}{16}$	$\frac{1}{16}$
Cut off...............	$19\frac{3}{8}$ (half ins.)	19 (half ins.)
Opening	full.	full.

Again, in a little different form, and more fully.

5°—Travel $1\frac{7}{8}''$, Stroke $12''=24$ half ins.

	Full Gear Forward.		Full Gear Backward.	
	Front End.	Back End.	Front End.	Back End.
Lead	$\frac{1}{16}''$	$\frac{1}{16}''$	$\frac{1}{32}$	$\frac{1}{32}$
Opening	$\frac{6\frac{1}{2}}{16}''$	$\frac{6\frac{1}{2}}{16}$	$\frac{6}{16}''$	$\frac{5\frac{1}{2}}{16}''$
Cut-off in $\frac{1}{2}$ ins...	18.2	17.6	16.9	16.7

6°—Travel $2\frac{1}{4}''$, Stroke $12''=24$ half ins.

	Full Gear Forward.		Full Gear Backward.	
	Forward Stroke.	Backward Stroke.	Forward Stroke.	Backward Stroke.
Lead	$\frac{1}{16}$	$\frac{1}{16}$	$\frac{1}{16}-$	$\frac{1}{16}+$
Opening	$\frac{9}{16}$	$\frac{9}{16}$	$\frac{9}{16}$	$\frac{9}{16}$
Cut-off..... $\frac{1}{2}$ ins.	19.	19.2	19.35	19.45
Compression $\frac{1}{2}$ ins.	22.4	22.5	22.6	22.6

7°—Partial or Mixed Gear. Travel reduced to $1\frac{9}{16}''$.

	Forward Motion or Gear.		Backward Motion or Gear.	
	Forward Stroke or End.	Backward Stroke or End.	Forward Stroke or End.	Backward Stroke or End.
Lead	$\frac{1}{8}''$	$\frac{1}{8}''$	$\frac{1}{8}-$	$\frac{1}{8}$
Opening	$\frac{5}{16}''$	$\frac{5}{16}''$	$\frac{5}{16}''$	$\frac{5}{16}''$
Cut-off..... $\frac{1}{2}$ ins.	12.″	12	12	12
Compression $\frac{1}{2}$ ins.	19.3	19.6	19.25	19.6

"*Throw*" is differently defined as the *radius*, or the *diameter* (208) of the circle made by the eccentric centre. Either way has its convenience. The question, "how far will the eccentric throw anything," gives the answer throw = the *diameter* named.

With this summary account, the reader is referred to the works of Auchincloss, Zeuner, Colburn, and others, in which this and other link motions are more fully treated than is possible or necessary here.

By now putting together the eccentric, with its straps and rods, Ex. XXIX.; the link, Ex. XXVIII.; the cylinder, Ex. X.; and valve, Ex. XLII.; with the valve stem and rockers, the student can draw a valve motion, from which he can learn much. The following may afford further data for practice in drawing, while the tables annexed are interesting as experimental determinations of the best relative positions of points of the link for making the main events of each stroke alike.

Example LX.

Data of Valve Motions.

I.—*Valve Motion of a* 15″ × 22″ *Cylinder Passenger Engine: Atlantic and Gt. Western R. R.*

Length of connecting rod .... = 6′–10¼″
Centre of shaft to centre of rocker .... = 5′– 9⅜″
" line of engine to centre of rocker, vertically = 5⅛″
" line of link to centre of eccentric rod pin .... = 3⅛″
" of tumbling shaft, from centre of driving axle, horizontally .... = 4′– 4¾″
" of tumbling shaft, above centre line of engine = 10¾″
Radius of link, centre arc .... = 5′– 8⅜″
Length of suspending link .... = 13½″
Distance between centres of eccentric rod pins ..... = 11½″
Saddle pin back of link arc .... = $\frac{7}{16}$″
Lower rocker arm out of centre, towards axle ...... = ⅜″
Length of eccentric rod .... = 5′– 5⅝″

Travel of valve .... =	4″	Rocker arms, each . =	9⅝″
Outside lap .... =	¾″	Bridges	1⅛″ × 15″
Inside lap .... =	0	Exhaust port	2 ″ × 15″
Steam ports	1 ″ × 15″	Drivers, diameter .. =	5′– 0″

II.—*Valve Motion of a* 16″ × 24″ *Cylinder Passenger Engine: New York C. and Hudson River R. R.*

Length of connecting rod	7′– $5\frac{1}{4}$″
Centre of shaft to centre of rocker, horizontally	5′– 3″
" " rock-shaft above centre line of Cyl.	$6\frac{5}{8}$″
Eccentric rod pin back of centre arc of link	$3\frac{1}{8}$″
Centre of tumbling shaft, from centre of main axle, horizontally	3′– $7\frac{1}{8}$″
Centre of tumbling shaft, below centre line of engine	1′– $3\frac{1}{2}$″
Radius of link arc	5′– 3″
Length of supporting link	1′– 2″
Lower rocker pin out of centre towards axle	$\frac{5}{16}$″
Saddlo pin back of link arc	$\frac{7}{16}$″
Length of eccentric rods, centre of eccentric to centre of eccentric rod, = knuckle joint, pins	4′–$11\frac{3}{4}$″

Eccentric rod pins, apart	13″	Rocker arms, each	9″
Travel of valve	5″	Bridges	1″
Outside lap	$\frac{7}{8}$″	Exhaust port	$2\frac{1}{2}$″ × $14\frac{1}{2}$″
Inside lap	$\frac{1}{16}$″	Four coupled drivers, Diam.	5′– 2″
Steam ports	$1\frac{1}{8}$″ × $14\frac{1}{2}$″		

III.—*Valve Motion of an* 18″ × 22″ *Cylinder Freight Engine: N. Y. C. and Hudson River R. R.*

Steam ports	$1\frac{3}{8}$″ × 15″	Outside lap	$\frac{3}{4}$″
Bridges	$1\frac{1}{2}$″	Inside lap	$\frac{1}{16}$″
Exhaust port	3″ × 15″	Eccentric, diam.	$14\frac{3}{8}$″
Valve travel	5″	Length of rockers	9″

Hor. dist., centre of main axle to centre of rock-shaft	55″
Centre of rock-shaft above centre line of engine	$8\frac{1}{2}$″
Radius of link, centre arc	50″
Saddle pin back of link arc	$\frac{1}{2}$″
Length of eccentric rods, as above (II.)	52″
Eccentric rod pins, = knuckle joints, apart	13″
" " back of link arc	$3\frac{1}{4}$″
Centre of eccentric to link arc	$55\frac{1}{4}$″
Hor. Dist. of lifting (tumbling) shaft from centre of driver	$36\frac{1}{8}$″
Centre of lifting shaft below centre line of engine	14″
Length of lifting arm	18″
Six coupled drivers, Diam.	4′– 9″

Determination of well adjusted [Valve Motion for a 15" × 22"

Forward Motion.										Back Motion.									
Forward Stroke.				Backward Stroke.						Forward Stroke.				Backward Stroke.					
Lead.	Opening.	Cut off.	Compression.	Lead.	Opening.	Cut off.	Compression.	Difference.	Valve Travel.	Lead.	Opening.	Cut off.	Compression.	Lead.	Opening.	Cut off.	Compression.	Difference.	Valve Travel.
1.	CENTRE OF SADDLE DIRECTLY OVER CENTRE LINE OF LINK.																		
	full				full			Cut off.											
1/16"	=1"	18 1/2"	1"	1/16	=1"	18 5/8	1"	1/8"	4"	<— T	umb	ling		sh	aft	arm	—	18"	—>
1/8	7/8	16 1/8	2"	1/8	3/4	15 3/4	2 1/8"	5/8	3 3/32										
3/16	5/8	14 1/8	3	3/16	9/16	13 5/8	3 1/8	3/4	2 11/16										
3/16	7/16	11 1/2	4 1/8	3/16	7/16	10 5/8	4 5/8	1 1/8	2 5/8										
7/32	3/8	9 7/8	5	7/32	3/8	8 5/8	5 5/8	1 1/4	2 1/4										
7/32	5/16	7 7/8	6	7/32	5/16	7"	6 3/4	7/8"	2 3/32										
2.	CENTRE OF SADDLE 7/16" FROM CENTRE LINE OF LINK.																		
1/16	1"=full	18 1/2	1	1/16	1"	18 5/8	1	1/8	4"	1/16	full.	19 1/8	3/4"	1/16	full.	18 3/4	7/8	3/8	4"
1/8	7/8	16 5/8	2	1/8	1 3/16	16 5/8	2	0	3 1/4	5/32	1 1/16	15 3/4	1 7/8	5/32	1 3/16	15 3/4	2 1/4	0	3
3/16	11/32	14 1/4	2 7/8	3/16	3/4	14 1/8	2 7/8	1/8	2 3/4	3/16	9/16	13 3/4	2 3/4	3/16	5/8	13 7/8	3	1/8	2 11/16
3/16	9/16	12 3/4	3 1/4	3/16	1/2	12 3/4	3 5/8	1/4	2 9/16	3/16	7/16	11 5/8	3 5/8	3/16	1/2	12 1/8	3 3/4	1/8	2 7/16
7/32	7/16	10 1/4	4 1/2	7/32	3/8	10	4 3/4	1/4	2 5/16	7/32	3/8	9 1/4	4 5/8	7/32	7/16	10 1/8	4 3/4	7/8	2 5/16
7/32	3/8	8 1/2	5 5/8	7/32	5/16	8 1/2	5 7/8	3/8	2 3/16	7/32	5/16	7 1/4	5 3/4	7/32	3/8	8 5/8	5 7/8	1 1/8	2 3/16
3.	CENTRE OF SADDLE 1/8" FROM CENTRE LINE OF LINK.																		
1/16	full.	18 1/2	1 1/8	1/16	full.	18 5/8	1 1/8	1/8	4"	1/16	full.	19	7/8	1/16	full.	18 3/4	1	1/4	4"
5/32	1 3/16	16 1/8	2	5/32	1 3/16	16 1/8	1 7/8	0	3 3/16	5/32	31/32	15 7/8	1 7/8	5/32	25/32	15 3/4	2 1/8	1/8	3
3/16	5/8	13 3/4	3	3/16	5/8	13 7/8	3"	1/8	2 11/16	5/32	9/16	14	2 5/8	5/32	3/4	14	3	0	2 11/16
3/16	1/2	12 1/8	3 3/4	3/16	1/2	12 3/8	3 5/8	1/4	2 1/2	3/16	7/16	11 3/4	3 5/8	3/16	1/2	11 3/4	3 7/8	0	2 7/16
7/32	3/8	9 7/8	4 7/8	7/32	3/8	10	4 7/8	1/8	2 1/4	3/16	3/8	9 7/8	4 5/8	3/16	7/16	10 1/8	4 5/8	1/4	2 1/4
7/32	5/16	7 7/8	6	7/32	5/16	8	6	1/8	2 1/8	3/16	3/8	7 7/8	5 5/8	3/16	5/16	8 3/4	5 7/8	7/8	2 5/32
4.	CENTRE OF SADDLE 9/16" FROM CENTRE LINE OF LINK.																		
1/16	full.	18 1/2	1	1/16	full.	18 5/8	1	1/8	4"	1/16	full.	19 1/8	7/8"	1/16	full.	18 3/4	1	3/8	4"
1/8	7/8	16 5/8	2	1/8	7/8	16 1/8	1 7/8	1/8	3 1/4	5/32	3/4	16	1 7/8	5/32	1 3/16	15 3/4	2 1/4	1/4	3
5/32	5/8	14 1/8	3	5/32	5/8	14 1/4	2 3/4	1/8	2 3/4	3/16	9/16	13 7/8	2 5/8	3/16	5/8	14	3	1/8	2 11/16
3/16	1/2	12 3/8	3 5/8	3/16	1/2	12 3/8	3 1/2	0	2 1/2	3/16	7/16	11 3/4	3 1/2	3/16	1/2	11 7/8	3 3/4	1/8	2 7/16
3/16	3/8	10 5/8	4 5/8	3/16	3/8	10 5/8	4 1/2	0	2 5/16	7/32	3/8	9 5/8	4 1/2	7/32	7/16	10 1/8	5	1/8	2 1/4
7/32	5/16	8 1/4	5 3/4	7/32	5/16	8 1/4	5 1/4	0	2 1/8	7/32	5/16	7 1/2	5 1/8	7/32	5/16	8 1/8	5 7/8	3/8	2 3/16
7/32	7/32	2 1/4	11	7/32	7/32	2 3/4	11	1/8	1 15/16	<	—M	id-g	ear	Sa	ddle	9/16"	etc.	—	—
1/4	1/4	2 1/4	11	1/4	1/4	2 1/8	11	1/4	1 7/8	<	—	"	"		"	7/16"	"		

Cylinder Engine,] *by successive experimental approximations.*

Forward Motion.										Back Motion.									
Forward Stroke.				Backward Stroke.						Forward Stroke.				Backward Stroke.					
Lead.	Opening.	Cut off.	Compression.	Lead.	Opening.	Cut off.	Compression.	Difference.	Valve Travel.	Lead.	Opening.	Cut off.	Compression.	Lead.	Opening.	Cut off.	Compression.	Difference.	Valve Travel.
5. CENTRE OF SADDLE 5/8" FROM CENTRE LINE OF LINK.																			
1/16	full.	18 1/2	1"	1/16"	full.	18 5/8	1"	1/8	4"	1/16"	full.	19 1/8	3/4	1/16	full.	18 5/8	1	1/8	4 1/32
1/8	13/16	16"	2	1/8	13/16	16 1/8	2	1/8	3 1/8	1/8	3/4	16	1 7/8	1/8	13/16	15 5/8	2 1/4	3/8	3"
5/32	9/16	13 1/2	3	5/32	9/16	13 3/4	3 1/8	1/4	2 11/16	3/16	9/16	14 1/8	2 1/2	3/11	5/8	13 7/8	3	1/4	2 11/16
3/16	1/2	12	3 7/8	3/16	1/2	12 1/2	3 3/4	1/4	2 7/16	3/16	7/16	12	3 3/8	3/16	1/2	12	4	0	2 7/16
3/16	11/32	9 1/2	4 7/8	3/16	3/8	9 3/4	4 3/4	1/4	2 1/4	3/16	3/8	10	4 1/4	3/16	7/16	10	4 3/4	0	2 5/16
7/32	9/32	7 5/8	6 1/4	7/32	5/16	7 3/4	6	1/8	2 1/8	7/32	5/16	8	5 1/2	7/32	3/8	8 5/8	5 3/4	3/8	2 3/16
6. CENTRE OF SADDLE 9/16" FROM CENTRE LINE OF LINK. TUMBLING SHAFT-ARM = 17".																			
1/16	full.	18"	1 1/4	1/16	full.	18 1/8	1	1/8	3 3/4										
1/8	13/16	16 1/8	2	1/8	13/16	16 3/8	1 3/4	1/4	3 1/8										
5/32	9/16	13 7/8	2 7/8	5/32	9/16	14	2 3/4	1/8	2 5/8										
3/16	1/2	12 3/8	3 5/8	3/16	1/2	12 1/4	3 1/2	1/8	2 7/16										
7. CENTRE OF SADDLE 9/16" FROM CENTRE LINE OF LINK. TUMBLING SHAFT-ARM, 17 3/4".																			
1/16	full.	18 5/8	7/8	1/16	full.	18 5/8	7/8	0	4	1/16	full.	19 1/8	7/8	1/16	full.	18 5/8	1	1/4"	4"
1/8	7/8	16 5/8	1 3/4	1/8	7/8	16 5/8	1 5/8	0	3 3/16	5/32	11/16	15 5/8	2	5/32	3/4	15 5/8	2 5/8	1/8	3
5/32	5/8	14 5/8	2 5/8	5/32	5/8	14 1/2	2 5/8	1/8	2 3/4	5/32	9/16	13 5/8	2 3/4	5/32	5/8	13 5/8	3 1/4	0	2 5/8
5/32	1/2	12 7/8	3 5/8	5/32	1/2	12 7/8	3 5/8	0	2 1/2	3/16	7/16	11 5/8	3 5/8	3/16	1/2	11 5/8	4	1/4	2 5/8
3/16	3/8	10 1/2	4 5/8	3/16	3/8	10 5/8	4 5/8	1/8	2 1/4	3/16	3/8	9 1/4	4 5/8	3/16	3/8	9 3/4	5	1/8	2 3/16
3/16	3/8	8 5/8	5 5/8	3/16	3/8	8 5/8	5 5/8	0	2 1/8	3/16	5/16	7 1/4	5 3/4	3/16	5/16	7 3/4	6	1/8	2 1/8
7/32	7/32	3"	10	7/32	7/32	3 5/8	10	5/8	1 7/8	At centre notch = Mid-gear.									

Inspection of this table shows that, as the link approaches its *mid-gear* position, *lead* increases; *port-opening* and travel diminish, and *cut-off* and *compression* occur sooner. The crank being at OD′, or OD″, lead occurs. The eccentrics will be at F′ and B′ and lead is greatest at mid-gear, because, see Pl. XXXIII., Fig. 4, shifting the link to mid-gear advances it a little, but withdraws it when the crank is at OD″, thus opening the ports more, at p and p', respectively.

"*Setting*" *the Valve Motion of a Locomotive.**

225. The engineer in charge first locates his cylinders and steam chests, places the valves on their seats, and the yokes (with their stems) over them. The drivers and axles (eccentric pulleys being on) are fitted to their boxes. Guide rods, one cross head, and connecting rod, tumbling shafts and springs, rocker boxes and rockers, located according to the drawings. The links, with their saddles attached, are swung from the tumbling shaft by their suspending links. The link blocks are then fastened to their respective rockers.

The Engineer, having attached the stub ends of the eccentric rods to the straps and links, prepares a trammel, and centre punch points, to indicate for the smith the proper length to which the rods should be "pieced out." They are subsequently put in position. The reversing lever is mounted, as well as the unslotted arches; the tumbling shaft is ready for its connecting rod to form its attachment with the reversing lever. This rod is "pieced out" to the length indicated, by dropping the link into extreme gear, and observing how far over the reversing lever is capable of motion without interference with the cab.

There is no occasion to place the pistons or their rods, but simply to take all dimensions which have reference to its motion, from centre punch marks on the cross head.

The engine is then "jacked up" under the boxes of the drivers, so that the latter clear the track. The first step in the process of alignment is to mark the quadrant points on the drivers, with reference to an arm clamped on the main frame which will so overlap the face of the driver that the passage of the point during the revolution of the wheel will be noted very distinctly.

The driver is now connected with the cross head; then it is revolved until the latter reaches the end of its stroke, and this extreme point is marked on the guide rods. Distant about 2″ from this point along the rod another centre punch mark is made. Then, after revolving the drivers, their motion is arrested when the end of the cross head arrives at this second point; while stationary, the "scriber" is drawn along the overlapping

* By W. S. Auchincloss, C. E.

arm, and a line marked on the face of the driver. Upon revolving the drivers, their motion is arrested when the points are a second time in alignment, and the face of the wheel is scribed as before. We thus have 2 points on the face of the wheel equally distant from the point when the crank-pin is on one of the "centres" or dead points. Therefore the bisection of this arc gives a point, which, placed immediately under the scribing edge of the arm, will place the crank pin on the "centre."

By repeating this process with the cross head when at the other end of the guides, we are able accurately to find the other "centre." Thus obtaining the "centres" on the face of one driver, we can readily "train" the quadrants. The strap of the connecting rod is next removed, and the rod allowed to rest on its yoke.

With these four points carefully determined, we are able at any moment to "pinch" the crank pins of either cylinder over to their "centres."

The eccentric pulleys are then placed with nearly the proper amount of lead (as nearly as the eye can judge), the reversing lever thrown into full gear forward, and the drivers "pinched" until one of the crank pins is on its centre.

It should here be observed, that instead of measuring the lead directly from the valve and the edge of its port with steam-chest bonnet off, it is customary after the valve is scraped to its seat, 1st, to place it so that on one side the steam port is *just* closed, and having made a centre punch hole on the face of the stuffing-box flange, place the small leg of the valve gauge in it; with the other leg scribe a line on the valve-stem, on which line make another centre punch hole. 2nd, place the valve so that the opposite port is *just* closed, and make a second centre punch hole on the valve stem, as will be indicated by the valve gauge. This gauge is made of a small piece of square steel bent thus, Fig. 108, and sharply pointed on the legs. One of these accompanies each engine. It is thus apparent that the engineer can (in the case of the eccentric pulley slipping) readjust his valve with the proper lead without removing the steam-chest bonnet.

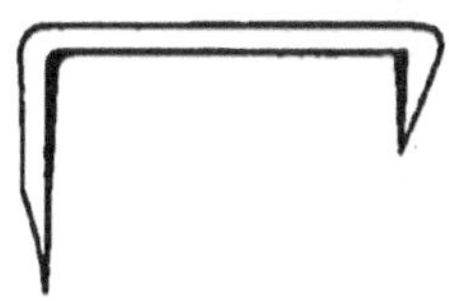

FIG. 108.

The lead the valve now has is noted in the forward and back stroke; if not equal, it is made so by adjusting the length of the valve stem, by turning the right and left nut, which connects its two parts, and then locking the check nuts. The forward eccentric pulley is then altered until it gives the required lead. This operation is performed on both sides of the engine until a perfectly "square" motion is obtained in forward gear.

After this, the point at which the reversing lever lock bolt strikes the arches is carefully marked with a scriber, and the lever is thrown into full gear back.

If, on trial, it appears that the motion is not a "square" one, it will be necessary to introduce a slip of sheet iron between the head of the backing eccentric rod and the eccentric strap; then draw the bolts tightly together. This slip should just equal in thickness half the difference between the leads. After this adjustment results in a "square" motion, the backing eccentric should be altered, until the same lead is produced in full gear back as was given in full gear forward.

Having repeated this process on both valves, the arches should be marked. In order to insure perfect accuracy, the lead and "squareness" of the valve in forward motion should be re-examined, in order to guard against any disarrangement which may have occurred while adjusting the back motion.

It now remains to mark the other "notches." These for a 24″ cylinder are usually 8″, 10″, 12″, 16″, 20″, 22″, and indicate the points of cut-off when the reversing lever is in either notch. They may be accurately described, by again attaching the connecting rod to the driver and cross-head, then laying off the points from the end of the stroke as shown on the guides; pinching over the driver until the cross-head mark corresponds with either of these and drawing back the reversing lever until the valve *just* closes the port. At this point, mark the arches and so continue to obtain the other points. The centre notch will be found at the point of bisection of the arches between the two points of shortest cut-off. Finally slot the "notches" in the arches.

It only remains to *rigidly* attach the eccentric pulleys to the driving axle. This is done by scribing the position of the feather on the pulley ribs and the axle; then sinking a $\frac{5}{8}$″ square feather—$\frac{1}{4}$″ into the axle $\frac{3}{8}$″ into the pulley, driving in solid. Also by letting the points of the two steel $\frac{3}{4}$″ set screws into the axle through each pulley.

REGULATORS.

Governors.

Elementary Principles.

226. A comprehensive idea of the governors used in equalizing the speed of steam engines, water wheels, etc., may best be had, at the outset, by considering the different principal ways in which they may be classified.

In respect to the essential governing member of the contrivance there are—

1. Ball governors, the most familiar kind.
2. Fan governors, in which the resistance of the air to an increased speed of revolving vanes, is made to diminish the steam passages.
3. Oil governors, in which the increased velocity of a paddle wheel or propeller, working in oil is made to act to produce the same effect.

In respect to the point at which the governor takes effect there are—

I. Throttle governors, acting to close a valve in the steam pipe.

II. Steam valve governors.

227. The popular idea of an engine governor is that it is a contrivance for rendering the speed of the engine uniform under a variable load.

It *is* true, that it will maintain an unvarying speed under a *uniform* load, and with a uniform steam pressure; and furthermore, that it will maintain a uniform *average* speed under a uniform *average* load. But it will *not* maintain an unchanged speed, if the load be permanently increased or diminished, though it will, by virtue of the consequent diminution or increase, respectively, of the speed, so increase or diminish the steam supply delivered to the piston, as to make the alteration of speed by the alteration of load less than would naturally result without a governor. In other words, it holds the engine from "running away," as it is called, if the load be greatly diminished; prevents it from stopping, if the load be correspondingly increased, and makes the speed more nearly, sometimes *much* more nearly, uniform, under a variable load, than it would be without one.

228. *Of the inherent defects in the simple ball governor, and of the methods of overcoming them.*

Fig. 109 represents a sleleton ball governor, where the lowest

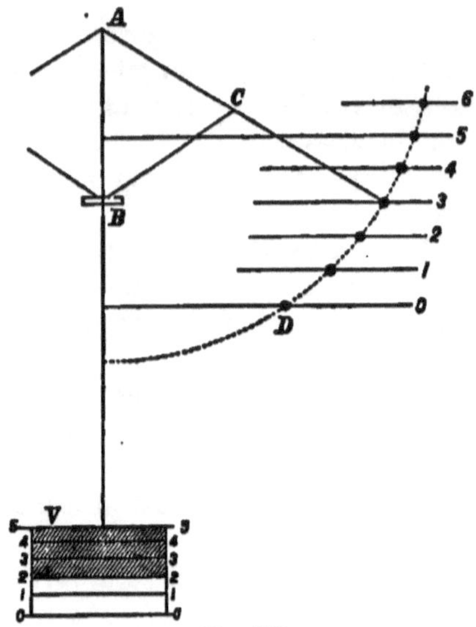

FIG. 109.

and highest positions are on the lines 0 and 6. B and C are fixed points, so that as the balls *rise*, a forked, or toothed top, at A, depresses the valve V and closes the steam pipe. Now if the balls rose through equal heights for equal increments of speed, the valve would be proportionally closed. But they do not, on account of the greater lever arm with which the weight of the ball acts to depress the ball from its higher positions.

229. *One method* of neutralizing this defect is, to give the balls a short range of action, as from line 4 to line 6, only, and a high velocity, say 60 revolutions to 30 of the engine, so as to keep them in a high position. Then, as the entire arc, through which they act, approaches to a vertical direction, equal increments of velocity will elevate the balls by nearly equal amounts.

A second method of compensation is to balance the governor balls by a weighted arm, and so to hang the governor by jointed arms, that they remain in a horizontal plane, and all work of *lifting* them ceases, and only the resistance of their inertia to an increased velocity in their plane remains.

An example of this construction will presently be given.

Finally, the principle of *graduation*, used in the celebrated Judson governor's, has been employed in connection with the first method of compensation just described. This principle consists in shaping the steam openings, which are regulated by the governor valve, not as rectangles, but as a tapered opening, so adjusted that *equal increments* of engine velocity will cause the governor through rising by decreasing increments of height, to shut off equal successive areas of steam port in the governor valve seat.

On the other hand, some governors abandon the ball regulator; examples of such will be described presently.

230. In regard, now, to the second classification (226). The *essential idea* of the class of *throttle governors* is, to deliver to the piston, by means of a variable steam-pipe opening, and at each instant of each stroke, until cut off takes place, a pressure of steam due to the work being done at the instant, the point of cut-off being invariable.

Whatever quantity of steam, more or less, is thus delivered to the piston before cut-off takes place, is used expansively after that point.

The *essential idea* of the *cut-off governor* is, to cut off the steam supply, which is of constant pressure, coming through an unvaried steam pipe opening, at such a point in each stroke that the total work of the steam for that stroke shall be equal to the resistance to be overcome during that stroke.

231. *The failing case of the cut-off governor, and its remedy.*—A cut-off governor is considered perfect, according to its sensitiveness, by which it may cut off at one-eighth, perhaps, of the stroke, or not at all. Now suppose the case of *frequent, sudden*, and *great* changes of load, as in a rolling-mill. In driving the empty rolls, it may therefore happen not very seldom that a piece of iron may be fed to the rolls, just *after* an early cut-off has taken place. In this case, the governor was a false prophet, not knowing the future beyond the point of cut-off, but, nevertheless, it must abide by its own doings, and no more steam can be had to do the work required, till the beginning of

the next stroke. But the difficulty is not necessarily a serious one, there being at least two remedies. First, to give out the steam power by quick strokes of a small cylinder, instead of slow strokes of a great one, so that the *time* before the steam will be ready to meet its work will be very short. Second, to provide a fly wheel, so heavy that its inertia will maintain a nearly uniform speed during the remainder of a single stroke.

232. With the throttle governor, the fixed cut-off occurs so late, comparatively, that there is a much smaller chance that the foregoing conditions will happen; so that a good throttle governor, placed directly on the steam valve chest, so as to quickly deliver to the piston the proper pressure, is an excellent regulator.

Without further general explanations, we will proceed to describe several governors; chosen from among a series of them, only with reference to having them as different from each other as possible, and each the best of its class, so far as could be ascertained, having reference also to novelty. If space allowed, it would have been interesting to have illustrated the Sickel's and other marine cut-offs; the Corliss and the Greene (of Providence, R. I.) variable cut-offs, and the Judson, Tremper, and Snow throttle governors.

Example LXI.

Chubbuck's Fan Throttle Governor.

Description.—Not many fan governors are made. This appears strange, in view of the apparent simplicity and delicacy of some of them, and the inherent defects of the unbalanced ball governor.

Pl. XXX., Figs. 9–13, represent a very interesting one, the figures being nearly facsimiles of the sketches and measurements taken directly from a governor which was taken apart. Fig. 9 is a side view; Fig. 11 is an end view, looking in the direction of arrow *q*; and Fig. 12, one, looking in the direction of arrow *r*.

A is a section of the driving pulley. B the end of a $\frac{5}{8}''$ spindle, II; solid with which is the spur wheel Ḥ. C is the hub of A, keyed to I, so as to turn the latter. JJ is a stationary *sleeve*, held by the set screw *c*, in the standard KD. The barrel, E, contains a coiled spring; one end fast to the inner surface of E, the other, to the sleeve P, solid with the sector GGF, and rotating on sleeve JJ. Fixed to G, is the stationary spindle, *oo*, on which the spur wheel M revolves loosely. The spindle, *oo*, also carries the fan, not shown, whose arms are fixed to a sleeve, *k*, sliding over *oo*, and solid with M. The fan carries four skimmer-formed vanes, 6½ inches diameter, and whose centres are 7 inches, from that of *oo*, Fig. 13.

The geared sector, G, actuates the sector *m*, which carries the spindle *n* of the wing valve, W, seated on the seat *p*. The opening at N, covered by the cap, Q, Fig. 10, affords access to the valve, W; which, when fully closed, comes edgewise against the *stud* or *stop*, *t*, Fig. 12. At L, steam enters from the boiler, and passes through the valve, and around its seat, which forms a partial partition within K; and passes out at the opposite opening, R, to the engine.

The parts on the valve spindle, *n*, are of brass. The others are iron.

The *operation* of this governor is as follows: The barrel E, is held to the sleeve, J, by a set screw. It can therefore be turned, to coil the spring to any degree of tension. The spring is also fastened to the sleeve P, carried by the sector, G, so that it tends to hold the valve W wide open, and the more forcibly so, the tighter it is coiled.

On the other hand, if M were solid with G, its connection with H would cause G to revolve about B as a centre. Hence, just in proportion as M resists rotation, does each successive radius of it act as a rigid arm, attached to G, on which H acts to revolve G about the centre B; while, when M turns with perfect freedom on *oo*, no motion is imparted to G. Now see what takes place in practice. If, by throwing off a part of the load of the engine, its speed is increased, the wheels A, H, M, and thence the fans, will revolve faster; and the resistance of the air to the increased velocity of the vanes will make M act as described, partially as if solid with G, so that G will turn; and thence, by turning *m*, partially close the valve, W, until the diminished steam supply reduces the speed.

But note: If the engine be designed to make *a* revolutions a minute, with a load L, with the valve open to a certain amount, the spring will be so set as to hold the valve at that opening, in spite of the vanes, until that speed is attained. If *then* the speed be increased, as supposed, the resistance of the vanes will be sufficient to overcome the tendency of the spring to keep the valve open, and it will be partly closed. *But the speed can not thus be permanently reduced to its former rate*, for at that rate, the valve must be open a certain amount, by reason of the mutual adjustment of the spring, and fan, and valve. To run at exactly the same speed with a less load, the boiler pressure must be reduced, or the hand valve in the steam pipe must be partly closed, or the spring must be relaxed, so that the fans will hold the valve at a given opening with less resistant effort.

Construction.—The three figures should be placed side by side, with B, B, B, on the same horizontal line; Fig. 11, to the left, and Fig. 12 to the right of Fig. 9.

Proper scales would be from one-*half* to one-*fifth* of the full size.

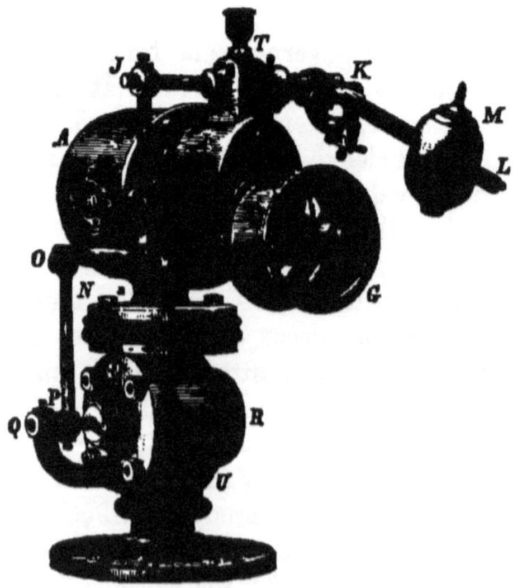

FIG. 110.

EXAMPLE LXII.

The Huntoon Oil Throttle Governor.

Description.—See Pl. XI., Fig. 3 and Fig. 110. Like

letters refer to like parts. A is a cylindrical reservoir of oil, within which works a small common screw propeller B, whose axis, C, slides freely in its bearings DD. A long pinion, E, is keyed to this axis, so as always to be in gear with its driver, F, which is of greater diameter. F is fast to the axis of the driving pulley, G, which is driven by a belt. H is a lever, actuated by the moving axis C, and keyed to the axis JK, at one end of which is the lever KL, weighted with the movable weight, M. From J proceeds the succession of levers JN, and OP, and their connecting link, NO; by which the axis, QR, of the cylindrical valve, SS, is oscillated within its concentric seat in the case, UU, into which steam enters at R, and leaves at V, for the steam chest on which the governor should stand.

Action.—Suppose an engine is to act at a certain speed, under a certain average load. *First,* suppose that load *uniform.* Then a definite aggregate opening of all the rectangular ports, whose sections appear in T, will be required, to which a certain position of the lever, KL, corresponds.

Now, as the engine is brought to the required speed, we find experimentally the position on the lever of the weight M, in order that it shall be sustained by the action of the propeller, whose operation is as follows. As the propeller works in the oil reservoir, it strives to propel the oil towards the end D, of the reservoir. There being no free escape for the oil, its reaction drives the propeller and its axis towards H, and thus shifts the lever H, and raises KL, which turns the valve SS till the required opening is obtained.

Second. Suppose the load, or the steam pressure, variable. A momentary increase of velocity of the engine, under decrease of load or increase of pressure, will instantly produce any reasonable required increase of velocity in the propeller, by means of suitable proportions between G and the wheel which drives it, and between F and E. Then, as the resistance of fluids to motion through them is as the square of the velocity, or more, perhaps, the reaction against a slight increase in the velocity of B will instantly raise KL, and close the openings in T. The contrary effect will result from a sudden increase of load or a decrease of pressure. If the load is to continue for some time, more or less than before, the weight M must be shifted till, at the same speed as before, the valve opening shall be more or less also, to the extent required. And as the lever is moved but a very short distance to close the ports T, its angular movement, as indicated by the difference of length of F and E, is so small for any given average load and speed, that it is raised with practically equal facility through all points of its small motion.

Finally, when, in case of a nearly uniform load and pressure, the weight M need not be shifted, it is hung by a chain wrapping on a curved sector, of which KL is the arm or spoke. It will then rise and fall vertically.

Construction.—No measurements have been placed on the figures, since the governor is made of various sizes. If the diameter of B be taken at 7½ inches, it may serve as a scale for the construction, and Pl. XI., Fig. 3, will afford all the data essential for transforming Fig. 110 into plans, elevations, and sections.

Example LXIII.

Wright's Variable Cut-off by the Governor.

Description.—See Fig. 111, giving a general view of the front portion of the engine, and Pl. XXXI., Figs. 6–10, showing the governor, with plan enlarged. Fig. 8 is a smaller plan view, showing the valve-stems and their heads.

The engine, to which this governor is attached, has a *variable cut-off*, and its connections with the governor are such, that the point of cutting off steam is made automatically variable to suit the requirements of the machinery driven by the engine, thereby measuring out just the amount of steam necessary to meet any variations in the power required, which are constantly occurring in all engines used for manufacturing purposes. The induction valves, which are of the balance poppet kind, are arranged in separate chests, II, Fig. 111, on the side of the cylinder, having a steam connection with the pipe cast with the cylinder. The engine has *independent exhaust valves* in the bottom of the cylinder, which are brought as close as practicable to the end of the cylinder to obviate waste of steam in filling the passages. These are slide-valves, worked by a rod taking hold of the valve from the underside, the rod being in the exhaust steam, thereby obviating the necessity of stuffing-boxes in the live steam, and they are worked with the minimum amount of friction. The maximum pressure on these valves is at the ends of the stroke, and they are relieved of the pressure

FIG. 111.

in proportion to the expansion of the steam in the cylinder during the stroke.

It may be the case with this class of engines, when used for manufacturing purposes, that when they are working with a minimum amount of power, the cut-off takes place so early in the stroke as to reduce the pressure of steam down to the atmosphere *before* the stroke is completed. This involves a loss of power during the balance of the stroke by producing a partial vacuum on the steam side of the piston; but by giving a due lead to the exhaust, the valves prevent this vacuum, and the consequent loss of power; which result cannot be produced without an independent exhaust.

Indeed, engines generally, with variable cut-offs, have independent exhaust-valves, as in the *Corliss*, *Greene*, and *Putnam* engines.

Both the eduction and induction valves are worked from a horizontal shaft, parallel to the cylinder, and driven by spur and bevel gearing from the crank shaft. Cranks on this parallel shaft actuate the exhaust valves transversely to the cylinder. The induction valves are opened in the direction of the arrows by a cam, F, on a hollow upright shaft, K, arranged in suitable fixed bearings between the heads, *h*,*h*, of the two valve-stems, *d*,*d*′. See also Fig. 8. The valves are closed to produce the cutting-off of the steam, by springs or by the pressure of the steam on the ends of their stems. The cam shaft, K, has a bevel gear at the bottom, through which it is driven by a bevel gear, G, on a horizontal shaft, H, which is arranged alongside the cylinder, and which derives motion through bevel gearing from the main shaft first mentioned. Fig. 7 shows a side elevation of the gear and cam boxes S and R.

The cam, F, is constructed with two pairs of sliding toes, *t*,*t*′, and *f*,*f*′, one pair for operating each valve, and as one of the toes for each valve operates during every half revolution of the shaft, K, the said shaft only makes *one* revolution for every *two* revolutions of the crank shaft. The toes are cogged on their inner edges, as shown in plan, to gear with long straight cogs, *n*, on a spindle, N,N′,N″, which passes through the hollow main spindle, MM′, of the governor, and which is so suspended from the governor at O as to be *raised* and lowered as the balls of the governor *rise* and fall. This spindle, N, has on its lower part a series of spiral cogs or threads, *r*, which fit and work like a many-threaded screw in a nut, P, formed or screwed within the lower part of the cam-shaft. The Governor is driven by bevel gear, *m*,*m*′ and *q*,*q*′, on the upper end of the shaft, K.

The spindle, N, rotates with the cam-shaft, and always rotates at the same velocity, so long as the speed of the engine and governor is invariable; but, whenever the speed of the engine varies, and causes a variation in the plane of revolution of the Governor balls, the spindle, N, rises or falls, and in so doing is caused to turn independently of the cam-shaft by the longitudinal movement of the spiral cogs or threads, *r*, in the nut within the lower end of the said shaft. By that means, the said spindle is caused to turn within the cam-shaft, and in this way the straight cogs, *n*, on the said spindle, are made to act upon the cogs of the toes, *tt'*, of the cut-off cam, and thereby to produce a greater or less opening of the induction valves, and to expedite or retard the closing or cutting-off movement, according to the requirements of the engine.

The toes are ribbed on their upper and under sides as seen at *t'*. These ribs guide the toes in and out, horizontally, in the grooves *aa* and *bb*. See the arrows in the plan.

Thus, the farther out the toes are thrown, by the rotation of the long pinion, consequent on the falling of the balls and the spindle, N', with the screw, *rr*, the wider will the valves be opened, and the longer they will stay open.

Construction.—By placing the figure lengthwise of the plate, or by making it on a folding plate it may be made on a scale of from *three-eighths* to *five-eighths*.

EXAMPLE LXIV.

Babcock and Wilcox Governor and Variable Cut-off.

Description.—With this remarkable example of the present group, there are coupled, by way of a summary of information, various points concerning steam engines; collected from descriptive articles in several scientific periodicals.

Most of the features of modern steam engineering originated in the fertile brain of James Watt. He found the steam engine in a very crude state, and left it in quite as perfect a condition (excepting only mechanical construction) as that of the ordinary engines at the present time. He invented separate condensation, expansion, steam jacketing, superheating, and the governor. The combination of the governor with a cut-off valve gear, was reserved to a later period, having first been published in the "Repertory of Patent Inventions" for 1826, as the invention of James Whitelaw. Since then the steam engine has advanced by improvement in details and construction, rather than by the development of new principles.

The engine herewith partly presented makes no pretension to radical improvements in the principle of using steam expansively, but it embraces a novel, simple, and highly effective method of operating and controlling the action of the valves for admitting and cutting off the steam.

There is no necessity at the present day to argue the superiority of an engine regulated by a good cut-off, so far as economy of fuel or regularity of speed are concerned.

This engine has a novel construction of the governor, by which the variation (228) due to the pendulum action of the ordinary governor is overcome, and a regulator produced, which will give the same speed whether the engine be lightly or heavily loaded, or the pressure of steam in the boiler be greater or less. The governor, as invented by Watt, and adopted by modern engineers with rare exceptions, gives only an approximation to equal speed, requiring a variation of from five to thirty per cent. between the extremes of motion. This we have seen.

In designing this engine, it has been the object not only to introduce peculiar ideas and improvements, but to combine therewith all those features which long practice has proved to be most conducive to economy of fuel, and the durability of all the working parts. The steam jacket has been much neglected in this country, though in almost universal use by the best engine makers of Europe; and so little are its theory and advantages understood here, that often where it has been introduced in this country it is filled with the exhaust steam, thus partly defeating the very object for which it was designed. This engine is jacketed with live steam from the boiler, in both heads, as well as around the cylinders, thereby keeping the metal of the cylinders as hot as the hottest steam which enters it.

The valves which affect the distribution of the steam in the steam engine, are the most important part of the machine, as upon their properly performing their functions depends the efficiency of the engine. They must not only admit, exhaust, shut off, and close, at the proper periods, but they must be perfectly tight when closed; and, when open, admit the steam with the least possible resistance. They should also permit of such a relation to the cylinder as to give the least practicable lost space or clearance. There are four distinct varieties of valves used for this purpose, viz.: the plug or cock, the piston, the seat valve or poppet, and the slide. The first variety is never used now by competent engineers, having but one good quality, viz.: the equal pressure of steam on its sides, to balance its many bad features, such as leakages, sticking from expansion, and unequal wear. The piston valve is also nearly out of use owing to the lost space inherent in its construction. The same objection applies to the poppet valve, with the additional ones of great liability to leakage and inability to open and close quickly, from the fact that it opens immediately on starting, and is not closed until brought to rest. It is impossible to start or stop the valve instantaneously; therefore the opening and closing must be correspondingly so slow as to be objectionable except on slow moving engines.

The universal experience in this country and in Europe, is in favor of the slide valve for opening and closing the ports of all quick moving engines. It is simple and easily fitted, admits of the least lost space, opens and closes the ports with the quickest possible motion, and is the least liable to become leaky from use of any form of valve. Of the two forms of slide valve the flat is preferable to the curved, from the greater facility of accurate fitting, and the more equal wear of two planes as compared to inner and outer cylindrical surfaces.

An important condition of equal wear in a slide valve, however, is a constant travel. Where the induction valve is made also to act as a cut-off valve, as in a link motion, this condition cannot obtain, and as a consequence we find that such valves are more apt to leak.

The adaptation of a cut-off mechanism, to act in conjunction with a plain slide valve, the latter to admit and exhaust the steam, and the former to close the port at any desired point in the stroke, has been a favorite pursuit of engineers for the past half century. Nine-tenths of all the expansion engines now built in Europe have some modification of this form of valve gear, and the engine of Messrs. Farcot & Sons, which received the Grand Prize at the late Paris Exposition, was of this class.

One of the points in which the Babcock & Wilcox engine differs from those which have preceded it, is the manner in which the cut-off valves are operated, viz.: by the action of the steam itself, independent entirely of the action of the main valve; thus insuring an instantaneous, positive, and easily controlled cut-off, at any desired point in the stroke of the piston. The distribution of the steam to the alternate sides of the piston, and its release from the cylinder when the stroke is completed, are performed in the manner most approved by experienced engineers, by means of a plain slide valve operated by the ordinary eccentric. But from the fact that the induction valve has in no case to act as a cut-off valve, and from the further fact that the cut-off is actuated independently of the motion of the main valve, the functions of "lead" and "cushion" can be adjusted to any desired degree, without in any manner affecting the action of the cut-off valve. This is an important distinction between the operation of the main valve of this engine, and of those which have preceded it. In the ordinary three-ported slide valve, or in any other arrangement where the several functions of *lead*, *cut-off*, *release*, and *compression*, or closing the exhaust, are dependent on the motion of one eccentric, the "exhaust" functions—*i. e.*, the release and compression—must always be subservient to the "steam" functions—*i. e.*, the lead and suppression, or cut-off. The cut-off mechanism consists of two cut-off slides, a miniature steam cylinder, and a valve for controlling the admission of steam to the same. This small cylinder, being enveloped in the steam, requiring no packing, and having only the weight of its piston to produce wear, is, for all practical purposes, indestructible. The cut-off slides are always balanced when they move, consequently they are not exposed to injurious wear.

The bed or framing which has been adopted for the horizontal engines is of the form first introduced by Horatio Allen, of the Novelty Iron Works, New York. It is bolted to the end of the cylinder, and extends to the pillow-block, and the metal is so disposed as to give the greatest rigidity with the least weight. The cross-head is upright, and is supported on flat slides, a drip cup cast on the bed serving to catch all drippings, not only from the slides, but from all the stuffing boxes.

The regulator or governor is driven by gearing, thus avoiding all danger of breakage or slipping of belts, and the consequent damage to the engine and machinery from the "running away" of the engine.

In addition to the steam jacket for preserving the temperature of the cylinder, a covering of felt is employed around all the exposed parts, and this in turn is covered by a casing of polished metal. The latter is the best possible protection against loss by radiation.

Pl. XXXII., Fig. 1, represents a *horizontal* section of the cylinder and valves, on X′X′ and YY′, Fig. 2, showing the peculiarities of the cut-off motion. A is a cylinder, which is steam jacketed, as are also the heads, at *aa*. B is a portion of the bed piece, which forms also the front head of the cylinder. C is the piston, and C′ the piston rod. D is the main valve, *ee* the induction ports, and F is the exhaust port. The body of the valve is hollow, and conveys the exhaust steam, from either end of the cylinder, alternately, to the exhaust port, F, whence it goes into the exhaust pipe. The steam passes through ports *e′* in each end of the valve, into the induction ports of the cylinder, alternately, as they are opened by the motion of the valve derived from an eccentric in the usual manner. On the back of the valve, at each end, is a slide, G, which can be made to cover the port at that end, and these slides are attached to a piston H, fitting in a small steam cylinder bolted to the back of the valve, and so adjusted so that when the port in one end of the valve is closed the other is open. Upon steam being admitted to either end of the piston H, the piston is shot over, and the corresponding slide closed, to cut off steam from that end of the main cylinder; while the port at the other end of the main valve is opened ready to admit steam to the other side of the main piston when the valve shall arrive at the proper position.

It will be observed that the cut-off slides, G, are always balanced when moved. The one about to close having steam of equal pressure upon each side; while the other one has been balanced by the main valve riding past the end of the valve face on the cylinder, thus admitting steam behind the slide, G. This condition obtains during the whole stroke of the piston until the steam is cut off, after which the cut-off slides G, remain stationary relatively to the main valve until ready to cut off steam on the return stroke, previously to which they have been balanced by the over-riding of the valve at the other end These slides experience, therefore, almost no wear, and, once fitted tight, they will remain so indefinitely. The piston H, in the small cylinder, is turned to fit, and has no packing, neither have the rods stuffing boxes, as the pressure is equal on both sides, except during the inappreciable time which intervenes between the exhausting of the cylinder, I, and the movement of the piston. The only tendency to wear in these parts is due to the weight of the piston and rods, which are supported on large surfaces. In fact, after twenty months constant use, none of these parts have worn sufficiently to obliterate the tool marks upon the surfaces.

Steam is admitted alternately to each end of the piston, H, at every revolution of the engine, causing the cut-off slides to move at every stroke, cutting off the steam at the point determined by the governor.

Fig. 2 shows a cross section on XX. The valve, *b*, *b'* of the cylinder, HH', is balanced by the plate, J, upon its back, and is operated by a toe, *t*, upon the rock shaft, L, carried upon the main valve, and extending through the end of the steam chest where it receives motion from a crank, *m*, on a shaft, *n*, which is oscillated by the governor. The exhaust ports, *f*, of the cylinder, I, are made upon the bottom, and are at a little distance from the end, while the steam ports, *g*, are upon the side and at the extreme end of the cylinder. By this arrangement the piston closes its own exhaust port, and cushions on the remaining distance, thus dispensing with all dash-pots or air cushions, and causing the valve to move without any noise.

The valve, *b*, being balanced, and the rod, L, carried through its stuffing box by the main valve, there is the least possible power required by the regulator to adjust the crank, *m*, thereby ensuring more sensitive action than can be attained where the governor has labor to perform.

The governor is peculiar, and is shown at Fig. 112. The balls, N, are hung upon arms in the usual manner, which arms are jointed at their upper ends to a head attached to the rod, *o*, which slides within the hollow shaft, *k*, that drives the balls; the motion being communicated through the radius rods, *p*, which are jointed at their lower ends to the gearing shaft, and at their upper ends to the centre of the arms, *n*. The rods, *p*, are half the length of the arms, *n*, measuring from the centre of the ball, and it will be readily seen that, in consequence of this arrangement, the arms, *n*, and rods, *p*, form a parallel motion, and compel the balls to move outward in a horizontal plane.

In the ordinary pendulum governor, the balls move in the arc of a circle, and rise as they extend. It therefore requires an increased speed to maintain them in their advanced position. The engine *must* consequently run faster when the load is light than when it is heavy, *and such is the case with all ordinary governors* (228). In this improved governor, it will be seen that the gravity of the balls has no tendency to move them in either direction, and exerts no influence whatever upon the speed of the engine. The centrifugal force causes them to diverge, and a weight, W, tends to bring them towards the shaft. When, therefore, these two forces are in equilibrium, the balls will remain in the same position; but, as either preponderates, they are moved in a corresponding manner, thus affecting the speed of the engine by varying the amount of cut-off. The weight, W, is supported upon a bent lever, *l*, which is so proportioned, that the centrifugal force, at any given speed, will just balance the weight in all positions. The speed of the engine will, therefore, remain at that fixed point with all variations of load or pressure of steam; for any increase or diminution will cause either the balls or weight to preponderate, and the point of cut-off to be changed, until the speed is again brought to the standard where the two forces are in equilibrium.

Any desired speed for a given load, can be obtained by altering the weight, W, and the action of the governor will be as perfect in one case as in any other. A spiral on the rod, *o*, serves to advance or retire the crank, *m*, relatively to the main crank, so as to cause the cut-off to occur earlier or later in the stroke, as the balls diverge or converge; and the amount of this adjustment is such that the cut-off may be varied from one eighth, of the stroke, to the end of the stroke.

The Balanced Governor is a radical improvement. The vices of the old governor are, that the extension of the revolving balls (by which the steam supply is shortened or lengthened as the speed is accelerated or checked) is resisted by their weights at a progressive leverage, and fails to represent truly the changes of speed: and, secondly, that a given force of steam can only be had at a given speed,—because a given position of the balls determines a certain opening of the governor valve,—whatever the load may demand, so that if the load is much lightened the steam is shortened only by running at a high speed, or if the load is heavily increased, the engine must run slow to get steam to meet it—*except as regulated by hand.* The balls of the new governor have nothing of the pendulum character, but extend in a horizontal plane, with equal ease at either extreme of their oscillation, faithfully representing all the fluctuations of the speed. Their centrifugal tendency, at the speed intended to be maintained, is accurately counterbalanced by a weight, drawing them inward with a progressive leverage responding to every change in their position and force. The consequence is that so long as the prescribed speed is maintained, the position of the balls is independent of the speed; they revolve far in or far out, indifferently, so as to give steam according to the wants of the load, with one and the same speed in all cases. If a change is made in the load, the speed *for the moment* suffers change, the equilibrium of forces in the balls is disturbed, and the preponderating force places the unresisting balls instantly in the proper position to meet the new demand for steam; and *there they stay* (the speed being righted and equilibrium thus restored) until another change of load summons them to a new position. The simple mechanism, by which the regulation of steam is perfected in this governor, will repay a more particular examination.

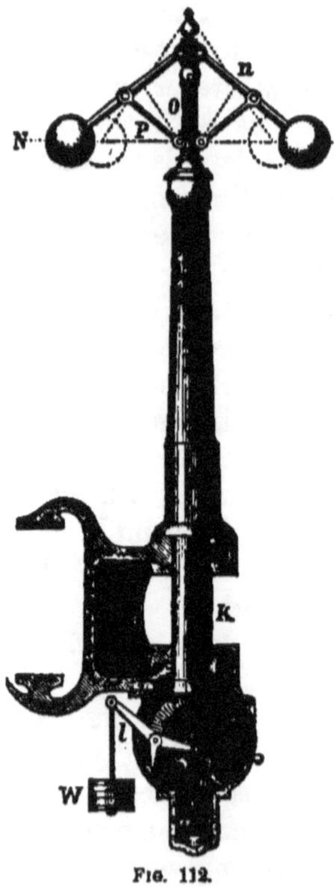

FIG. 112.

Construction.—The student can profitably make Plate XXXII. on double the present scale, or from one-*fourth* to one-*sixth* of the full size.

Indicator Diagrams.

233. The only means we have of tracing the action of the steam within the cylinder—the time and rapidity of its entrance,

the point of cut-off, its action in expanding, the time of release, the amount of back pressure, compression, etc.—is by diagrams drawn by an "Indicator,"—an instrument actuated by the pressure of the steam and the motion of the piston. These two motions, acting at right angles, produce a curve which indicates the exact pressure of steam at each portion of the stroke of the piston, in and out. The line thus drawn, during a complete revolution of the engine, encloses an irregular figure, the shape of which varies with every different condition in the elements which form it, and by it we are enabled, not only to determine the actual power exerted by the steam, but also the relative perfection of the valve motion, and the effect of different proportions between the piston and passages.

The importance of the Indicator as a means of studying the action of any given engine, and of comparing the relative values of different constructions and proportions, though known from the time of Watt, has but recently been fully appreciated by engineers; and, in fact, not until within a very few years has there been an instrument manufactured, capable of being used with any satisfactory degree of accuracy upon the quick moving engines now employed for most stationary purposes. To its employment the world is indebted for its most satisfactory practical knowledge of the action of steam, and the best means of obtaining the highest economical results.

But in order to compare one engine with another, they should be in precisely similar circumstances. As, however, this rarely occurs, it is necessary to have some standard by which all engines may be compared and their relative performance determined. The best means for doing this is to compare each engine with a theoretically perfect engine of the same size under the same circumstances.

The expansion of steam follows certain laws, and the quantity of steam being known, as well as the space which it occupies, it is possible to tell the correct pressure for each variation in the space occupied. A curve can thus be calculated which is of hyperbolic form, and which will give a diagram of the theoretical action of a given amount of steam in a given size of cylinder. The diagrams taken from any engine may thus be compared with a theoretical diagram, *for the same quantity of steam* used in the same sized cylinder, and the ratio existing between the actual and the theoretical diagrams will serve as a measure of the perfection of the engine and valve mechanism.

For the illustration of this subject, several diagrams follow, which have been taken from different engines, with the theoretical diagram, for each case, the latter allowing for *no losses* from any source. It is impossible to construct an engine in which there shall be no loss from friction of the steam in the pipes and passages, or from clearance.

In these diagrams, the highest line, AB, represents the pressure of steam in the boiler, and, with the exception of those from condensing engines, the lowest line, CD, that of atmospheric pressure. The scale marked upon each diagram is the fraction of an inch which represents one pound of steam pressure in the *vertical* lines of the diagram. The horizontal length of the diagram represents the length of stroke of the engine, plus an amount of space at the end (exterior to the heavy outline), which is called the "clearance," and represents, in the same scale as the stroke, the amount of space included between the end of the cylinder and the piston at the extreme of motion of the latter, and also the contents of the passage ways. It will be seen that the length of stroke is represented by no particular scale, but each of the divisions is one-tenth of the full stroke. The *heavy* outline is the diagram formed by the indicator, and represents truly the pressure in the cylinder at each fraction of the motion of the piston. Where the line commences to fall abruptly, as at F, is the "point of cut-off," and shows the portion of the stroke in which the steam is admitted. During the remainder of the stroke, the steam expands, reducing the pressure, and forming a curve, called the "expansion curve." At, or just before, the end of the stroke, the steam is released, and "exhaust" commences. The returning line shows, by its distance from the base line, the amount of "back pressure." In a properly constructed engine, the exhaust closes a little before the termination of the return stroke, thus confining the remaining steam, compressing it, and forming a "cushion" to stop the momentum of the piston, and prepare it for the return stroke. This is shown by the rounding of the corner and the rising of the pressure at the termination of the stroke.

The *dotted* outline represents the theoretical power of the amount of steam exhausted from the cylinder in each instance, when used in a cylinder of the same size, with no losses from friction in the passages, back-pressure, or clearance. The proportion of the area of the actual to the theoretical diagram represents the *relative* efficiency of the given engine, and is stated in percentage of the theoretical efficiency in connection with each individual diagram.

Nos. 1 and 2 were taken from a 12-inch engine, 3 feet stroke,

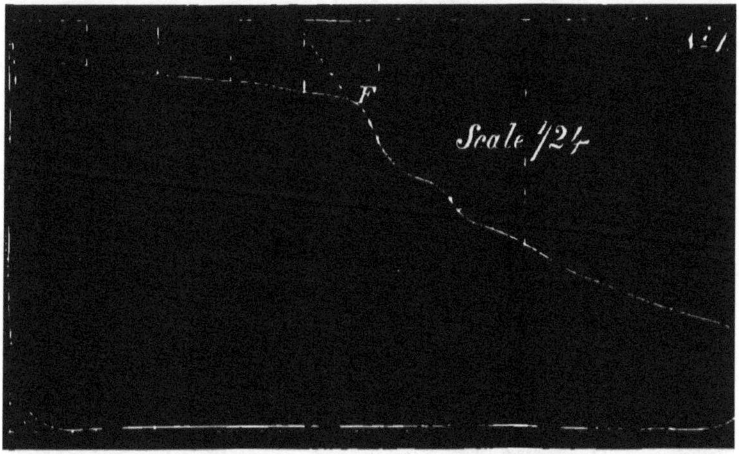

making 64 revolutions. In No. 1 the actual is 89 per cent. of the theoretical diagram, and in No. 2, 82 per cent. The losses in this case are due to the friction of the steam in the pipe and passages.

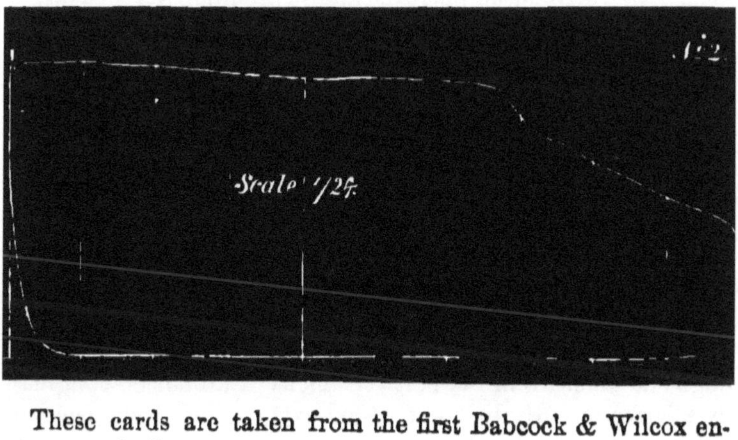

These cards are taken from the first Babcock & Wilcox engine ever built.

No. 3 is from a 14 by 42 inch engine. The engine makes 65 revolutions per minute or 455 feet of piston speed. This is one of a large number of equally good cards taken from this engine, and the actual diagram is $92\frac{3}{4}$ per cent. of the theoretical.

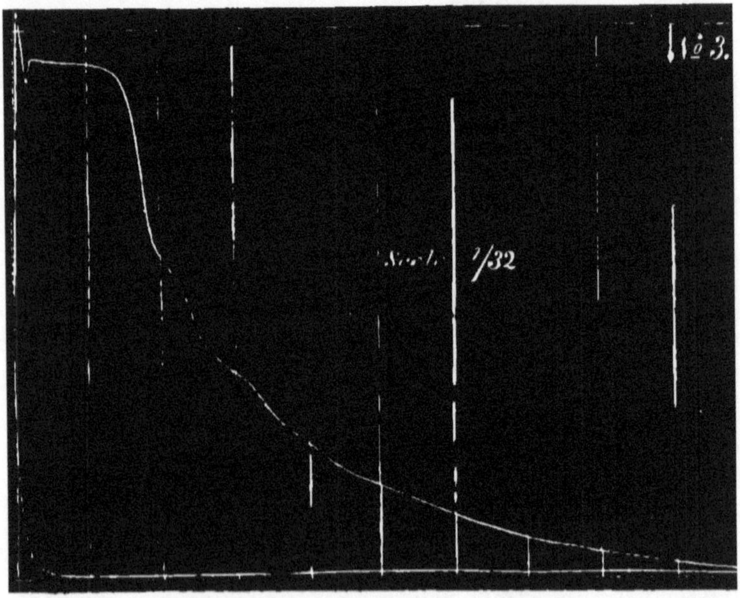

233. Leaking valves, or pistons, ill proportioned steam passages, an improper amount of compression, back pressure arising from a too long, circuitous, or otherwise hindered exhaust, a general bad condition of the engine, or an overestimated boiler pressure, given by inaccurate gauges, may all make an abnormal difference between the actual power given out, as represented by the *indicator* diagram, correct as that may be, and any alleged corresponding theoretical diagram.

With these explanations, which the rigorous impartiality of a text book imperatively demands, the following vicious diagrams are given in illustration of them; and without implying that better ones could not be taken from perfect engines of the designs which they represent.

Indeed, it is much to be desired that a series of diagrams might be taken from the best specimens of all our best engines, by a board of entirely disinterested experts, and from each engine by each of a number of indicators.

No. 4 is from a plain slide valve engine, 18 inches bore and 2 ft. stroke, of good construction and apparently in good order. It shows but 55.33 per cent. of the theoretical diagram. It will be noticed in this case, that the theoretical compression curve

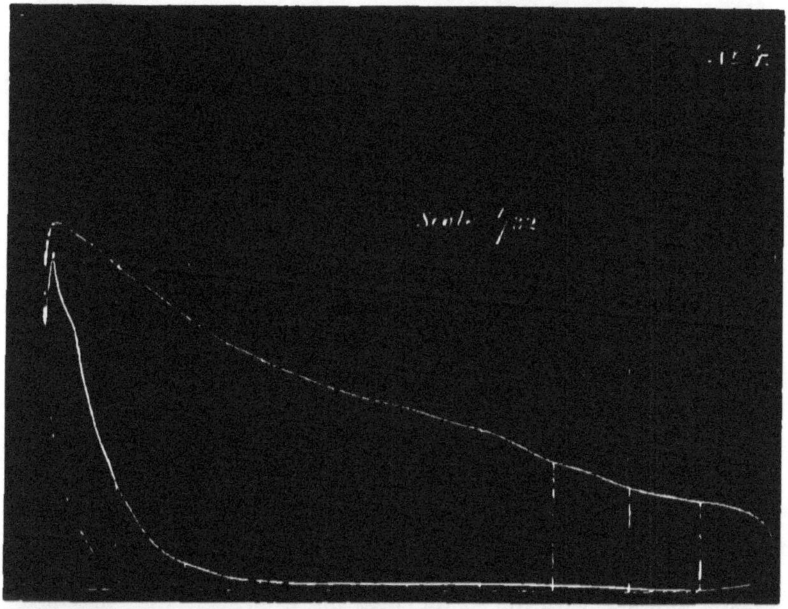

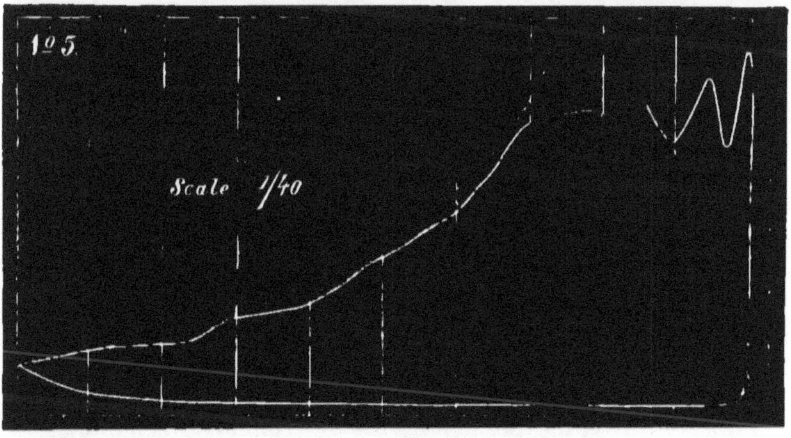

does not touch the actual line, except at the point of the peculiar hook, which is probably owing to a leakage of the piston.

No. 5 is from an engine of 23 inches bore and 4 ft. stroke. It figures 71½ per cent. of the theoretical diagram.

No. 6 was taken from a condensing steam engine, fitted with poppet valves, actuated by toes, and shows 71⅝ per cent. of the theoretical effect.

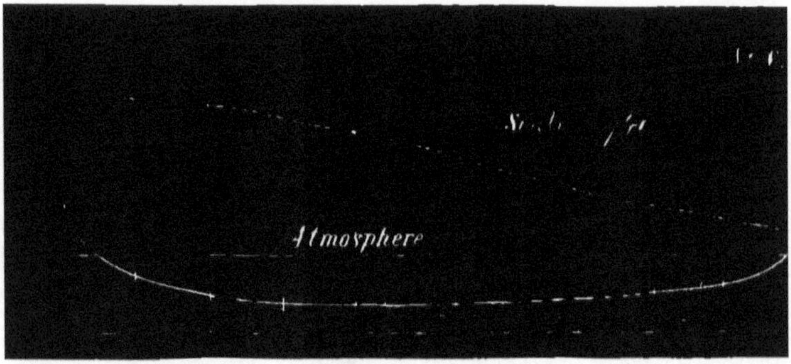

Example LXV.

The Putnam Machine Co.'s Variable Cut-off.

Description.—Pl. XVI., Figs. 11, 12. Let the observer be supposed to be facing the front end of the cylinder of a horizontal engine, with the steam chest on what will then be the right hand side of the cylinder. In this chest are the poppet valves opening upward; and the parts shown in the figure will be under the steam chest.

A is a section of the cam shaft, moved by gearing on the main shaft, and lying parallel to the cylinder.

The manner of lifting the valves is designed to overcome all shock in their rising and falling. The actual construction varies somewhat, but the principle is shown in the figure. The cam shaft, in this case, has the same angular velocity as the engine shaft, and when connected as shown in Fig. 11, is adapted for large engines. The auxiliary lever, D, has a horizontal and vertical motion, the corner, 7, describing nearly a semicircle and returning on the diameter, operating against the lever, B, to raise the valve, the position of the fulcrum of B being under control of the governor. The rise and fall of the governor spindle oscillates the lever, F, by means of an arm, keyed to H, or in any convenient manner; and thus determines when the corner, 7, shall pass the corner 8, and thus let go of the valve; which will then drop to its seat. The lever as in Fig. 11, is operated by a double cam—GE, Fig. 12—the part *c* of the lever rests upon the G part of the cam, during the entire revolution, while the lever receives horizontal motion from the eccentric E, acting against the straight sides, *mn*, of the opening of the lever within which it plays. It will be observed that the corner, 7, never falls below the under side of the lever B, but slides along in contact with it.

FIG. 113.

EXAMPLE LXVI.

The Rider Cut-off.

Description.—Fig. 112 clearly illustrates this novel and simple variable cut-off, with the valve chest removed. The rise and fall of the governor spindle is made to rotate the obliquely truncated half-cylindrical cut-off valve, CO, and its stem, T, by means of an arm from T; or a toothed sector, gearing with a rack, formed by teeth on the lower end of the governor spindle. The angular position of the valve thus determines the time of cutting off steam from the oblique ports, seen in the main slide valve. These ports pass spirally through the valve, SS, so that on its plane under side they are rectangular and perpendicular to the axis of the steam cylinder in the usual way.

Construction.—After the practice from measurements, which the student has thus far had, it will be sufficient to assume them for this example. As a key to the relative position of the piston and the two valves, in making a section as in Pl. IV., Fig. 1, note that when the piston is at either end of its stroke, the main valve, having no lap, is exactly at midstroke, except by the amount of the *lead;* and the cut-off valve is in a position to have the port on top of the main valve wide open. Also, the main valve having no lap, its *travel* need be but twice the width of its steam port.

The following more definite data from a model, will serve as well as if from actual practice, for the purpose of locating the eccentrics, and thence the position of both valves, for any piston position. They are all in 60ths of an inch.

Stroke of piston = 168. Width of steam ports, 8½. Do. over all the ports, 43.
Travel of main valve, constant = 17, = 8½ each way from midstroke.
" " cut-off " " = 21, = 10½ " " "
" " cut-off " on main valve 13, = 6½ " of centre of cut-off from centre of main valve.

These half travels, 8½ and 10½, give the throws of the eccentrics; and then the last item will show what angle they should make with each other.

Example LXVII.

Sibley and Walsh's Water-Wheel Governor.

Description. Pl. XXXIV., and Fig. 114. In the plate, Fig. 1 represents a sectional elevation of the governor. Fig. 2 is a sectional plan, on the plane *xx*. Fig. 3 shows a separate view of the stop-plate which operates to detach the pawls, LL′.

Like letters indicate like parts on all the figures. A is the bed. B is the frame. C is the governor head. D, the governor balls, and E, the governor spindle. F is a band pulley, by which the governor is rotated through the action of the bevel gears, G and H, as shown on Fig. 114. I is a frame—or bar, Fig. 114—sliding on the bed, and moved by the eccentric, J, on the spindle, E, by means of the rod, K.

This frame, I, is provided with four pawls, LL′—two only in Fig. 114—which engage with the ratchet wheel, M, through the reciprocating motion of the frame, or bar, I. The pawls LL turn the wheel M in one direction, and the pawls L′L′ turn it in the opposite direction; but they engage with M only when there is a variation in the speed of the governor. N is a shield on the shaft, or spindle, O, to which shaft the ratchet M is attached.

The shield N extends in the form of guard-plates, P—the concentric shell segment, M′, in Fig. 114—which so cover some of the teeth of M, as to prevent the pawls from engaging with M, except when opening or closing the water-wheel gate by the action of the bevel gears, Q, Fig. 114. The positions of N and P are controlled by the governor.

The vertical motion of the governor spindle, E, is transmitted by the grooved thimble, R, through the forked bell-crank, S, oscillating on the shaft, T, and through the rod, U, to the shield N; which is thus made to oscillate on the shaft O, and to vary the position of the guards, P.

The pawls are kept in contact with the guards, when not engaged with M, by means of the small springs, V, attached to the studs, W, as shown. X is the stop, which consists of a bar in which the shaft, O, works by a screw thread, so that the rotation of O carries X in the direction of the length of O.

As the wheel M is turned, X will finally be brought to the shoulder on the screw-shaft, O; and then the water gate will be fully open, and the stop, X, will turn with the wheel M; its motion being limited, however, by the stud X′. The pin, *y*, in the bar, X, will be in contact with the pin, *z*, in the shield N, and will hold the latter, with the guards, P, in such a position as to prevent the pawls, L, L′, from engaging with M. At the same time, the governor, by the action of its varied velocity in putting the guards, P, out of the way of the pawls, LL′, is left free to close the water gate as may be required.

Fig. 3, *z*′, shows a plate which turns on the shaft O, and, by means of the rod, *l*′, may be made to disengage the pawls from the ratchet wheel M, when it is desired to close the gate by the hand wheel Y. The plate, O, is recessed as at $a'a'$, to allow the pawls to engage with M at other times.

FIG. 114.

The wheel M may be horizontal, as in the plate, or vertical, as in Fig. 114, and with two or four pawls.

Operation.—When the water-wheel revolves too slowly, the balls fall, E rises, S swings to the right, L′ is let into the teeth of M, and turns it as I makes its stroke to the left, so as to *open* the gate.

When the water-wheel revolves too fast, the balls rise, E falls, S swings to the left, L is engaged with the teeth of M and actuates it during the stroke of I to the right, so as to *close* the gate; I being constantly actuated back and forth by the eccentric J.

Construction.—These governors are made of various sizes, and of partly different proportions, to suit different cases. A scale for Pl. XXXIV. can be determined by assuming the balls D to be four inches in diameter.

With a little care, especially where an example of the governor is accessible, Fig. 114 can be transformed into plans and elevations. Also an end elevation could be added to Pl. XXXIV. In the elevation the top of the bed, A, and the centre lines of E and O, will be convenient lines of reference to work from; and in the plan, a line joining the centres of O and E will be a useful line of reference.

MODULATORS.

Example LXVIII.

Compound Speed and Feed Motions.

Among compound modulators forming trains of mechanism, *feed motions* may first be mentioned.

The utmost number of plates being now full, the skeleton, Fig. 115, may represent, in plan, the feed movement of a grand shaping machine by the celebrated Joseph Whitworth, of Manchester, Eng., for finishing up propeller blades. F is a tapering "*mandril*," so made for the purpose of taking up all wear, by slipping it further into its bearings, so that it will revolve without play in its bearings, or *journals*, at F and F; whose massive common support is called a headstock. FL is the radius of the circular face-plate, the many rectangular holes in which allow work to be clamped to it in any position. FM, FN, and FO are three concentric spur-toothed rings, bolted to the back of the face-plate, to allow it to be driven at various speeds. P is the band pulley, from which all parts take motion, and which may be attached to either of the other speed shafts, 1, 2, or 3. The spur wheel, R, sliding by a feather in a longitudinal groove in the shaft P*g*, may thus be in or out of gear with the spur-wheel S, on the shaft 1; which, by a pinion, *b*, at M, carries the face-plate. Pinion T is on the same hub, or "*boss*," with S, and is in gear with U on shaft 2. Now U can be put in gear with pinion V, on shaft 3, carrying the pinion X, to gear with O for driving the face-plate at its quickest speed. Shaft 2, carrying wheel V, has also upon it a pinion Y, in gear with wheel Z, on shaft 1. The pinion *a*, gearing with N, gives the intermediate face-plate speed, and pinion *b* on shaft 1, acting with ring FM

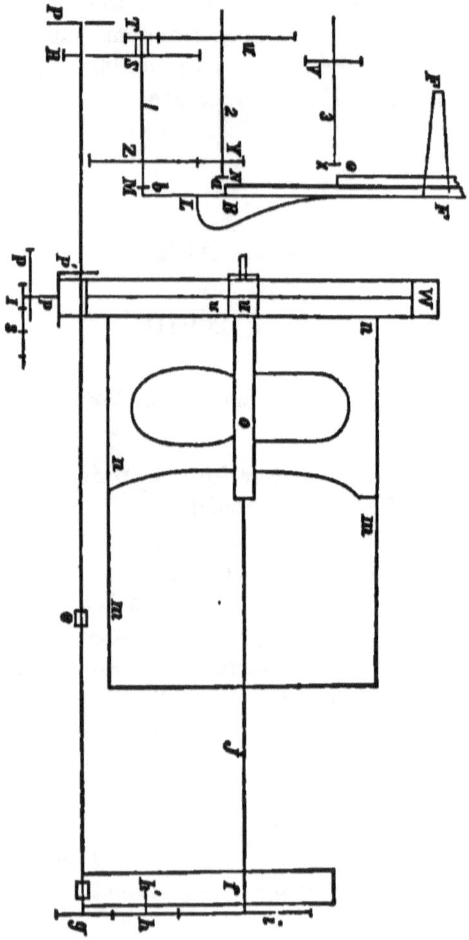

FIG. 115.

gives the slowest speed. The shaft P*g* is supported by intermediate bearings, as *e*, and one on the standard *f*, where it carries the spur-wheel, *g*, which gears into *h*, which turns *loosely* on a fixed stud *h'*, and gears with *i* on the screw shaft *j*. At *m m*, are guides, on which slides the great slide rest *n*, by the working of the revolving screw *j*, in the long fixed nut *o*, on the under side of *n*.

This part of the movement advances the tool, *t*, against the propeller blade B, according to the pitch of the latter, as the face-plate revolves.

Again: a bevel pinion *p'* with a long feathered boss, on P*g*, revolving with P*g*, moves another pinion *p*, on the axis of which is *r*, gearing with *s*, loose on a stud, and so driving a larger wheel, *c*, not shown, on the screw shaft, *u*, which, by working through the slide *v*, gives transverse motion to the tool, to enable it to make its concentric parings from the blade B.

Wheel q, by gearing with a pinion on the shaft of *s*, then put in gear with *c*, will give a quicker transverse feed.

Various sets of the wheels *g*, *h*, and *i*, provide for different pitches of the screw to be trimmed.

Construction.—This example, by the very rude illustration given of it, may serve, more fully than any previous one, as one in mechanical design.

Example LXIX.

Whitworth's Quick Return Motion.

Description.—Quick return motions are contrivances for withdrawing a tool across the surface of its work for a new cut, faster than it works in making a cut, so as to save a part of the time in which it is idle.

Pl. XXXI., Fig. 11, shows a very ingenious and admired form of quick return, for a shaping machine; that is, a planing machine for irregular, or curved work. An arm at A, takes hold of the stock, which carries the cutting tool, and leads it back and forth, as actuated by the connecting-rod, G, proceeding from the crank-pin, H; whose distance from the *fixed* centre of motion, O, is adjustable. H is clamped in the radial groove of the crank piece, I, which is carried by the spur-wheel J, which revolves in the very stout bearings K*k*. This being hollow, as shown by the dotted circles, K and *k*, the spindle, C, of the crank piece, I, passes through it. There is a radial slot in that side of I which rests against J, and in the contiguous side of the wheel J is a pin, L, carried in the square bush D, which slides in this slot. The pin, L, with its bush, D, imparts the motion of J to I. Now LO is a fixed distance, and so is CO, but LC is variable, being now = LO – CO; but when L is on the opposite side of O, we shall have LC = LO + OC. And this variable distance of L, from the centre of motion, C, of the crank, makes the return motion, in the sense of the arrow, quick, and the advance motion, when LC = LO + OC slow.

Construction.—This example may well be drawn twice as large as in the figure.

FIG. 116.

EXAMPLE LXX.

Mason's Friction Pulleys and Couplings or Clutches.

Description.—In the above cut, the friction pulley shown, consists of the two main parts, the loose pulley B, and the shaft

A, on which is keyed the part second, which consists of a plate or disk, D, and two segments, EE, and the sliding sleeve or thimble, F. The two segments, EE, are fitted to slide radially on the face of the disk, between ribs or guides, cast on the plate DD, and are operated by means of the adjustable toggle joints shown. It will be seen that when the thimble, F, is moved towards the plate, the segments EE are forced outwards in contact with the inside of the pulley A, thereby producing friction between the two surfaces sufficient to drive the machinery to which it is applied. The ball and socket joint used, is more plainly seen in the figures 117 and 119 of the friction coupling, shown by the letters *e*, *f*, and *g*.

Belts running over the friction pulleys will run much longer without necessity of tightening, than when they are constantly shipped from one pulley to another.

In some cases friction pulleys may be placed on the main line and used to stop and start machines driven directly from the main line.

Method of Adjusting the Friction.

The friction pressure may be nicely adjusted, by placing the centres of the segments in a horizontal position, and then taking two strips of stiff paper, and placing one strip between the centre of each segment and the inside of the pulley, when unshipped, then slowly ship the thimble towards the plate, and note which strip of paper tightens first; the adjustment is then easily effected by screwing the connecting arms out or in until the pressure is alike on each segment, and sufficient to drive without slipping; then tighten the check-nuts, FF, firmly against the joint to prevent the screw from loosening. Care should be taken not to set out too hard, so as not to prevent the thimble from always shipping up closely against the plate, as the joints are so arranged that the centres of the joints of the thimble will just pass by a line drawn through the centres of the joints of the segments, thereby holding itself in when the thimble is fully shipped up against the plate.

This adjustment should always be attended to by a good mechanic, on the first starting up, as when properly adjusted they will generally run one and two years, and often longer, without any readjusting.

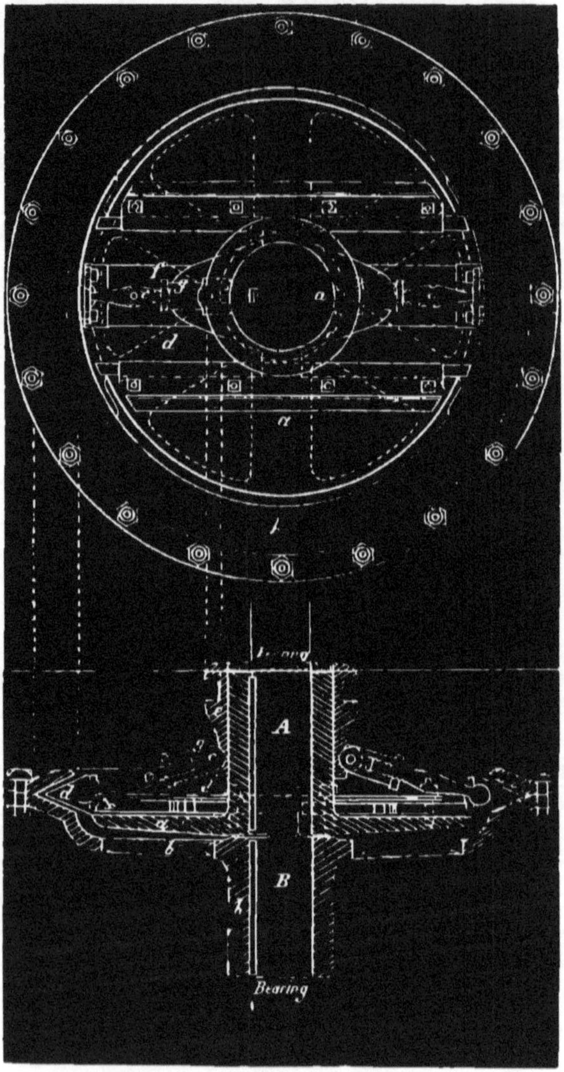

FIG. 117.—Thirty-six inch Friction Coupling, one-twelfth size.

Figure 117 represents a view of the friction coupling as applied for connecting shafting. The shafts A and B, are made one to enter the other, so as to help keep them in line; some times it is more convenient to drill into the end of each shaft, and put in a steel pin, instead of letting one shaft enter the other. The segments of this friction coupling, have a V, or wedge form, by which a very powerful friction is produced, for connecting heavy shafting and gearing.

FIG. 118.

Fig. 118 shows the slides or ribs on the plate, also, the ball joint and screw and check nuts, and T joints of the thimble, in their relative positions.

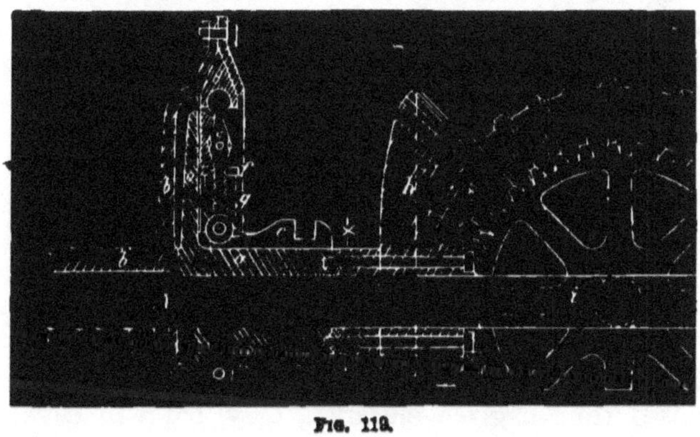

FIG. 119.

Fig. 119 shows one method of applying the friction for starting and stopping gears, as used for force pumps, grinding mills

and other machinery, which can by this method be driven directly from the main shafting; stopping both gears when the friction is unshipped. By this method, there is *no danger of breaking the gear teeth on starting, nor any necessity for slacking speed*, on starting any machinery which they may be employed to drive; as, in case of sudden unusual strain being brought on the machine, the friction may be so adjusted as to slip a trifle, thereby saving the breaking of shafts or teeth of gears. In cases of fires in factories, force pumps have frequently been disabled and rendered useless by the breaking down of the gears or shafts; and, in places *where frictions are not in use*, it is necessary to stop the engine, or water wheel, before the pump gear can be thrown in; while by the use of friction, any force pump can be started up promptly in case of fire, without stopping or hindering the motive power.

These friction clutches are also very useful for hoisting, as at coal, copper and gold mines, and in tunnel shafts and rolling mills, and have been furnished for use at a number of different copper mines in Chili, South America and elsewhere. They are applied between the engine and large winding drum, for reversing the motion of the drum, which they accomplish in the most perfect manner. Among the largest yet made are two employed to reverse the motion of heavy rolls for rolling sheet lead. The two weighed about five thousand eight hundred pounds, yet are light in comparison with the ponderous gears to which they are applied. By simply working the lever the power is arrested or transmitted in either direction almost instantaneously, without any shock or noise, or slacking speed of engine. They transmit sixty horse power, and are shipped probably two hundred times daily.

EXAMPLE LXXI.

Reversing Gear for the Compound Rolling-Mill Engine.

Description.—Pl. XI., Fig. 2. In this figure, the plan was improvised from a given elevation, only, but will answer to illustrate the character of the movement, and as a basis, from which the student can proceed to make modifications.

SS′ is the main shaft, on which is mounted, loosely, the

eccentric, EE′, solid with the bevel wheel, GG′; the collar, CC′, and the sleeve, AA′, which slides on the long feather *ff′*. CC′ and AA′ revolve with the shaft, by reason of the feather, and EG′—E′G′ revolves by reason of the hold which the teeth of the bevel sectors LL′ and FF′ have upon GG′.

To reverse the engine, it is only necessary to revolve the common eccentric, EE′, of the three valves, upon the shaft. For this purpose, RR′ is a loose ring on the sleeve, AA′. By operating a handle which proceeds from it, either by hand, or by power, AA′ is shifted to the right or left, and thence the arms, *aa′*, and *bb′*, rotate the bevel sectors, which will rotate GG′ as desired. These arms are centred on the studs, *cc′*, and *dd′*, on the sleeve, and the sectors are centred at *gg* on studs on the collar CC′.

Construction.—Let this figure be made on a scale of one-twelfth, and with the sectors in some other relative position, so that the arms *aa′* and *bb′* will not be parallel. Also some at least, of the teeth on the sectors can be added, and an end view made.

Escapements.

233. *Escapements* are of comparatively small immediate interest to the civil or mechanical engineer, as such; but are intrinsically attractive, on account of the refined ingenuity of their design, and by association with astronomical clocks; certain uses of which in engineering practical astronomy, the engineer should be acquainted with. Escapements are also indirectly interesting to the engineer as forming a part of some of the dividing engines, by which the degree circles of his field instruments are graduated.

Escapements are, finally, interesting as being the only means within the range of pure mechanism, that is by motion only, without employing inertia as in case of a fly-wheel carrying a crank over its "centre," for converting rotary into reciprocating motion.

A few examples therefore are here given; as a proper conclusion of this work, with the most refined constructions that can be found.

Example LXXII.

Bond's Escapement, No. 2.

Description.—What follows is in the words of the inventor and makers.

"In giving the description of my Isodynamic Escapement, No. 2, it will be necessary, in order to explain its advantages fully, to give some account of those now in use, and of the obstacles still remaining to be overcome in order to obtain one that is perfectly exact.

"The advantages to be derived from the possession of a clock of perfect accuracy (were such a thing possible), could hardly be over-estimated. The science of astronomy, in particular, would receive important benefit from such an instrument. But as no timekeeper has ever yet been constructed that could be relied upon as being absolutely free from error, it is evident that there is still room for improvement. The sources of irregularity have long engaged the attention of many able scientific investigators, and very numerous contrivances for counteracting them have been suggested. These remarks refer, of course, only to such timekeepers as have the pendulum for their regulator, and indeed, no other natural principle is practically so well adapted to produce regularity.

"The end to be attained is, to keep the pendulum vibrating always in the same *arc*, always encountering the same amount of *resistance*, and of motive *power*.

"The principal errors may be divided into two classes, those arising from the mechanical intervention necessary to maintain the vibratory motion of the pendulum, and those arising from such causes as would still affect it, provided its vibration could be maintained by a uniform force; for instance, changes of the themometer, barometer and hygrometer, magnetic influences, and, to a slight degree, the position of the moon. By far the most important of these errors are those arising from the friction influencing the pendulum through the escapement, as they are of such a nature as renders it almost impossible to ascertain beforehand what their influence will be, while the residuary errors of the second class are not only smaller in amount, but can, by close and accurate observation, be tabulated and the corrections applied. It is the first class, therefore, of these errors, that it is most important to remedy, and the power of doing this lies almost entirely in diminishing, or *equalizing*, the friction of the escapement. The mere *diminishing* of the friction is not the only, or even the principal desideratum in such improvement; the *equalizing* of the force is of far more consequence, as is sufficiently proved by the single instance of the deservedly high position which Graham's escapement has so long maintained, notwithstanding the great amount of friction which it involves. One reason of its superiority to many others which boast far less actual friction, is its principle of compensating a slightly varying power upon the pallets, by a corresponding variation in the arc described by the pendulum; this has given it its practical utility. This escapement, though invented nearly a century ago, met with no successful rival, until within a few years. It has, however, recently been replaced, in some astronomical clocks, by Denison's three-legged gravity escapement. The superiority of this latter consists in the impulse to the pendulum being given either by the force of gravity, influenced by a small amount of friction, or by the force of a spring without such friction; but in either case there is a certain amount of variable resistance, increasing or diminishing as the wheel-work of the clock carries more or less power to the escapement, and consequently, if a heavier driving weight is applied, the pendulum encounters more friction in unlocking the escapement, without gaining any additional impulse, as it would in the case of Graham's. The vibration of the pendulum is thus affected, and its rate changed, by a varying cause, dependent upon the freedom with which the wheelwork transmits the power, and which it is impossible to calculate. Notwithstanding this defect, this method has, upon the whole, smaller causes of error than any hitherto known.

"The Isodynamic Escapement, recently invented, overcomes entirely the difficulty of the varying power transmitted by the wheelwork, and thus obviates most of the objections to other escapements. In comparing it with previous ones, I refer only to Denison's and Graham's, they combining to a greater degree than any others the various requisites essential to *a good escapement.*

"Besides the difficulties already enumerated, which are to be overcome in making a perfectly reliable gravity escapement, there is one which has been exceedingly troublesome, namely, the necessity of guarding against what is called *tripping*, or the danger of two or more teeth of the escapement wheel passing the pallets at once, when one only is intended to do so. This causes the hand of the clock to gain by jumps, and of course, in a most unreliable manner, while the pendulum may be vibrating with perfect regularity. Mr. Denison, in his book on 'Clock and Watch Work,' speaks of guarding against this difficulty as first among the essential mechanical conditions. Even in his own gravity escapement, to which I have already referred, the possibility of *tripping* still exists, though rendered reasonably slight by the introduction of a fan upon the escapement wheel, but in the Isodynamic Escapement, it will be seen that this danger is completely removed."

The escapement, of which Pl. XXXIII., Fig. 2, is a drawing, is of the class known as *gravity escapements*, and has proved thoroughly good and given extremely good results. The extreme variation in the *hourly* rates of a clock with this escapement, for a considerable length of time, was only $0^s.020$.

In the figure, *g* and *g*, are the gravity arms hung on delicate pivots at *a*, *p* and *p'* the pallets, *f* and *f'* friction rollers. P the pendulum, and S the scape wheel, revolving five times a minute, having six arms, and six pins *d d'* projecting from the face, to act on the friction rollers, *f*–*f'*. At the moment shown in the figure, the pendulum has completed its swing to the left, and has just began to move to the right, the gravity arm, *g*, being in contact with it, and assisting the vibration by its weight, until the adjusting screw, *b*, comes in contact with the stop, *c*. The gravity arm, *g'*, has in the mean time been raised slightly by the pin *d'* in the scape wheel, coming in contact with the friction roller, *f'*, and the arm, *m'*, of scape wheel has locked on pallet, *p'*. As the pendulum continues to move to the right, it comes in contact with adjusting screw *n'*, and carries gravity arm *g'*, with it, to the right. This unlocks the arm *m'*, the scape wheel moves forward until the pin *d'* comes in contact with friction roller, *f*, raises it a little, and arm, *m*, locks on pallet *p*, until it is again unlocked. It will be seen from this, that the impulse is constant, being the weight of the gravity arm, acting on the pendulum through the distance it is raised by the pin in the scape wheel, since it falls back in contact with the pendulum that much more than the distance through which the *pendulum* has raised it.

EXAMPLE LXXIII.

Bond's Auxiliary Pendulum Gravity Escapement.

Description.—This again is nearly in the words of the makers. Pl. XXXIII., Fig. 3. This was the last invention of the artist, Mr. Richard F. Bond, perfected indeed on the morning of his death. Its mechanical beauty is shown even more clearly in the drawing, than it appears in the clock, where it is necessarily somewhat confused with the other details.

It is an entirely new escapement, nothing like it ever having been made before, and with the exception of the Remontoir clocks, which are entirely different in principle, this is the first clock which has ever been made, with a *perfectly detached* escapement. The border of the plate represents the back plate of the clock, and all the work shown is on the outside of this plate. The wheel, b, is in connection with the train of the clock, and is *constantly* revolving with a *uniform velocity*, making one revolution in little less than one second, and is regulated and controlled by the conical pendulum, not shown in the drawing, which also revolves once a second.

S is the scape wheel, with the scape arm, s_1, secured firmly to its axis, and moving freely on delicate pivots, working in polished jewel holes; or, as made in the first two clocks, sent by the makers to Paris and Liverpool, the pivots of the scape wheel worked on small friction rollers.

The scape wheel, s, gears in with the constantly moving wheel b; but, at the moment shown in the drawing, it is *entirely detached*, a few teeth being omitted in the scape wheel, so that when the arm s_1, locks on the pallet, p, in the gravity arm g, it rests there merely by its own weight, the wheel b continues to revolve, but the scape wheel s remains at rest. There is also a jewelled cam, e, on the axis of the scape wheel, which, just before the arm s_1 locks on the pallet p, comes in contact with the friction roller f on the gravity arm, thereby raising it slightly, in order to give the necessary impulse to the pendulum P, which is represented as moving to the right. At the lower end of the gravity arm is the screw *k*, which carries a jewel *o* on its end, and as the pendulum comes in contact with it, it moves the gravity arm also to the right, thereby releasing the arm $s_{,,}$, which falls by its own weight. This causes the wheels *s* and *b* to gear together again. The scape wheel is caught up by the constantly moving wheel, *b*, and is carried over, until the cam, *e*, raises the gravity arm, the arm, s', locks again on the pallet *p*, and the scape wheel, disengaged from wheel *b*, again comes to rest. There is also another jewelled cam on the axis of the scape wheel, which at the instant that arm $s_{,}$ drops from the pallet and the scape wheel is caught up by wheel *b*, engages with a tooth of wheel *c* and moves it forward one tooth. The axis of wheel *c* carries the *seconds* hand; and a pinion on the same arbor, under the dial, in connection with other wheels and pinions, moves all the hands at once. M is a stop screw, shaded dark behind screw k, which regulates the movement of the gravity arm to the left, and so adjusts the amount of impulse which the pendulum shall receive.

It will therefore be seen that the whole timekeeping part of the clock consists of the pendulum P, the scape wheel, and arms *s* and $s_{,,}$, and the gravity arm, *g*; the scape arm falling from the pallet *p* at regular intervals of two seconds, measured by the vibrations of the pendulum P. The rest of the clock may therefore be regarded as an auxiliary machine, to carry the scape wheel round until it locks, to raise the gravity arm, and to move the hands; all heavy work, which has nothing to do with the time-keeping qualities of the clock, and it might be used to give motion to any number of instruments or machines, for recording meteorological observations, or anything else that might be desired, and the time of the clock would not be influenced in the slightest degree.

The great difficulty to be overcome, was to make the two wheels *b* and *s* gear together properly, without having the points of the teeth jam together, which would stop the clock, and this was effected at last by making the first two teeth in wheel *s* movable, they pass through the rim of the wheel as shown, and rest on delicate springs *v v*, which yield to the slightest pressure and permit the teeth to slip into their proper place.

www.ingramcontent.com/pod-product-compliance
Lightning Source LLC
Chambersburg PA
CBHW031425230426
43668CB00007B/442